ISSUES AND TRENDS IN INTERDISCIPLINARY BEHAVIOR
AND SOCIAL SCIENCE

PROCEEDINGS OF THE 6TH INTERNATIONAL CONGRESS ON INTERDISCIPLINARY BEHAVIOR AND SOCIAL SCIENCES (ICIBSoS 2017), 22–23 JULY 2017, BALI, INDONESIA

Issues and Trends in Interdisciplinary Behavior and Social Science

Editors

Ford Lumban Gaol
Bina Nusantara University, Indonesia

Fonny Hutagalung & Fong Peng Chew
University of Malaya, Malaysia

CRC Press is an imprint of the
Taylor & Francis Group, an **informa** business

A BALKEMA BOOK

CRC Press/Balkema is an imprint of the Taylor & Francis Group, an informa business

© 2018 Taylor & Francis Group, London, UK

Typeset by V Publishing Solutions Pvt Ltd., Chennai, India
Printed and bound in Great Britain by CPI Group (UK) Ltd, Croydon, CR0 4YY

All rights reserved. No part of this publication or the information contained herein may be reproduced, stored in a retrieval system, or transmitted in any form or by any means, electronic, mechanical, by photocopying, recording or otherwise, without written prior permission from the publisher.

Although all care is taken to ensure integrity and the quality of this publication and the information herein, no responsibility is assumed by the publishers nor the author for any damage to the property or persons as a result of operation or use of this publication and/or the information contained herein.

Published by: CRC Press/Balkema
Schipholweg 107C, 2316 XC Leiden, The Netherlands
e-mail: Pub.NL@taylorandfrancis.com
www.crcpress.com – www.taylorandfrancis.com

ISBN: 978-1-138-55380-4 (Hardback)
ISBN: 978-1-315-14870-0 (eBook)

Issues and Trends in Interdisciplinary Behavior and Social Science – Lumban Gaol et al. (Eds)
© 2018 Taylor & Francis Group, London, ISBN 978-1-138-55380-4

Table of contents

Preface	ix
Committees	xi
The opportunities and strengths of Thailand's MICE industry *Y. Sudharatna & P. Chetthamrongchai*	1
Evaluation of the differentiating factors in the avocado value chain *J.A. Martínez-Arroyo & M.A. Valenzo-Jiménez*	7
Competitiveness in the handicraft sector in Mexico *M.A. Valenzo-Jiménez, J.A. Martínez-Arroyo & L.A. Villa-Hernández*	15
Higher-order thinking skills in teaching the Malay language through questions and questioning among the teachers *F.P. Chew, Z.H. Hamad & F. Hutagalung*	25
Using poems to increase phonological awareness among children *C.T. Lim & F.P. Chew*	33
Web-based catering marketplace business plan: Cateringku.com *B. Susanto, F. Lumban Gaol, S. Si & M. Kom*	41
Practice of mobile guide and learning system by GPS positioning technology *K.-Y. Chin, H.-Y. Hsiao & Z.-W. Hong*	57
Mathematical models for the spread of rumors: A review *M.Z. Ndii, E. Carnia & A.K. Supriatna*	65
Relationship between emotional intelligence and job performance among early childhood professionals: A preliminary study *N. Ghanieah, F. Hutagalung, N.A. Rosli & C.H. Leng*	75
Russian partnership banking integration in Islamic financial system *N.F. Yalalova & A.A. Daryakin*	81
Influencing factors of ethnic tolerance among multiethnic youths *A.R. Ahmad, A.A.A. Rahman, M.M. Awang & F.P. Chew*	89
Analysis of innovation cluster formation in industrially developed countries *I.M. Ablaev*	97
An approach to developing a programmer competency model in Russia *L.V. Naykhanova, N.B. Khaptakhaeva, N.N. Auysheeva & T.Y. Baklanova*	103
Organizational behavior norms as a factor boosting the emotional stability of staff at housing and utility services enterprises *M.D. Mironova, R.M. Sirazetdinov & M.Y. Virtsev*	111
Methodology for the technical and economic analysis of a product at the projection stage *V.N. Nesterov & N.N. Kozlova*	117

Income inequality in Russia *R.S. Timerkhanov, N.M. Sabitova & C.M. Shavaleyeva*	125
Parental challenges in the filial therapy process: A conceptual study *C.C. Han, D.-L. Baranovich, L.P. Li & F. Hutagalung*	131
Report on the results of a monitoring study conducted in the Russian Federation on the prevention of addictive and suicidal behavior and internet risks among adolescents *A.V. Khydyrova, N.A. Mamedova & E.G. Artamonova*	139
Assessment of the quality of language education of future kindergarten teachers *I.N. Andreeva, T.A. Volkova, M.R. Shamtieva & N.N. Shcheglova*	145
Developing design and construction of backspin serving skill tests to assess the learning outcomes for table tennis serving skills *Sumaryanti, Tomoliyus, Sujarwo, B. Utama & H.M. Jatmika*	149
Marketing management of Madrasah Ibtidaiyah (MI) *E. Munastiwi*	155
Development of competitiveness as a condition of professional demand for graduates of pedagogical directions *K.Yu. Badyina, N.N. Golovina, A.V. Rybakov & V.A. Svetlova*	159
Development of students' fire safety behavior in the school education system *V.V. Bakhtina, Zh.O. Kuzminykh, A.G. Bakhtin & O.A. Yagdarova*	165
Development and enhancement of communication for junior school children with vision impairments *E.Y. Borisova, O.V. Danilova, T.A. Karandaeva & I.B. Kozina*	171
Neuropsychological approach to the correction of developmental deviations of preschool children with health limitations *E.Y. Borisova & I.B. Kozina*	175
Active and interactive methods for teaching psychological counseling *I.E. Dremina, L.V. Lezhnina, I.A. Kurapova & S.V. Korableva*	181
Superstitions as part of the frame "pregnancy and birth" in the English, Russian, and Mari languages: Postliminary level *E.E. Fliginskikh, S.L. Yakovleva, K.Y. Badyina & G.N. Semenova*	185
Functional problems of the tourism development institutions *G.F. Ruchkina, M.V. Demchenko, N. Rouiller & E.L. Vengerovskiy*	193
Formation of students' readiness for safe behavior in dangerous situations *M.N. Gavrilova, O.V. Polozova, S.A. Mukhina & I.S. Zimina*	197
A poetics of Mari verbal charms *N.N. Glukhova, E.V. Guseva, N.M. Krasnova & V.N. Maksimov*	203
The trends in the dynamics of Mari ethnic values *N.N. Glukhova, G.N. Boyarinova, G.N. Kadykova & G.E. Shkalina*	211
On the importance of comparative phonetic analysis for bilingual language learning *I.G. Ivanova, R.A. Egoshina, N.G. Bazhenova & N.V. Rusinova*	219
Resource of a public debt in forecasting the development of the economy *M.V. Kazakovtseva & E.I. Tsaregorodtsev*	225
Conceptual theories of forming leadership qualities of a competitive university student *S.Y. Lavrentiev, D.A. Krylov & S.G. Korotkov*	231

Pragmatic aspect of implicit addressing means in different spheres of communication T.A. Mitrofanova, T.V. Kolesova, M.S. Romanova & E.V. Romanova	237
The problem of early development of creative abilities via experimentation in productive activities O.A. Petukhova, S.L. Shalaeva, S.O. Grunina & N.N. Chaldyshkina	245
A negative assessment of the concept insolvency in French and English paroemias T.A. Soldatkina, S.L. Yakovleva, G.N. Kazyro & G.L. Sokolova	251
Training of the Russian Federation's national guard officers in information security R.V. Streltsov, T.A. Lavina, L.L. Bosova & E.I. Tsaregorodtsev	257
Models of the learning styles within the adaptive system of mathematical training of students V.I. Toktarova	261
The problem of realizing administrative responsibility for an offence provided in Article 19.21 of the administrative code of Russian Federation A.V. Vissarov, A.M. Gavrilov, V.V. Timofeev & A.A. Yarygin	265
Grant-supported collaboration between specialists and parents of children with special psychophysical needs O.P. Zabolotskikh, I.A. Zagainov, M.L. Blinova & O.V. Shishkina	269
Preparing university educators for tutoring adult learners in distance education N.A. Biryukova, D.L. Kolomiets, V.I. Kazarenkov & I.M. Sinagatullin	275
Theoretical analysis of the rehabilitation study of children with limited health capacities J.A. Dorogova, O.N. Ustymenko, E.A. Loskutovan & N.V. Yambaeva	283
Analysis of tourism management development of the Bunaken National Marine Park S.B. Kairupan & R.H.E. Sendouw	289
A three-stage data envelopment analysis combined with artificial neural network model for measuring the efficiency of electric utilities P.M. Kannan & G. Marthandan	297
Emotional change of immigrant youths in Korea after collage art therapy: A case study G. Wang, Y. Kim & Y. Oh	305
A study on the "individual accountability" of undergraduate students in the cooperative learning-centered liberal education class Y. Kim, G. Kim & H. Choi	313
Effect of competency on the job performance of frontline employees in Islamic banks: The moderator effect of religiosity M.F.B. Hamzah & M.N.B. Md. Hussain	321
Role of communication in IT project success K. Shasheela Devi, G. Marthandan & K. Rathimala	329
Pupils' self-regulated skills in relation to their perceived position in formal and informal school life processes K. Hrbackova	337
Author index	345

Preface

The theme of the 6th International Congress on Interdisciplinary Behavior & Social Science 2017 was "The Role of Social Sciences in Cross-Cutting Issues in Humanities".

In cross-cutting issues of broad relevance, Social Sciences and Humanities are fully integrated to enhance industrial leadership and to tackle various societal challenges. ICIBSoS 2017 provided the economic and social analysis necessary for reforming such Humanities issues as education, sociology, anthropology, politics, history, philosophy and psychology as well as food security.

Contributions to ICIBSoS 2017 give necessary insight into the cultural and human dimension of such diverse research areas as transport, climate change, energy and agriculture. ICIBSoS 2017 also analyses the cultural, behavioural, psychological, social and institutional changes that transform people's behaviour and global environment.

ICIBSoS 2017 proposes new ideas, strategies and governance structures for overcoming the crisis in the global perspective, innovating the public sector and business models, promoting social innovation and fostering creativity in the development of services and product design.

One example that was discussed at ICIBSoS 2017 was the evolution from the concept of Ecosystem that adapts the framework to the Millennium Ecosystem Assessment Nature's Contributions to People classification assessments. Issues such as sharing economics and the economics of disruptive technology were also discussed at ICIBSoS 2017.

Hence, at ICIBSoS 2017, there was a discussion about new paths that have to be forged, including revisiting basic ontological and epistemological considerations, such as how we understand the world, what knowledge is, and the role of science. Constructive interdisciplinary dialogues in support of the development of innovative frames and terminologies have become the goal of and contribution by ICIBSoS 2017.

Finally, we hope that this book will make a significant contribution to the social sciences and humanities.

Ford Lumban Gaol
Bina Nusantara University, Indonesia

Fonny Hutagalung
University of Malaya, Malaysia

Fong Peng Chew
University of Malaya, Malaysia

Issues and Trends in Interdisciplinary Behavior and Social Science – Lumban Gaol et al. (Eds)
© 2018 Taylor & Francis Group, London, ISBN 978-1-138-55380-4

Committees

CONFERENCE CHAIR

Fonny Hutagalung—*University of Malaya, Malaysia*
Wayne Marr—*University of Alaska, USA*
Ford Lumban Gaol—*IEEE, IAIAI, & SERSC, Indonesia*

PROGRAM COMMITTEE CHAIR

Validova Asiya Faritovna—*Kazan Federal University, Russia*
Varlamova Julia Andreevna—*Kazan Federal University, Russia*
Larionova Natalia—*Kazan Federal University, Russia*
Malganova Irina G.—*Kazan Federal University, Russia*
Odintsova Julia L.—*Kazan Federal University, Russia*
A.J.W. Taylor—*Victoria University of Wellington, New Zealand*
Maria de Lourdes Machado-Taylor—*Taylor Center for Research in Higher Education Policies (CIPES), Portugal*
Seifedine Kadry—*Lebanese University, Labanon*

PUBLICATION CHAIR

N. Panchuhanatham—*Annamalai University, India*

PUBLICITY CHAIR

N. Panchanatham—*Annamalai University, India*
T. Ramayah—*Universiti Sains Malaysia, Malaysia*
Yousef Farhaoui—*Moulay Ismail University, Morocco*

PROGRAM COMMITTEES

Valentinas Navickas—*Kaunas University of Technology, Lithuania*
Hsin Rau—*Chung Yuan Christian University, Taiwan*
Panos M. Pardalos—*University of Florida, USA*
Siham El-Kafafi—*Manukau Institute of Technology, New Zealand*
T.C. Edwin Cheng—*The Hong Kong Polytechnic University, Hong Kong*
Baldev Raj—*Indira Gandhi Centre for Atomic Research (IGCAR), India*
Yuosre Badir—*Asian Institute of Technology, Thailand*
Maria Fekete-Farkas—*Szent István University, Hungary*
Hong Yan—*Hong Kong Polytechnic University, Hong Kong*
Qi Yu—*Rochester Institute of Technology, USA*

Tatsiana N. Rybak—*State Economic University, Republic of Belarus*
Lalit Mohan Patnaik—*Indian Institute of Science, India*
Janardan Nanda—*Indian Institute of Technology, Delhi, India*
Nazmi Sari—*University of Saskatchewan, Canada*
Md. Ghulam Murtaza—*Bangladesh Khulna University, Bangladesh*
Wan Khairuzzaman Wan Ismail—*International Business School (UTM IBS), Malaysia*
Ha In Bong—*School of Economics and Trade, Korea*
Hui Tak Kee—*National University of Singapore, Singapore*
Andrew Rosalsky—*University of Florida, USA*
Jennifer Chan Kim Lian—*University Malaysia Sabah, Malaysia*
Celso Ribeiro—*Universidade Federal Fluminense, Brazil*
Sajid Anwar—*University of the Sunshine Coast, Australia*
Pradyot Jena—*Institut für Umweltökonomik und Welthandel, Germany*
Wayne Marr—*University of Alaska, USA*
Siti Zaleha Abdul Rasid—*International Business School (UTM IBS), Malaysia*
K.L. Mak—*Hong Kong University, Hong Kong*
Maria de Lourdes Machado-Taylor—*Taylor Center for Research in Higher Education Policies (CIPES), Portugal*
A.J.W. Taylor—*Victoria University of Wellington, New Zealand*
Yuelan Chen—*Economist Consultant, Australia*
Jens Graff—*SolBridge International School of Business, Woosong Educational Foundation, South Korea*
Chandana Withana—*Charles Sturt University, Sydney, Australia*
Will Hickey—*SolBridge International School of Business, Woosong Educational Foundation, South Korea*
Constantinos J. Stefanou—*ATEI of Thessaloniki, Greece*
N. Panchanatham—*Annamalai University, India*
Binnur Yeşilyaprak—*Ankara University, Turkey*
Athanassios Vozikis—*University of Piraeus, Greece*
Marina Riga—*University of Piraeus, Greece*
Haretsebe Manwa—*North West University, South Africa*
Javier de Esteban Curiel—*Rey Juan Carlos University, Spain*
Arta Antonovica—*University Rey Juan Carlos, Spain*
George M. Korres—*University of the Aegean, Greece*
Zhou Xu—*The Hong Kong Polytechnic University, Hong Kong*
Sola Fajana—*University of Lagos, Nigeria*
Kate Daellenbach—*Victoria University of Wellington, New Zealand*
Brij Mohan—*Louisiana State University, USA*
Ian Hunt—*Flinders University, Australia*
Ramadhar Singh—*Indian Institute of Management Bangalore, India*
Raymond K.H. Chan—*City University of Hong Kong, Hong Kong*
Anek R. Sankhyan—*President Palaeo Research Society, India*
T. Wing Lo—*City University of Hong Kong, Hong Kong*
Jerzy Gołosz—*Jagiellonian University, Poland*
Leonid Perlovsky—*Air Force Research Laboratory, USA*
Sheying Chen—*Pace University, USA*
Antonio Marturano—*Catholic University of the Sacred Heart, Italy*
Eric Chui—*The University of Hong Kong, Hong Kong*
Wenceslao J. Gonzalez—*University of A Coruña, Spain*
Maduabuchi Dukor—*Nnamdi Azikiwe University, Nigeria*
Rajendra Badgaiyan—*University at Buffalo, USA*
Robert J. Taormina—*University of Macau, China*
Rabia Imran—*Dhofar University, Oman*
Validova Asiya Faritovna—*Kazan Federal University, Russia*

Varlamova Julia Andreevna—*Kazan Federal University, Russia*
Larionova Natalia—*Kazan Federal University, Russia*
Malganova Irina G.—*Kazan Federal University, Russia*
Odintsova Julia L.—*Kazan Federal University, Russia*
Chew Fong Peng—*University of Malaya, Malaysia*
Aishah Rosli—*University of Malaya, Malaysia*
Zulkifli Md Isa—*University of Malaya, Malaysia*
Kusmawati Hatta—*University of Ar-Raniry-Aceh, Indonesia*
Zulkefli Mansor—*University of Kebangsaan Malaysia, Malaysia*
Mohd Rushdan Abdul Razak—*Open University, Malaysia*

The opportunities and strengths of Thailand's MICE industry

Y. Sudharatna & P. Chetthamrongchai
Kasetsart Business School, Kasetsart University, Bangkok, Thailand

ABSTRACT: The purpose of this paper is to explore the ways in which Thailand is one of the best places to organize MICE events. In terms of academic analysis, there are several respectable reasons for why Thailand is ideal, and these are called 'opportunities'; this industry also has many strong points, which are labelled 'strengths'. A mixed method of qualitative and quantitative research is conducted. Focus groups as well as in-depth interviews with the snowball technique are applied to this research study. Additionally, the triangulation method is used to verify the qualitative data. The fact-finding is concluded, and then a questionnaire survey distributed to key informants in this industry is used to confirm the qualitative data collected once more. The findings state that Thailand's MICE industry embraces the strength of natural and cultural endowments, a high standard of service quality, outstanding infrastructure, and public and private organizational cooperation, and opportunities that support this industry are the location of the country at the centre of the Asian region, the reasonable costs and benefits for investment in events and the support of the industry's policies.

1 INTRODUCTION

This paper is presented as part of 'The Study of Thailand MICE Capability' research that was granted by the office of the National Research Council of Thailand (NRCT) and the Thailand Research Fund (TRF). The goal of the research is to clarify the capability of Thailand MICE in order to advance the industry to a higher level of competitive advantage. The research question is to what extent the country is capable in the MICE industry. However, this paper's highlight is simply on the 'strengths' and 'opportunities' of this industry.

The purpose of this paper is, therefore, to demonstrate the opportunities and strengths of Thailand's MICE (Meeting, Incentive, Convention, and Exhibition) industry, and other issues will be presented in a subsequent paper.

According to marketingoops.com, Bangkok, Thailand, is claimed to be one of the top destinations for tourists to visit in 2016 and therefore brings 21.47 million travellers to this destination and contributes to 5.45% GDP to the country. To this amount, MICE contributed more than a million travellers in 2016 as well as in 2014. The study of Frost & Sullivan suggested that Thailand's MICE industry has contributed to 0.77 percent GDP (Frost & Sullivan by TCEP, 2015).

The study suggested that the contribution of this industry should be rising up to 1.2 percent of the GDP. The gap is significant for the industry's stakeholder to fill in. Moreover, the spending of travellers of the MICE industry is quite interesting, since it claims to be 3.5 times that of travellers from the tourist industry. Hence, the opportunities and strengths of this industry will be clarified in this paper in order to reinforce the capability of this industry.

2 MICE AND THAILAND MICE INDUSTRY

'MICE' stands for 'Meeting, Incentive, Convention and Exhibition'. The industry is composed of 4 businesses altogether (shown in Figure 1), and each one is certainly different from one another in terms of structure, occurrence processes, and customers as well as impact to the country.

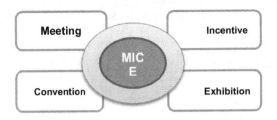

Figure 1. MICE industry structure (Sudharatna, Y, 2016).

One meeting in this research paper is the overseas private organization corporate meeting, which refers to specific activities that have been planned in advance, involving at least 10 participants and lasting at least 4 hours for a meeting. The participants for a meeting could come from any organization, any country or any continent.

Incentive: According to SITE (Society of Incentive Travel Excellence), Incentive is defined as "a global management tool that uses an exceptional travel experience to motivate and/or recognize participants for increased levels of performance in support of organizational goals." Normally, the majority of traditional incentive trips include a group of people/organizations' employees for whom an activity and entertainment programme is tailored.

A convention is defined as the international meeting of a group of people in either similar or matching professions who come to join together, aiming towards the exchange of ideas.

An exhibition, according to The Global Association of the Exhibition Industry (UFI) refers to a show that is organized to occur over a certain period of time. The main objective is to provide a platform for manufacturers or service providers to show their products and sell directly to businesses, consumers and the public in order to promote the company's products.

Thailand's MICE Industry is at the growth stage and plays an important part in the tourism scene. Additionally, it is expected to further enlarge Thailand's GDP. The country is located in central Asia, and it is the prime location for MICE activities such as corporate meeting of organizations, incentive travel which rewards the employees, conferences for some professional and academic areas and exhibition for various potential markets in the region. Although the MICE industry shares some of tourism's facilities, for example, airlines, international airports, transport infrastructure routes and networks, freight delivery stations, local transportation, accommodation, restaurant, local transportation, communication, tourist resources and others, the MICE industry is for real business trips that demand specific details for each 'M' 'I' 'C' and 'E' for taking place.

The MICE industry not only has contributed to the country in terms of economic benefit but also has an influence on the community and environment, which include employment, training and development of staff in related industries, knowledge transfer, international awareness, and creating of network opportunities. Moreover, the MICE industry also contributes to and magnifies the business ecosystem of the industry, for example, airlines, accommodation, food and beverages, sightseeing and attractions, and ground transportation. Therefore, many countries in this region are recently re-focusing on this industry, including Thailand.

3 THEORY BACKGROUND

This paper justly presents the SWOT to discover the opportunities and strengths of Thailand's MICE industry.

3.1 *SWOT matrix*

This research applied SWOT Analysis for scrutinizing the Thailand MICE Industry, which focused on the analysing of the strengths, weaknesses, opportunities and threats of the country. The strengths and weaknesses are internally focused while opportunities and threats are external considerations. This theory is argued to be best at conducting a study of a group of people with different perspectives and stakeholders of the industry. However, this paper presents merely the strengths and opportunities of the industry. Another part presenting weaknesses and threats analysis will be present in a following additional paper.

Strength, which is outlined as all the internal positive characteristics of the unit of analysis, (http://articles.bplans.com/how-to-perform-swot-analysis/) is shown in the below checklist:

- What does the country do well (in this industry)?
- What internal resources does the country have? For example: people, knowledge, skills, education, credentials, network to other countries in the region, reputation, and level of technology in the country.
- What advantages does the country have over the competitor?
- Does the country have strong research and development capabilities, manufacturing facilities?
- What other positive aspects (internal factors of the country) add value that offers competitive advantage to the country?
- What is the uniqueness or lowest cost that the country can draw upon which other countries cannot?
- What is seen as the country's strengths in terms of customers' point of view?

Opportunities, which are the external significant factors that indicate the country is likely to succeed (http://articles.bplans.com/how-to-perform-swot-analysis/) or exploit the advantage. The sample of questions illustrate as follows:

- What opportunities exist in the market or the environment that the country can benefit from?
- Is the perception of the country positive?
- What are the current ongoing trends?
- Has the recent market been in the growth stage or has there been any change in the market that creates any chances? Or, what external changes will bring you opportunities?
- Which interesting trends is the country aware of?
- Will natural causes such as weather and climatic changes give you the competitive edge?
- Is your brand name helping you to get finance easier?

4 METHODOLOGY

This research study used a mixed method of qualitative and quantitative research methodology to study the overall environment and current situations within the MICE industry through an inductive approach. Data are collected from both primary and secondary data, with the primary data gathered through focus groups (The key informants are representative from ICCA, APT Showfreight (Thailand) limited, Dusit Thani Bangkok, Bi-teach Management Limited, Kingsmen C.M.T.I. Ltd., Thai Airways International Public Company limited, TCEP, IMPACT Management Ltd., NCC Management & Development Ltd., and Bangkok Airways) and the in-depth interviews of the MICE industry's key stakeholders using the snowball method.

Triangulation is applied to verify the qualitative data. Furthermore, a weight-rate questionnaire survey is answered by some key persons in the industry and is used to confirm the analysis once more.

Moreover, for each sub-research project, the questionnaire survey is also used to verify the relevancy of qualitative research results to key industry stakeholders both from the supply and the demand sides. However, this paper reports solely the integrated project.

5 RESEARCH RESULTS

The analysis shows Thailand's MICE value chain as Figure 2. It is composed of three parties named as follows: MICE owner—the one who would like to organize 'M' 'I' 'C' or 'E'; service organizers—hotels, airlines, local transportation, venues, and other related services; and, last

Figure 2. The value chain of Thailand's MICE industry (Sudharatna, Y, 2016).

but not least, the business supporters—TECP, TICA, government, TAT, MOTS, professional associations, etc. However, this paper presents the background of the industry, showing simply the opportunities and strengths of Thailand's MICE industry.

Thailand's MICE industry strengths display the following issues:

The country has a high quality of supported facilities to nurture the MICE industry such as accommodations, venues, supported services, and airlines as well as international airports. The MICE industry shares some of these tourism facilities, therefore creating more value to be added to the country; for example, an international airport that could support the MICE industry, accommodations/hotels with high standards, and superior restaurants.

The Thai Government highlights the industry by setting up a Government Bureau for Public organization labelled Thailand Exhibition and Convention Bureau or TECB, which directly reports to the prime minister and is in charge of the overall MICE industry of the country.

The private sectors have great capabilities to attract 'M' 'I' 'C' 'E' to the country, such as the association in various area, e.g., cane and sugar, dentistry, and surgery. Additionally, organizers such as PCO (professional conference organizers), PEO (professional exhibition organizers), and DMC (destination management companies) have an important role in providing service management for the organization when companies would like to organize 'M' 'I' 'C' and 'E'.

Thailand's MICE Venue Standard is developed to certify that customers will receive an international standard (MICE Review, 2015) when events occur in the country.

The focused industry is distinguished in exhibitions. The country noticeably emphasizes 5 industries, which are food and agribusiness, automotive industries, energy, infrastructure, and healthcare and wellness. Therefore, the related stakeholders are conveniently focused.

This industry has a high rate of incorporation among shareholders/business ecosystem. The nature of the MICE industry requires a long-length plan, normally more than a year or longer. Therefore, good networking among stakeholders, the ability to have an adjustable and flexible plan of events is necessary for activities to occur.

One of the strengths of Thailand's MICE industry arises from the cost of living in the country, that is, the value for money spending, which brings about the competitive advantage to the industry.

The analysis shows that the opportunities of Thailand's MICE industry are shown below:

Thailand is located at centre of the region; consequently, it is the gateway to the ASEANs. Additionally, the growth rate of the region is high.

To hold MICE events in the country, the cost, price and benefit are reasonable and acceptable. Moreover, the ability to organize MICE activities is satisfactory to the customer, e.g., the wealthy bride and groom from India who request to celebrate their marriage in Thailand.

The country has a variety of natural and cultural resources which circuitously support the MICE industry, such as the dressing style among sub-Thai groups, well-known Thai food, service mind and others' Thai-ness. These bits and pieces could facilitate 'Incentive' as well as attract pre- and post- 'M' 'C' and 'E'.

The Thai government supports both directly and indirectly this industry. For example, there is a clear policy for enlarging the competitiveness of the industry. The National Economic and Social Development Plan is unblemished to support the MICE industry.

The global trend of Asia is raising, and, therefore, so is the growth of MICE industry.

The study shows that Thailand's MICE industry holds strengths and opportunities that could support the industry to an higher growth level; however, this study finds that there are some weaknesses and threats that need to be simplified in another paper in order to enhance the industry.

6 DISCUSSIONS

The study shows that Thailand's MICE industry has strengths in various ranges: a government organization, TCEP, that directly supports the industry named, supportive government policies, MICE venue standards, the network of the MICE business ecosystem, and the value of money spending. Meanwhile, the opportunities supporting Thailand are providential to some extent, for example, location of the country, the costs and benefits of hosting an event in Thailand being reasonable and acceptable, the variety of cultural and natural resources, and the uniqueness of the country.

Although the strengths give a competitive advantage to the industry itself and to the country, the weaknesses must be keep in mind, such as data to support the industry, the long-term policy for convention events, the positioning of the country when competing with others, and the difference in understanding issues between public and private sector. Moreover, other threats that need to be brought into consideration include anxiety about political issues in the past period and some regulations that are not as easy as some other countries.

7 CONCLUSIONS

This research finds that Thailand's MICE industry has many strengths and opportunities to boost up to the target at which it is aimed. Additionally, this research is congruent with the study of Dwyer and Mistilis (1999), who suggested that to strengthen this industry, the country should be committed to the cooperation between stakeholders in the industry or good networking with the business ecosystem. Moreover, the government needs to support either some policy or infrastructure to the industry. Additionally, public and private sectors have to work together to facilitate the industry. Further, the MICE service standard should be developed. Last but not least, the training of staff in the industry must be strengthened. Therefore, having confidence in this research, Thailand is on the right path to strengthening the Thailand MICE industry, and those opportunities are supported as claimed.

8 IMPLICATIONS FOR THIS RESEARCH

Thailand's MICE industry embraces some strengths – natural and cultural endowments, high quality of service standards, outstanding infrastructure, public and private organizational cooperation – and opportunities – its location at the centre of the region, reasonable costs and benefits for investment in events and strong support of the industry's policies.

Although the country seems to be favourable to having those physical appearances, the weaknesses of the industry also need to be clarified and corrected. Studies focus on both positive and negative aspects of the industry; nonetheless, only the positive aspects are presented in this paper. Additionally, this research gives the background of this industry from an academic point of view. This approach could help encourage any interested person to join the industry either in terms of supply side (chance to join the business) or demand side (organize the MICE activities in Thailand).

ACKNOWLEDGEMENT

The authors would like to give sincere thanks to financial support from the office of the National Research Council of Thailand (NRCT) and the Thailand Research Fund (TRF).

REFERENCES

Bauer, T.G.; Lambert, J. and Hutchison, J (2001) Government Intervention in the Australasian Meetings, Incentives, Conventions and Exhibitions Industry (MICE), Journal of Convention and Exhibition, Vol. 3(1).

Cameron, R. (2009). MICE is not about tourism. *TTG mice Magazine*. Retrieved 3 July 2011 from http://ttgmice.com.

Dwyer, Larry and Mistilis, Nina (1999) Development of MICE Tourism in Australia: Opportunities and Challenges, Journal of Convention & Exhibition Management, Vol. 1(4).

Frost & Sullivan by TCEP, 2015. Study on the Economic Impact of MICE in Thailand.

http://articles.bplans.com/how-to-perform-swot-analysis/ BY TIM BERRY "What Is a SWOT Analysis?" 20 Feb 2017.

http://pestleanalysis.com/swot-analysis-questions/.

http://www.mots.go.th/ewt_dl_link.php?nid=7533.

http://www.seebtm.com/definition-of-the-incentive-travel/?lang=en.

https://www.marketingoops.com/news/brand-move/mastercard-global-destination-cities-index-2016/.

https:www.mintools.com/pages/article/newtmc)05.htm. "SWOT Analysis Discover New Opportunities, Manage and Eliminate Threats" by the Mini Tool Editorial Team 20 Feb 2017.

Porter M. E. (1990). The Competitive Advantage of Nations. New York: Free Press.

SITE International Foundation. (2010). 2010 Study of the German Incentive & Motivational Travel Market. Site Germany.

Sudharatna, Y. et al. (2016). The Study of Thailand MICE Capability. The Research Report of the National Research Council of Thailand (NRCT) and the Thailand Research Fund (TRF).

TCEB Annual Report 2007.

Thailand's MICE Industry Report: The gave way to ASEAN's MICE (2014). Thailand Convention & Exhibition Bureau (TECB).

Thailand Convention & Exhibition Bureau (Public Organization). (2015). Thailand's MICE Industry Report, 2015; MICE Review 2015.

Evaluation of the differentiating factors in the avocado value chain

J.A. Martínez-Arroyo & M.A. Valenzo-Jiménez
Universidad Michoacana de San Nicolás de Hidalgo, Morelia, Michoacán, México

ABSTRACT: The relevance of this research lies in evaluating environmental management and innovation as sources of competitive advantages in the value chain of avocado exporting companies located in Uruapan, Michoacán México. It was applied to the nursery producers and packers through the application of a questionnaire with 70 questions fully measuring the model of variables this analysis allows to know the origin of the competitive advantages. The hypotheses were verified. The Spearman correlation shows that innovation has a greater relationship with the dependent variable and in the indicators, the process innovation is the one of greater relation, the environmental management the evidences show very little relation between the dimensions and indicators with the dependent variable, Inferring that the actions that the members of the value chain carry out to preserve the environment are few or none. It is important to mention that the collection of information was hampered by the environment of insecurity that exists in this region.

Keywords: Value chain, competitive advantage, environmental management, innovation

1 INTRODUCTION

This research work is aimed at analyzing the actions in environmental management and innovation carried out by the organizations that are part of the chain of the avocado value chain, located in Uruapan, Michoacán México, activities that are carried out with the purpose of adding value to the Final product, with the aim of better competitive advantage, and the research goals are to identify the sources of competitive advantages in environmental management and innovation activities in the avocado value chain.

Today the business environment is more competitive and hostile, so that each competitive advantage of the company will be quickly eroded and overcome by the rapid pace of competition (Grimm, Lee, & Smith, 2006). Markets are in a constant state imbalance ie the current business environment is far from stable and predictable. However, currently most emerging economies have a comparative advantage in the supply of labor and land and the exploitation of certain natural resources and climatic advantages over the more developed countries.

A widely competitive sector in Mexico is agroindustrial, with emphasis on avocado, which has a high demand in the domestic and international market mainly in the North American market. The development of the avocado industry in Mexico in recent years has increased significantly and with great development opportunities especially with the diversification of markets and final presentation of the product. Mexico is the main producer, exporter and consumer of avocado in the world, with a production of more than one million tons per year and produces 42 percent of the avocado that is cultivated worldwide. Michoacán is the first Mexican state with approximately 83% of the total national production. In addition, since 2008, the export exceeded 200 thousand tons of fresh avocado (Agropecuaria, 2008), and in the next period 2009 exported more than 300 thousand tons to the United States, by 2016 avocado exports were 779.613 thousand tons (HassAvocadoBoard.com, 2017) the main Mexican avocado-consuming countries are the United States, Japan and Canada. There are currently 20 avocado-producing municipalities in Michoacán corresponds to the municipalities of Tancítaro, Uruapan, Peribán, Ario de Rosales, Tacámbaro, New Parangaricutiro and Salvador Escalante, This research was carried out mainly in the municipalities of Uruapan, San Juan Nuevo and Tancitaro Michoacán, with nurseries, producers and packers installed in these municipalities.

In studies carried out by Bonales and Sánchez, the michoacano avocado sector is characterized by its poor organization (Bonales & Sánchez, 2003), this being one of its main weaknesses, it is also undeniable that some comparative advantages translated into competitive advantages of avocado producers and exporters have been emulated and sometimes improved by some producing countries and competitors in the international markets of avocado (Sánchez, 2007). For example, according to the avocado industry in Israel and according to the information presented at the World Congress held in Viña del Mar, Chile on November 13, 2007, this producing country is expected to achieve yields of almost 20 tons per hectare for the following years (Naamani, 2007), while avocado producers in Michoacán barely manage to produce ten tons per hectare. In analyzing in detail the successful export performance of Mexican avocado producers, we find weaknesses in relation to other foreign competitors, especially in the North American market, especially they are more notable in terms of technological development and the technification of production. Likewise, competition for the US market has increased with the increase in the volume of avocado introduced in this market in countries such as Peru, Colombia, New Zealand and the Dominican Republic.

2 THEORETICAL FRAME

2.1 *Value chain*

The analysis of the value chain is a method used to decompose the chain in each of the activities that comprise it, and in which the activities that add value to the final product for the client are sought. The value chain approach analyzes the particularities between the different links that make it up and its purpose is to know the factors that are affecting the competitive advantages, assessing their relative incidence, in order to be able to define priorities and strategies of action agreed between the different actors. It is therefore very important to consider the identification of the bases of the competitive advantages of the avocado value chain that facilitate the performance of the different economic agents. It is not enough that a link in the chain reaches the desired competitiveness, since it is required that the whole chain or system achieves it (Venegas & Loredo, 2008) in other words, value chain analysis is essentially a value creation system, it is an analytical tool that facilitates the identification and evaluation of strategic alternatives (Walters & Rainbird, 2007). Likewise, the value chain is an important unit of analysis to understand the competitive advantages of the company (Nations, United, 2007).

2.2 *Temporary competitive advantage*

Competitive advantage in the literature is synonymous of value creation (Rumelt, 2003). The competitive advantage that some companies have achieved through the adoption of the strategy has its beginnings in the basic concept of the end of 1930, called "competitive adaptation" (Alderson, 1937) in which intellectual activities and relations with suppliers are the main sources of competitive advantage. This is one of the first literature on competition in which competitive adaptation is the specialization of suppliers. The competitive advantage is "sustained profitability" above normal (Peteraf M. A., 1993). Therefore, competitive advantage is not something that is "got", but is "reached"; is not simply something that makes us different from the competition, but to obtain a higher profitability and it can be created in many ways, for example, by size of company, location, access to resources (Ghemawat, 1986). In other words, competitive advantage can be created by combining the resources available to the company, with an adequate strategy in which these resources are involved, the knowledge and skills of the owners and workers, as well as the opportunities of the environment. (Morales & Pech, 2000).

2.3 Sustained competitive advantage

Understanding the sources of sustained competitive advantage for companies has become the largest area of research in the field of strategic management (Porter, 1985). In such a way that the sustained competitive advantage almost all the organizations look for and try to develop it (Cheney & Jarrett, 2002). Y is defined by Bar-Eli, Galily & Israeli, (2008) as "that which the competition cannot copy or simulate". In addition, they must possess four attributes: rarity, value, inability to be imitated, and their inability to be replaced. The company's resources from this perspective include: all the assets, capabilities, organizational processes, company attributes, information, knowledge, controlled by the company that allow the organization to implement strategies to make it more efficient (Daft, 1983). However, not all resources have the potential to create sustained competitive advantage. The dynamic nature of the business environment, especially in relation to the influence of competitors, customers, regulation, technology and the supply of financing, is such that the achievement of competitive advantage is somewhat dynamic. Holding a competitive advantage permanently is very difficult, particularly in the era of uncertainty, the crisis and the impact of the Internet on consumer behavior and transaction capabilities.

2.4 Environmental management

For organizations of any commercial activity it is very important to measure and disclose their environmental performance is substantial for firms, because if a company performs well in this area, or if it does better things than its competitors, this can be a source of competitive advantage. On the other hand, if a company does not meet the regulatory concerns that the market and the public demand, it could "affect" it's financial result (Montabon, Sroufe, & Narasimhan, 2009). The degradation of the environment now reaches unprecedented levels in history and poverty affects a significant part of the world's population. It is not surprising, therefore, that these problems are today among the major challenges facing the international community (Tortella, 2006). Contemporary degradation has reached such a magnitude that even the most serious studies warn that if the future generations are to continue, the future may be threatened (Comisión Mundial del Medio Ambiente y del Desarrollo, 1992).

Both the high-consumption societies of developed countries and the poverty of developing countries they are a threat to the environment. As a result of growing pressure from non-governmental organizations, local, national and international authorities, consumers, competitors and other stakeholders, the environmental responsibility of a company has acquired unprecedented importance. Environmental management is defined as "the techniques, policies and procedures a firm uses that are specifically aimed at monitoring and controlling the impact of its operations on the natural environment (Montabon, Sroufe, & Narasimhan, 2007). An EM can be a powerful tool for organizations to improve their environmental performance and enhance their business efficiency (Chavan, 2005).

This trend has led an increasing number of companies to introduce the dimension of environmental care into their competitive strategy. This is a reason of enormous weight and truth, due to the fact that environmental issues are important today and clearly representative of the current conditions of business and in the future (Berchicci & King, 2007). Some research has argued that a proactive environmental strategy creates entry barriers for competition and is a source of competitive advantages in the markets (Aragon-Correa & Sharma, 2003).

2.5 Innovation

Technological changes increase the need for companies to innovate (Brown & Eisenhardt, 1995), companies that generate innovations will be more successful in responding to changes in the environment and developing new capabilities to achieve better performance (Montes, Moreno, & Fernández, 2004). Innnovation is "the intersection of invention and insight, led to the creation of economic and social value" (Competitiveness, 2005). Innovation plays a more crucial role in today's global competition, and is the main source of competitiveness of a country. That is, it has become the biggest differentiator in the competitive race (Roberts, 2007). One of the sources of innovation is knowledge, which is a strategic asset that enables companies to maintain distinctive competencies and discover innovation opportunities (Chen & Lin, 2004). Innovation is generated in an interactive process in the exchange, absorption and assimilation of knowledge shared in a social and physical context (Autio, Hameri, & Vuola, 2004), it is obvious that organizations learn and create innovations through sharing and combining knowledge (Kogut & Zander, 1992). Thus, knowledge sharing contributes to innovation because it generates a collective knowledge and generates beneficial synergies, which improves the stock of knowledge available in the company (Nonaka & Takeuchi, 1995). In addition, successful innovation requires a combination and association of recent knowledge and existing knowledge. Innovation is a social process where strategic choices are not simple as it involves the exercise of control of knowledge communication (Scarbrough, 1995). In such a way, that the acquired knowledge, will allow the individuals then to respond to the demands of the environment with new and innovative performances (Wang, Wang, & Horng, 2010). At the same time the literature mentions that the innovation strategy helps companies in three ways: Provides new offers or experiences that stimulate the customer, stay at the forefront of competition in the market and enter new market segments or the creation of new business (Anthony, Johnson, & Eyring, 2004). Continuous innovation is mentioned as the only endogenous factor that offers the possibility for all companies to obtain and maintain a sustained competitive advantage, even in a resource-constrained environment (Ren, Xie, & Krabbendam, 2010). Therefore innovation is a constant process of search and exploration. However, most of the researchers have focused on the technological and technical innovations that correspond to the innovations of products and processes (Lundvall, 1992). In the same way, innovation is an interactive process of generation, diffusion and use of knowledge (Bottazzi & Peri, 2003). There is no doubt that innovation plays a key role in gaining crucial competitive advantages (Becheikh, Landry, & Amara, 2006). In addition, the most innovative and successful companies have adapted their approach to innovation management due to changes in the external environment in an attempt to protect or foster their competitive advantage (Ortt & Smits, 2006).

3 METHODOLOGY

This article comes from scientific research and has a correlational cross, inductive and deductive qualitative and quantitative descriptive-design hypothetical, since, described the object of study and second because it determines the relationship with the independents variables with the dependent variable Competitive advantages in the value chain, a questionnaire was applied to the three elements of the value chain, nurserymen, producers and packers of avocado which contains 70 questions fully measuring the model of independent variables, dimensions and indicators.

3.1 *General research questions*

What is the impact of environmental management and innovation on the competitive advantage of the avocado value chain located in Uruapan, Michoacán?

3.2 *General objective*

Determine the impact of environmental management and innovation on the competitive advantage of the avocado value chain located in Uruapan, Michoacán.

3.3 *General hypothesis*

Environmental management and innovation are the variables that influence the competitive advantage of the avocado value chain located in Uruapan, Michoacán.

3.4 Sample

In this scientific research the sample of study is formed by the links of the value chain, a questionnaire was applied to 57 nurseries, 316 producers and 26 packing companies, represented by the owners, managers, administrators or the head of production of the export companies of avocado located in Uruapan, Michoacán. The representative sample was calculated with a confidence level of 95% and a maximum error level of 5%.

4 STUDY RESULTS

The questionnaire is reliable when measured with the same precision, it gives the same results in successive applications made in similar situations (Santillana, 1998). The results are shown in Table 1.

This table shows the Cronbach's alpha of questionnaires used in this research, reliability, which is inferred to be reliable instruments and stable in their training and application containing relative absence of measurement errors, this is more explicitly expressed in percentages observed, so that this measurement may show a precision, homogeneity and internal consistency of the measurement instrument used.

Table 1. Reliability.

Element	Questionnaire	V.I Innovation	V.I Environmental management
Alfa cronbach	0.962	0.942	0.940

Source: Research data.

Table 2. Spearman correlation measurement.

Dependent variable	Independent variable	Dimensions	Indicators		
Competitive advantage in the value chain	Environmental Management 0.685 0.469	Supplies 0.432 0.214	Requirement	0.580	0.334
		Legislation 0.514 0.276	Obligations	0.356	0.126
			Policies	0.345	0.119
			Updating of personnel	0.317	0.100
			Benefits	0.278	0.077
			Training	0.345	0.119
		Processes 0.576 0.298	Evaluation	0.297	0.088
			Care policies	0.307	0.094
			Planning	0.276	0.076
	Innovation 0.662 0.438	Product 0.456 0.207	1-Frequency innovations	0.376	0.141
			2-Frequency propose	0.398	0.158
		Process 0.495 0.245	3-Most frequent type	0.345	0.119
			4-Frequency invested	0.374	0.139
		Marketing 0.568 0.322	5-How often E. M.	0.207	0.042
			6-Considers innovation Advantage	0.398	0.158
		Organization 0.501 0.251	7-Innovation is planned	0.234	0.054
		Design 0.614 0.316			

Source: Own elaboration.

Table 3. Hypothesis testing.

Variable	Test	Significance	Decision
Environmental management	Independent-Samples Kruskal-Wallis test	0.314	Retain the null hypothesis.
Innovation	Independent-Samples Kruskal-Wallis test	0.179	Retain the null hypothesis.

Asymptotic significances are displayed. The significance level es .05.

Source: Measurement data.

4.1 Spearman correlation index

This calculation serves to determine the relationship between variables and how strong is.

In this table the total measurement of the research model concerning the correlation of Spearman, in black and the coefficient of determination shown in red, this allows to observe the impact of each indicator, dimensions and independent variables on the dependent variable, determining the origin of the greatest impact and thus more accurately identify the sources of competitive advantages in the value chain.

Table 2 shows the measurement of the correlation of Spearman (black numbers) and the coefficient of determination (red numbers), so that according to these results show the poor relation between the indicators and dimensions as well as the independent variable with the dependent variable, being inferred that the actions or activities of innovation and environmental management that the Agents of the value chain of avocado in Michoacán, resulting in the low possibility that these activities are sources of competitive advantage for these organizations, causing weakness for them mainly in the image of companies or organizations with environmental responsibility.

4.2 Hypothesis verification

Hypothesis testing was performed for each of the independent variables, Environmental Management and Innovation with the SPSS software for non-parametric ordinal variables and Independent samples, using the test Kruskal-Wallis and the results indicate retaining the null hypothesis, therefore the hypothesis raised is rejected the results are shown in Table 3.

5 DISCUSSION

Avocado producing organizations face massive internal and external competition and compete for a better position in the market and for survival, employ known techniques, routines are time-honored with habits based on experience. Decision making comes from processes based on past experiences that gives work, trust and instinct. There is a substantial body of knowledge regarding the concept of innovation. Innovation and environmental management are considered in current literature as sources of competitive advantage. It is important for companies in the avocado value chain to infer that innovation and environmental management must be successfully managed at a time when these strategies are required to survive and compete. It is clear that companies have different capacities to manage and generate innovations, which are transformed into the delivery of benefits to customers, and companies in the chain intuit that when you innovate a product or service, you need to have clear knowledge Of what the client wants, is interested and what he expects, so that mutual benefits are obtained. The results show clear evidence of the lack of interest of these organizations to promote environmental management and innovation as two pillars of competitive advantage, weakening in a way these organizations since every day consumers are more demanding in their tastes and needs and the Care for the environment is a concern that is gaining momentum among consumers.

6 CONCLUSIONS

After observing the results can be interpreted as a lack of interest in the implementation of activities such as innovation and environmental management that would help in building the competitive advantage of the elements of the value chain, nurseries, producers and packers. However, when analyzing the information in a particular way, it is the link of the packers who carry out more activities aimed at taking care of the environment as well as encouraging innovation among their employees, thereby establishing a clear differentiation between the three elements, this link is the strongest of the three in relation to capital, infrastructure, technology and the preparation of its leaders. Likewise it is important for each one of the links in the value chain to improve its competitiveness through the implementation of innovation and environmental management, to face growing competition in the North American market.

Future lines of research are related to the creation and dissemination of knowledge as well as quality and economies of scale.

REFERENCES

Agropecuaria, I. (12 de Febrero de 2008). *iMAGEN Agropecuaria.Com*. Recuperado el Marzo de 2008.
Alderson, W. (1 de January de 1937). A marketing view of competition. *Journal of Marketing*.
Anthony, S., Johnson, M., & Eyring, M. (2004). A Diagnostic for Disruptive Innovation. *Harvard Business School-Working Knowledge*.
APEEAM. (28 de Enero de 2006). Marca referente mundial. *Cambio de Michocán*.
Aragon-Correa, J., & Sharma, S. (2003). A contingent resource-based view of proactive corporate environmental strategy. *Academy of Management Review, 28* (1), 71–88.
Arellano, R. (6 de Junio de 2014). Sufre "oro verde revés". *La Voz de Michoacán*, pág. 3.
Autio, E., Hameri, A. p., & Vuola, O. (2004). A framework of industrial knowledge spillovers in big-science centers. *Research Policy, 33*(1), 107–26.
Becheikh, N., Landry, R., & Amara, N. (2006). Lessons from innovation empirical studies in the manufacturing sector: a systematic review of the literature from 1993–2003. *Technovation, 26*, 644–64.
Berchicci, L., & King, A. (2007). Postcards from the Edge: A Review of the Business and Environment Literature. *ERIM REPORT SERIES RESEARCH IN MANAGEMENT*, 60.
Bonales, J., & Sánchez, M. (2003). *Competitividad de las empresas exportadoras de aguacate*. Morelia, MICHOACÁN, MÉXICO: UMSNH.
Bottazzi, L., & Peri, G. (2003). "Innovation and spillovers in regions: evidence from European patent data". *European Economic Review, Vol. 47* (4), 687–710.
Brown, S. L., & Eisenhardt, K. M. (1995). Product development: past research, present findings. *Academy of Management Review, 20*(2), 343–78.
Chen, C., & Lin, B. (2004). The effects of environment, knowledge attribute organizational climate, and firm characteristics on knowledge sourcing decisions. *R&D Management, 34*(2), 137–46.
Cheney, S., & Jarrett, L. (2002). Up-front excellence for sustainable competitive advantage. (E. Host, Ed.) *Trainning and Development*, 4.
Comisión Mundial del Medio Ambiente y del Desarrollo. (1992). *Nuestro Futuro Común*. Madrid: Alianza S.A.
Competitiveness, C.o. (2005). *National Innovation Initiative*. Washington, DC: Council of Competitiveness.
Daft, R. (1983). *Organization theory and design*. New York: West.
Grimm, C., Lee, H., & Smith, K. (2006). *Strategy as action: Competitive Dynamics and Competitive Advantage*. Oxford New York: Oxford University Press Inc.
HassAvocadoBoard.com. (13 de Marzo de 2017). *Avocado Shipment Volume Data*. Obtenido de Avocado Shipment Volume Data: https://www.hassavocadoboard.com/shipment-data/historical-shipment-volume/2013.
Hernandez Sampieri, R., & Fernandez, C. (2010). *Metodología de la investigación*. México: Mc Graw Hill.
Kogut, B., & Zander, U. (1992). Knowledge of the firm, combinative capabilities, and the replication of technology. *Organization Science, 3*(3), 383–398.
Lundvall, B. (1992). National Systems of Innovation:Towards a Theory of Innovation and Interactive Learning. *Frances Printer*.

Montabon, F., Sroufe, R., & Narasimhan, R. (2009). An Examination of Corporate Reporting, Environmental Management Practices and Firm Performance. *Journal of Operations Management, 25*(5), 998–1014.

Montes, F. J., Moreno, A. R., & Fernández, L. M. (2004). Assessing the organizational climate and contractual relationship for perceptions of support for innovation. *International Journal of Manpower, 25*(2), 167–80.

Morales, M., & Pech, J. L. (Abril-Junio de 2000). Competitividad y estrategia: el enfoque de las competencias esenciales y el enfoque basado en los recursos. *Revista de Contaduría y Administración* (197), 48–50.

Naamani, G. (2007). Agrexco Tel-Aviv, Israel.

Nations, United. (19 de February de 2007). Global value chains for building national productive capacities. *Trade And Development Board*, 4.

Nonaka, I., & Takeuchi, H. (1995). *The knowledge-creating company; how Japanese companies create the dynamics of innovation.* New York: Oxford University Press.

Ortt, J., & Smits, R. (2006). Innovation management: different approaches to cope with the same trend. *International Journal of Technology Management, 34,* 296–318.

Peteraf, M. A. (1993). "The cornerstones of competitive advantage: A resource-based view. *Strategic Management Journal, 14,* 179–191.

Ren, L., Xie, G., & Krabbendam, K. (2010). Sustainable competitive advantage and marketing innovation within firms A pragmatic approach for Chinese firms. *Management Research Review, 33*(1), 79–89.

Roberts, E. B. (2007). Managing invention and innovation. *ResearchTechnology Management, 49*(1), 35–54.

Rumelt, R. P. (2003). What in the World is Competitive Advantage? *Policy Working Paper*, 5.

Sánchez, G. (2007). *El Cluster del Aguacate en Michoacán.* Uruapan, Michoacán, México: Fundación Produce Michoacán.

Santillana. (1998). *Diccionario de las Ciencias de la Educación* (Undécima edición ed.). México|: Editoreal Santillana.

Scarbrough, H. (1995). Blackboxes. Hostages and Prisoners. *Organization Studies*, 991–1019.

Tortella, G. (2006). *Los orígenes del siglo XXI: Un ensayo de la historia social y económica contemporánea* (2da. ed.). Madrid: Gadir l.

Venegas, B., & Loredo,. N. (2008). El empleo de la cadena de valor en la búsqueda de la competitividad. Recuperado el 2008

Walters, D., & Rainbird, M. (2007). Cooperative innovation: a value chain approach. *Journal of Enterprise Information Management, 20*(5), 595–607.

Wang, Y.-L., Wang, Y.-D., & Horng, R.-Y. (2010). Learning and innovation in small and medium enterprises. (E. G. Limited, Ed.) *Industrial Management & Data Systems, 110* (2), 175–192.

Competitiveness in the handicraft sector in Mexico

M.A. Valenzo-Jiménez & J.A. Martínez-Arroyo
Universidad Michoacana de San Nicolás de Hidalgo, Morelia, Michoacán, México

L.A. Villa-Hernández
The Morelia Institute of Technology, Morelia, Michoacán, México

ABSTRACT: In this study, we investigate the competitiveness in the handicraft sector in Michoacán, Mexico. To determine the competitiveness levels of this Mexican group, we used independent variables, namely quality, the pricing strategy, the technology used, logistics, and marketing strategies, while the dependent variable used was competitiveness. The methodology that we used consisted of a survey which was divided into three sections: the first section considers age, gender, and type of products; the second section includes a SWOT analysis; and the last section consists of 57 Likert-type items. We included 155 handicraftsmen for the study, in which the fieldwork processing showed that Cronbach's alpha reached up to a value of 0.896, indicating high levels of reliability. The results obtained was located in the range of "regular competitiveness", which means that the handicraftsmen analyzed reached only up to 62% of the expected competitiveness levels.

Keywords: Competitiveness, Handicraft, Mexico, and SWOT analysis

1 INTRODUCTION

Competitiveness of a company shows its competitive advantage; therefore, competitiveness, as perceived by customers, most likely indicates customer loyalty, far beyond traditional isolated service/satisfaction quality measures (Baumann, Hoadley, Hamin, & Nugraha, 2017).

At present, the handicraft sector is considered privileged, which represents an element of collective identity, as well as one of the ways of conserving cultural heritage. Moreover, it contributes significantly to the local, regional, and national economies. In Mexico, 250 micro-regions are dedicated to producing handicrafts, employing approximately 9 million people (Duarte, 2012).

Throughout the world, production and marketing of crafts are crucial to the local and national economies and represent a group, a place, or a country (MacDowell & Avery, 2006).

Mexican handicrafts are valued and recognized potential attractions in the international market. Also, they are highly demanded in countries like Spain, Canada, the United States, Colombia, Germany, Italy, and Australia (PROMEXICO, 2011).

Therefore, it is unsurprising that Mexico has the potential to compete in the international market; however, factors such as decreasing sales due to the economic crisis, insecurity and therefore lack of tourism, as well as the presence of Chinese products have affected this sector (Gómez Prado, Romo de Vivar y Mercadillo, Dávila Vázquez, Arellano Espinoza, & Vega Cano, 2010).

1.1 Problematic situation

In Latin America, Peru, Mexico, Colombia, Honduras, Guatemala, Bolivia, and El Salvador are the main exporters of handicrafts. Market experts have identified Mexico, Peru, Columbia, Honduras, and Guatemala as countries with particularly high export capacity on the basis of the number of exporters able to handle large order volumes in their own facilities or through subcontracting to smaller producers (Barber & Krivoshlykova, 2006).

Handicrafts in Mexico, like in many developing countries, represent a lifestyle, because artisans dedicate long workdays to each piece. Besides, the profits are only sufficient to meet the basic needs of their families (housing, education, and clothing). Nonetheless, the handicraft sector is one of the most important in Mexico and plays a significant role in the economy and in the employment of many artisans of the country (Duarte, 2012).

According to INEGI, there are more than 3 million business units (where the artisanal economic units are inserted), of which 99.7% are MSMEs (micro, small, and medium-sized enterprises): 95.7% are micro, 3.1% small, and 0.9% medium-sized enterprises. The number of workers employed in these organizations range from 1 to 500, contributing to 42% of the GDP and 64% of the national employment (Duarte, 2012).

Another problem of artisans is exposed by Suzuki (2005), who states that artisans tend to be conservative toward the adoption of new technologies, which would allow them to produce the same products with greater efficiency and higher quality.

Despite the economic role of MSMEs, the handicraft sector has suffered lower sales volume, growth, and development in the last years. Artisanal work generates employment and income for artisans living in rural areas, and hence, it is necessary to seek ways to increase the competitiveness of the handicraft sector. Thus, the following research question arises: What are the main factors that affect the competitiveness of the handicraft sector in Mexico?

1.2 Literature review

In general, "competitiveness" is used by the government, enterprises, and the media as an ambiguous concept. However, even among researchers, there is a lack of consensus on the concept of competitiveness, which has led scholars to approach this concept from different theoretical perspectives (Valenzo, Martinez, & Bonales, 2010).

Ambastha and Momaya (2004) define competitiveness as the ability to compete, i.e., the ability to design, produce, and offer products superior to those offered by competitors, with regard to price. Therefore, for a customer, a competitive organization is the one that can deliver better value compared to its competitors, achieving lower prices with benefits equivalent to or higher than those of its competitors.

In the literature review, some models were identified to measure the level of competitiveness in the handicraft sector.

The Spanish government identified that business competitiveness is a multidimensional and dynamic concept that refers to a company's ability to maintain and increase its market share and that it is closely linked to its competitive advantage, a concept that is specified in costs lower than their competitors or product differentiation (Gobierno de España, 2009). It also states that, a company needs to excel in the following six key indicators in order to be competitive in the craft sector: (1) human resources; (2) incorporation of ICT; (3) innovation; (4) improvement of marketing channels and internationalization; (5) financing; and (6) cooperation.

Dominguez, Hernandez, and Toledo (2004) observed that competitiveness is increased with the innovation of artisans, the generous environment, and the strategies of differentiation. Innovation influences the perception of generous environments to generate competitiveness. In a generous environment, safety is perceived in executing the operations undertaken by the artisan. This situation favors the possibilities of business growth because it increases the level of sales revenue. If the artisan has a presence in the market with improved products or new products, with quality, with price strategies based on product innovations, then the technological change in craftsmanship would allow him to survive in any environment, whether generous, dynamic, or complex.

Several authors have studied the handicraft sector in Mexico. For example, Sánchez-Medina, Corbett, and Toledo-López (2011), focusing on the relationship between environmental innovation and sustainability, analyzed 168 Mexican handicraft businesses in Oaxaca, Puebla, and Tlaxcala, and Bouziane & Hassan (2016) considered the handicraft industry as an important asset to support and help develop the tourism sector of a country, taking Algeria as an example. Such conventional products appear as part of the tourism industry of any country.

The results obtained by Venkataramanaiah and Ganesh-Kumar (2011) show a direct, positive relationship between environmental innovation and sustainability in three dimensions: economic, social, and environmental. In this study, we present various drivers affecting the competitiveness of handicrafts manufacturing units of a cluster through empirical study. The recommendations were categorized into the following four major areas: productivity improvement, financial resources, marketing, and logistics and support services.

Programs such as AL-INVEST IV support small and medium-sized enterprises (SMEs) in Latin America, where they seek to improve the competitiveness and internationalization of companies by promoting innovation, knowledge, and economic relations with their European counterparts during 2009–2013 that have benefited 648 SME's of handicrafts as well as decoration and gift products in Mexico and Cuba (AL-INVEST, 2014).

1.3 *Quality*

Parking (2004) defines the quality of a product as the physical attributes that make it different from the products of other companies. The quality includes design, quality in the service provided to the buyer, and the ease of access of the buyer to the product. Quality can be measured in a spectrum that goes from high to low.

Gutiérrez (1997) affirms that the quality of a product or service is the coefficient of perception of customers in which that good meets their expectations, as well as indicates that quality is determined by the performance of the product, features attractive to the customer, reliability, durability, appearance, safety, and service.

1.3.1 *Price*

Price is the medium of exchange to acquire a good or a service. It is typical that the price is the money exchanged. Moreover, price can be related to anything with perceived value, not just money. When goods and services are exchanged, the deal is called barter (Lamb, Hair, & Mc Daniel, 2005).

1.3.2 *Technology*

Morita and SONY (1990) define technology as an organized knowledge for production purposes, which is incorporated into the workforce (skills).

1.3.3 *Logistics*

Lambert (2007) states that logistics is the process of effectively and efficiently planning, implementing, and controlling the flow and storage of raw materials, products in process, and finished products—with related information—from the point of origin to the point of consumption in order to adapt to the client's requirements.

1.3.4 *Marketing*

Lamb, Hair, & Mc Daniel (2005) define marketing is the process of planning and executing the conception, pricing, promotion, and distribution of ideas, goods, and services to create exchanges that meet individual and organizational goals. They also define the promotional mix as the combination of promotional tools, including personal sales, advertising, public relations, and sales promotion, used to reach the target market and meet the overall goals of the organization.

Table 1. Competitiveness in the handicraft sector.

Independent variable	Dimensions	Indicators
Quality	Product	Quality Characteristics Design Innovation
	Service	Tangibles Reliability Answer time Security Empathy
Price	Quality control systems	ISO 9000 norms Quality assurance
	Pricing strategies	Pricing oriented with utilities Pricing with sales orientation
	Pricing strategies	Pricing status quo Pricing with surcharge Price fixing with break-even point
	Government support	Financial sources Subsidies Promotion and dissemination Training
Technology	Research and technology development	Machinery and equipment Production logistics
Logistics	Logistic chain	Distribution logistics (wholesalers) Commercial distribution logistics
Marketing	Mix marketing	Personal sales Advertising Public relations Promotion
Dependent variable	Dimensions	Indicators
Competitiveness	Sales	Sales participation Fulfillment of planned objectives in the last 3 years Sales performance Relationship of sales with economic conditions
	Competence	Quality with respect to competition Price of the competition Technology with respect to competition Logistics with respect to competition Marketing with respect to competition

Source: Own elaboration based on the literature review.

Furthermore, Noîlla (2007) exposes that marketing must have proper training needs to be given in marketing techniques. Packaging, labeling, sizing, presentation, and color schemes are a few topics that need to be discussed with the craftspeople.

After reviewing some theoretical references of the craft sector, the following model is proposed to carry out this investigation:

Table 1 presents the research model used to measure competitiveness in the artisanal sector in Mexico and we can show how each of the independent and dependent variables is shaped and structured with their respective dimensions and indicators.

On the basis of the above data, the research question and the general objective that are addressed in this study are the following:

How are quality, price, technology, logistics, and marketing associated with the competitiveness of the artisanal sector of Paracho, Michoacán?

The general objective is to determine how quality is associated with handicrafts, proper pricing, a technology used in the manufacture of handicrafts, logistics, and an adjusted marketing strategy that may be the main factors affecting the competitiveness of the handicraft sector of Paracho, Michoacán.

2 METHODS

This article contains a descriptive-relational design, which includes statistical measurements, that is to say, the competitiveness of the handicraft sector of Paracho, Michoacán, with the purpose of answering the inquiry question and, in turn, carrying out hypothesis testing.

Conformation of the model used in this study was mainly obtained from the literature review, which allowed a greater theoretical understanding of the proposed variables. Quality, price, technology, logistics, and marketing were considered as independent variables and their relation to the dependent variable competitiveness.

2.1 Universe and sample

According to Casa de las artesanías (CasArt), in a census carried out in 2009, the number of artisans in the interior of the state of Michoacán was 20,462, distributed in the 113 municipalities that make up the entity, of which 987 are craftsmen from Paracho, becoming the universe of study (Paz Vega, 2012).

The sample was measured with a confidence level of 93% with a maximum error of 0.7%. The sample size was 147 artisans who should be surveyed. However, although only 147 surveys were collected at the end of the study, 155 questionnaires were applied, which shows us the interest of the artisans to participate in this study. Once an appropriate design of the research has been carried out and a sample based on our problem has been collected for investigation, the next stage is the collection of the variables involved, for which a measurement instrument was designed according to the needs.

2.2 Measurement instrument

The survey was divided into three parts. The first part is related to general aspects that cannot be included in the Likert-type scale. These items are related to age, gender, type, or product, belonging to an association, brand, etc. In fact, some of the questions were related to the company personnel.

The second part includes open questions about the strengths, weaknesses, threats, and opportunities of the business, to which a Likert-type scale was not included, not to influence the answers. Finally, the third part contains 61 questions.

2.3 Data collection

The data obtained in the first part of the survey, which were derived from closed-ended and multiple-choice questions, are not allowed to be processed or standardized, and thus were grouped in the program SPSS 20. In this section, the main results obtained in the research will be shown, where there is no group to identify if they are competitive; however, it will allow us to know some important aspects of the handicraft sector of Paracho, Michoacán.

In the second part of the survey that includes a SWOT analysis, a statistical count of the answers was made by calculating their frequency using Microsoft Excel 2007.

The third part of the survey included multiple-choice questions, using a Likert-type scale, where it was possible to identify the level of competitiveness of the dependent and independent variables. Subsequently, the responses to the statistical program SPSS 20 were entered, to make the corresponding analysis.

3 RESULTS

The results of the field research were analyzed to derive the results of the dependent and independent variables, in which all the answers were processed statistically. Due to the characteristics of the research, Cronbach's alpha coefficient was used, which only requires the administration of the measuring instrument to all the artisans located in the State of Michoacán.

The results of the reliability test for each variable are given in Table 2.

Table 2 presents the reliability of the instrument used using Cronbach's alpha technique, where it is observed that the variables studied, including quality, price, logistics, marketing, and competitiveness, show reliability levels accepted by the theory, where average and high levels of reliability are presented, as well as the complete questionnaire indicating an internal consistency of 0.896 confidence, from which it is inferred that its application is very reliable.

The independent variable technology was not considerate in this analysis because the artisans did not have tools or cutting-edge machinery.

The first part of the survey shows the following results: the artisans were aged between 15 and 84 years; the level of schooling of the owner, manager, or administrator corresponded to high school (36%), university (30%), primary school (18%), secondary school (15%), and master's degree (only 1%); the average educational level of the personnel corresponded to primary (63%), secondary (23%), and other (14%) – who have another occupations, most commonly merchants; the learning methods were familiar tradition (61%), training (23%), empiricism (14%), and school-based (2%); respect for the current organization yielded the following response: no (89%) and yes (11%).

At present, the Mexican handicrafts sector faces several problems in the search to be competitive in the local, national, and international markets. Hence, the objective of this study is to identify the problems of the sector more clearly and accurately. To achieve this goal, the SWOT analysis was used, with the aim of understanding the current situation of this sector.

The SWOT analysis is based on the key findings in relation to the competitiveness of the industry. It is important to note from the SWOT analysis that a distinction can be made between endogenous factors (strength/weaknesses) and more exogenous factors (opportunities/threats) (ECORYS, 2015).

To apply this technique, artisans and businessmen were questioned, the results of which are given in Table 3. This diagnostic represented the first approach to the problems of the handicraft sector of Paracho, Michoacán, Mexico.

Table 2. Reliability test.

Variable	Cronbach's alpha
Survey (61 items)	0.896
Quality (12 items)	0.623
Price (10 items)	0.834
Logistics (12 items)	0.862
Marketing (14 items)	0.581
Competitiveness (10 items)	0.700

Source: Own elaboration based on the fieldwork.

Table 3. SWOT analysis in the handicraft sector in Paracho, Michoacán, Mexico.

Strengths	Weaknesses
Quality (59%)	Lack of capital (25%)
Service (27%)	Manufacture of low-quality products (17%)
Design (21%)	Logistics (15%)
Price (19%)	High costs and prices (21%)
Fast production (17%)	No credits (10%)
Low costs (17%)	Cheap labor (not well-paid work) (5%)
Grant credits (17%)	Low production capacity (5%)
Creativity (6%)	Obsolete machinery (5%)
Innovation (5%)	Design (3%)
Ability (5%)	Does not use advertising (2%)
Responsibility/commitment (5%)	Retail sales, only selling in the local market (3%)
Training by Mexican and foreign teachers (4%)	Do not use government support (1%)
Durability (2%)	Some have gone abroad and have made their work known (1%)
Knowledge of wood treatments (2%)	Do not use the training (1%)
Unique products (2%)	There is no support for the artisan (1%)
Variety (1%)	Years on the market (1%)
Experience (1%)	Not exported (1%)
Availability for sale (1%)	Less diversity in products (1%)
Production in series (1%)	Less experience (1%)
Playing various musical instruments (1%)	I do not belong to any brand or collective group (1%)

Opportunities	Threats
Explore new markets, both national and international (23%)	Competition (44%)
By seasons (17%)	Insecurity/crime (33%)
New and improved products (25%)	Economic crisis (32%)
Increase participation in fairs and exhibitions (10%)	Taxes (15%)
Government support (3%)	Shortage of raw materials (mainly timber) (10%)
Promotion of its products (1%)	Absence of tourism (8%)
Increase in population (1%)	Low sales (7%)
Increase production (1%)	Increased prices of raw materials (3%)
It is produced with quality (1%)	Lack of market (2%)
Knowledge of wood treatments (1%)	New technologies (1%)
Creativity (1%)	Advanced age (1%)
With suppliers, when buying by volume (1%)	Immodest rate (1%)
Recognition of the mark of origin region (1%)	Espionage (try to copy what we produce) (1%)
Have client portfolios (1%)	Devaluation of handicrafts (1%)
Training to improve (1%)	
Have the best product on the market (1%)	
Grow business, more staff (1%)	
Make the products known through the Internet (1%)	
Meet customer's tastes (1%)	
Financing credits (1%)	
Sell directly to the public (1%)	

Source: Own elaboration based on the fieldwork.

Table 4 presents the correlation between the independent variables and the dependent variable using the statistical technique of Spearman's correlation applied for nonparametric and ordinal variables.

Table 4. Spearman's correlation coefficient and its incidence in the competitiveness variable.

Independent variable	Spearman's coefficient	Dependent variable
Quality	0.685	Competitiveness in the handicraft sector
Price	0.201	
Logistics	0.357	
Marketing	0.289	

Source: Own elaboration based on the fieldwork.

4 DISCUSSION

Analysis of the dependent variable competitiveness, where sales dimensions and the perception of the competition were taken into consideration, shows that the artisanal sector only reached 62%. The analysis of the craft sector that was carried out using 155 questionnaires shows that 2.6% have a very low competitiveness, 12.9% have low competitiveness, 50.3% have regular competitiveness, and 34.2% have high competitiveness.

Furthering the competitiveness analysis, 36% of artisans have seen a decrease in their sales since 2008, 28% acknowledge that sales have remained stable, and only 23% had increased sales in this period. Regarding the targets set for sales in the last three years, 52% say that only a few times they have been fulfilled, 30% say that they are regularly met, and 15% indicate that the objectives set for sales have never been reached.

Regarding the annual sales volume of 2012, results show that 74% of the respondents consider themselves in the regular range and 26% low or very low, and it is important to emphasize that, in any case, there was a high or very high volume of sales.

In addition, questions that allow respondents to self-evaluate their sales performance against the competition were used, with 77% responding that they are perceived in a regular range. Regarding the pricing strategy, 60% believe that it does. As for the perception of the use of technology, they are perceived in the low range with 52%. In terms of logistics, 37% indicate to have a high performance. With regard to marketing, only 38% say they are at a high level of performance compared to their competition.

Regarding the independent variable quality, 77% was reached, which is considered by artisans as their greatest strength to be more competitive. The effect of applying the 155 questionnaires was that about 8% of artisans rated a very high quality, 79% a high quality, 12% a regular quality, and 1% a low quality, according to the ranges in which they were designed for each of the levels of quality.

The effect of applying the 155 questionnaires was that 32% of artisans rated a regular price fixing strategy, 43% had a low pricing strategy, and 7% had a very low strategy in pricing. The price was fixed according to the ranges in which each of the price levels was classified.

Technology, although several authors consider it as a determinant of competitiveness, in this case, in the craft sector, largely does not apply, because precisely the manufactured products are handmade using tools such as screwdrivers, hammers, razors, woodcutters, and brushes, resulting in the production of unique products that are not mass-produced. Therefore, an analysis of the variable was done independently and it was decided to exclude it as a determinant variable of the artisanal competitiveness

Regarding the variable logistics, the effect of applying the 155 questionnaires was that 6% have very low logistics, 20% low logistics, 53% regular logistics, 3% high logistics, and 17% have very high logistics, according to the ranks in which each of the levels were classified. Regarding this variable, a general performance of the variable was reached at 60%, which allows the marketing of artisanal products and is classified as a regular performance.

Regarding the marketing variable, the effect of applying the 155 questionnaires was that 31.6% have a very low marketing, 65.8% a low marketing, 1.9% regular marketing, and 0.6%

have a high marketing according to the ranks in which each level of marketing was classified. According to Lambert (1997), with regard to the proportion of competitiveness in marketing, it has an average score of 28.2, that is, it reaches 40.2% of competitiveness in marketing, placing it in a low marketing.

5 CONCLUSION

This study investigated the handicraft sector to determine the way in which the variables in the model are studied. However, we believe that actions should be taken to improve the competitiveness of this sector, which are as follows: (1) include crafts initiatives in development plans; (2) improve access to credit to the word to gain access to the purchase of better materials and machinery to raise quality levels; (3) know that there is a greater link between government agencies and artisans; (4) establish strategic alliances with universities, schools, and the craft sector; (5) implement training, which involves (A) creating centers for crafts and trades, (B) educational programs, (C) design training workshops, (D) development of craft products; (6) use marketing to push the creation of promotion of craft villages; and (7) improve the road infrastructure so that artisans can access new customers.

REFERENCES

AL-INVEST. (17 de 08 de 2014). *Facilitating the internationalisation of Latin American SMEs*. Recuperado el 23 de 03 de 2017, de https://ec.europa.eu/europeaid/sites/devco/files/brochure-latin-america-al-invest-iv-2013_en.pdf.
Ambastha, M., & Momaya, K. (2004). Competitiveness of Firm: Review of Theory, Frameworks, and Models. *Singapore Management Review*, 1 (26), 45–61.
Barber, T., & Krivoshlykova, M. (2006). *GLOBAL MARKET ASSESSMENT FOR HANDICRAFTS*. USAID. Accelerated Microenterprise Advancement Project.
Baumann, C., Hoadley, S., Hamin, H., & Nugraha, A. (2017). Competitiveness vis-à-vis service quality as drivers of customer loyalty mediated by perceptions of regulation and stability in steady and volatile markets. *Journal of Retailing and Consumer Services*, 36, 62–74. http://doi.org/http://dx.doi.org/10.1016/j.jretconser.2016.12.005.
Bouziane, F., & Hassan, A. (2016). Strategic Determinants for the Development of Traditional Handicraft Industry of Algeria. *International Journal of Managing Value and Supply Chains*, 7 (1), 1–11.
Dominguez, M.L., Hernández, J. d., & Toledo, A. (2004). Modelo de competitividad y ambiente en sectores fragmentados. El caso de la artesanía en México. Cuadernos de Administración, 127–158.
Duarte, R. (2012). Tesis: "Artesanías de exportación, desarrollo local y regional, 1980–2007: un estudio de caso comparativo entre localidades michoacanas y peruanas". Morelia, Michoacán.
ECORYS. (24 de 11 de 2015). *Study on the competitiveness of the recreational boating sector*. Recuperado el 25 de 03 de 2017, de http://ec.europa.eu/DocsRoom/documents/15043/attachments/1/translations/en/renditions/native.
GOBIERNO DE ESPAÑA, M.D. (2009). Plan de competitividad.
Gómez Prado, C.A., Romo de Vivar y Mercadillo, M.R., Dávila Vázquez, I., Arellano Espinoza, S.A., & Vega Cano, R. (2010). Artesanías y Desarrollo local.
Gutiérrez, H. (1997). Calidad total y productividad. Mac Graw Hill.
Lamb, C.W., Hair, J.F., & Mc Daniel, C. (2005). Fundamentos de Marketing. Thomson.
Lambert, D. (1997). Supply Chain Management: More than a new name for logisitics. *The International Journal of Logistics Management*, 8 (1), 1–14.
MacDowell, M., & Avery, J. (12 de 2006). *Crafts Work Michigan*. Recuperado el 15 de 03 de 2017, de www.craftworksmichigan.org: https://www.michigan.gov/documents/hal/CraftWORKS_184765_7.pdf.
Morita, A., & SONY. (1990). Hecho en Japón (1ra Ed. ed.). México: Lasser Press Mexicana.
Parking, M. (2004). Economía. México: Pearson.
Paz Vega, R. (09 de Octubre de 2012). Director General de la Casa de las Artesanías del Estado de Michoacán.
PROMEXICO. (2011–12-agosto). promexico.gob. Retrieved 2012–20-julio from www.promexico.gob: http://www.promexico.gob.

Sánchez-Medina, P., Corbett, J., & Toledo-López, A. (2011). Environmental Innovation and Sustainability in Small Handicraft Businesses in Mexico. *Sustainability*, *3* (7), 984–1002. http://www.mdpi.com/2071-1050/3/7/984/pdf.

Suzuki, N. (07 de 2005). *Effective Regional Development in Developing Countries through Artisan Craft Promotion*. Recuperado el 18 de 02 de 2017, http://mitizane.ll.chiba-u.jp/metadb/up/thesis/Suzuki_Naoto.pdf.

Valenzo, M., Martínez, J., & Bonales, J. (2010). Lacompete tividad logística en Latinoamérica: comparativo entre elíndice logístico y la propuesta metodológica. *Mercados y Negocios,* 20 (10), 85–106.

Venkataramanaiah, S., & Ganesh-Kumar, N. (2011). Building Competitiveness: A Case of Handicrafts Manufacturing Cluster Units. *Indore Management Journal*, *3* (2), 27–37.

Higher-order thinking skills in teaching the Malay language through questions and questioning among the teachers

Fong Peng Chew, Zul Hazmi Hamad & Fonny Hutagalung
Department of Language and Literacy Education, Faculty of Education, University of Malaya, Malaysia

ABSTRACT: This study was conducted to investigate the issues related to the level of knowledge acquisition in Higher-level Thinking Skills (HOTS) among the Malay language teachers in Malaysian national primary schools. In this study, we applied quantitative approach using Bloom's taxonomy (1956, 2001) and intellectual enhancement and upbringing model (Philips, 1996). A total of 60 teachers from one district in the Negeri Sembilan were selected for the HOTS test. In addition, we also explored significant differences between the level of knowledge acquisition in HOTS questioning with teachers' social background. The collected data were analyzed using SPSS version 21.0. The findings showed that the level of knowledge acquisition of HOTS questioning in the Malay language was at a moderate level. The Mann–Whitney and Kruskal–Wallis tests showed no significant differences between the level of knowledge acquisition in HOTS questioning based on teachers' specialization, attendance to relevant courses, and teaching experiences. In other words, the social background of teachers did not determine the level of knowledge acquisition in HOTS questioning technique among the Malay language teachers in national primary schools. Therefore, Ministry of Education should plan effective HOTS courses in order to ensure success in the implementation of HOTS questioning technique in all schools.

Keywords: Higher-order thinking skills, Malay language, national primary schools, teacher

1 INTRODUCTION

Together with the development of Malaysia civilization and the effort to improve the national image, education is crucial. The transformation in improving the quality of the education system has been designed in the long term by Ministry of Education through the National Education Blueprint (2013–2025). One of transformational shifts in the Malaysian Education Blueprint (NEB) is to provide equal chance of access to quality education of international standards, and examinations and assessments prior to this will be revamped and devoted to the higher-order thinking skills (HOTS).

2 STATEMENT OF PROBLEM

The Assessment and Teaching of 21st Century Skills Consortium (AT21CS) organized knowledge, skills, and attitude into the following four categories: ways of thinking, ways of working, tools for working, and living in the world (Schleicher, 2012). Similarly, one of the wishes of national education in Malaysia is to educate people to think. However, Malaysia was ranked 55th in the assessment of reading skills for PISA test with 443 mean score (PISA 2015 Result in Focus, 2016). This directly shows that the position of the best students in Malaysia is far behind compared to those of other countries.

This finding was further evidenced by the report submitted by Kestrel Education (UK) and 21st Century School (USA) in November 2011, according to which the level of higher-order thinking among teachers and students in Malaysia is very low (Teachers Education Division, 2012), and by report by UNESCO in 2011, which stated that the teachers are bound by the examination-oriented method of teaching. With regard to this situation, Harmi and Mahamod (2013) were of the opinion that the dependence on examinations have had been raising issues that hinder the initiative in shaping the generation envisaged by the National Education Philosophy.

Teaching students higher-order cognitive skills, including critical thinking, can help individuals improve their functioning in multiple circumstances (Tsui, 2002) and promote higher level of language proficiency (Chapple & Curtis, 2000; Tarone, 2005). Thus, learners need to think critically and creatively when using the target language in order to achieve high proficiency.

Researchers and educators are generally aware of the importance of teaching critical thinking skills. However, they pose a question that how such skills could be promoted through instruction (Tsui, 2002). One controversial issue is whether critical thinking is discipline-specific or not. Some experts argue that critical thinking instruction is only effective when it is integrated in teaching subject-specific knowledge and skills. The lack of insight into this topic suggests that a review of recent empirical studies on the topic of instructional teaching method of critical thinking skills may yield important and timely findings.

Among the teaching methods, HOTS questioning technique is highly appropriate in the teaching and learning of the Malay language. Thus, it is clear that the thinking skills in the learning process will affect the development of a pupil from the aspect of learning speed and effectiveness of learning (Radovic-Markovic, M. & Markovic, D. 2012; Yee et al., 2013).

From the aspect of teachers, language teachers do not use appropriate questioning in their teaching skills as routine. It is also found in the study that teachers rely mainly on textbooks (Sarsar, 2008) and doing the exercises provided in the textbook only (Supramani, 2006). The problem of the dependence of teachers on textbooks also led to the conclusion that student thinking has not improved to a higher level including aspects of analysis, evaluative or even creative.

Rajendran (2001; 2010) also that teaching methods of the Malay language teachers nowadays are not blending elements of higher-order thinking skills. This issue is fundamental to the non-application of thinking skills in the teaching process in viewing the teacher's role as a model to lead students chase toward excellence.

Furthermore, it is more serious that Malay Language teachers use memorizing techniques in teaching writing skills to students. The ability of students to memorize will be an indicator of the success or failure of the process of writing skills. This negates the principles of teaching writing skills and the importance of thinking skills in essay writing. This fact is clearly supported by Brookhart (2010), who stated that higher-order thinking in problem solving cannot be met with a memorized solution. To recall something, students have to identify it as a problem.

On the basis of the issues mentioned above, it is clearly demonstrated that HOTS should be a culture of questioning techniques among the teachers during the teaching of the Malay language to produce quality students, as specified by Mansor (2009), according to whom questions and questioning play very important role in the national education curriculum to build the thinking skills more effectively to students.

Therefore, this study aimed at answering the following research questions:

i. What is the level of Malay Language teachers' acquisition of knowledge in HOTS questioning?
ii. Is there any significant difference between the level of knowledge acquisition HOTS questioning with social background among Malay language teachers?

Here, social background is defined as the profile that affects the HOTS questioning technique of Malay language teachers. The scope included their specialization, teaching experiences, and attendance in HOTS courses.

3 CONCEPTUAL FRAMEWORK

Figure 1 shows the conceptual framework for the teaching of higher-order thinking skills in the Malay language in national primary schools. This framework is based on the domain of cognitive thinking as revised by Anderson to Bloom's taxonomy existing model. The idea of the model is then divided into models of teaching thinking skills by Philips (1996) on the rationale that the thinking skills is a process that can be taught, learned, and can be implemented in a subject.

According to Phillips (1996), all subjects in school can be customized with thinking skills. He also noted that intellectual students will be formed through the teaching that leads to thinking and learning skills. However, he also sought support teachers through the confidence, skills, knowledge, adaptability, and willingness to disseminate skills to students, as he was aware of the importance of teacher as a facilitator in the application of thinking skills.

Teaching model is featured next to focus questions and questioning that became a milestone in teaching and learning. The form of a question and the process of entry into force of this questioning were focused on the methods and characteristics of the questioning and the question of the high order of thinking skills' nature as the adopted cognitive domain model.

3.1 Research instrument

The data pertinent to this study were collected by administering a set of testing questions to assess teachers' achievement in HOTS application. This instrument consists of six items drafted based on Bloom's taxonomy and Philip's model (1996). Content validity of the instruments was confirmed by a group of experts on the construction of each item in the test paper, including the Head of Malay Language, Madam T; principal trainer for Malay language in Negeri Sembilan, Mr. M, who has more than 10 years' experience in teaching the Malay language, and Madam A.

Before data collection, the test was checked for reliability. A pilot study was conducted among 30 teachers at three national primary schools. Results of the analysis showed that the test has consistency, with a Cronbach's alpha value of 0.85. This indicated that the instrument was robust and can be implemented in the study.

3.2 Respondents of the study

The sample size of this study was 60, including teachers from three primary schools, namely National School, National-type Chinese School, and National-type Tamil School. Purposive sampling method was used in this study.

The results showed that female teachers (n = 50, 83.3%) are more in number than male teachers (n = 10, 16.7%). This is inevitable in view of the fact the number of female teachers who teach the Malay language is more than the male counterparts in the location of this study. This study used same number of respondents from the three types of schools (20 each,

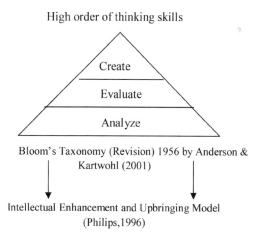

Figure 1. Conceptual framework of the study.

33.3%), and the total breakdown of respondents by ethnicity, namely Malays, Chinese, and Indians, are also the same for the three types of schools (20 each, 33.3%).

The number of teachers with more than 11 years' experience form the majority (38, 63.3%), followed by those with 6–10 years' experience (12, 20.0%), and the lowest was novice teachers, that is, new teachers with experience less than 5 years (10, 16.7%). Total non-graduate respondents were more (36, 60.0%) than graduated teachers (24, 40.0%).

Data also showed that respondents who are specialized in the Malay language were minorities in this study (18, 30.0%) compared to the non-specialized ones (42, 70.0%). In addition, the experience of attending HOTS courses was also taken into account in this study and it has been found that the number of teachers attending the related course (36, 60.0%) were higher those not attending the course (24, 40.0%).

4 FINDINGS

Research Question 1: What is the level of Malay language teachers' acquisition of knowledge in HOTS questioning?

To determine the level of Malay language teachers' acquisition of knowledge in HOTS questioning, frequency, percentage, and grade were used as shown in Table 1. The range of score in the table is based on the rubric for national schools set by Examination Board of Malaysia.

Table 1 shows the results of testing the level of knowledge and abilities of teachers to carry out the HOTS questioning techniques during the teaching process in Malay language class. The results showed that from the total of 60 respondents, 26 teachers (43.33%) achieved a grade C in the score range of 40–59. This showed that majority of Malay language teachers have a satisfactory level of skills. A total of 24 teachers (40.0%) were at a good level with average score of 60–79.

The findings also showed that 10 teachers (16.7%) scored A for the HOTS test, with marks in the range of 80–100. The results of this table showed that the level of knowledge and abilities of Malay language teachers were satisfactory, while the number of teachers who were outstanding level was low.

Research Question 2: Is there any significant differences of the level of acquisition of HOTS and social background among the Malay language teachers?

To answer this question, the Mann–Whitney test was performed using SPSS software, and the results are shown in Tables 2 and 3.

Analysis of the results obtained from the Mann–Whitney U-test showed the relationship between Malay language teachers' specialization and their score of HOTS. It was found that there was no significant difference in the scores between the specialized Malay language teachers (Md = 3, n = 18) and non-specialized teachers (Md = 3, n = 42), $U = 274$, $Z = -1.68$, $p > 0.05$, $r = 0.21$. This showed that the specialization of Malay language teachers was not related to their score in HOTS test. Z value was not significant at $p < 0.05$. Therefore, the null hypothesis, "there is no significant differences between the level of acquisition HOTS questioning with teachers' specialization", was accepted. In other words, teachers with or without specializing in the Malay language did not differ in their score in HOTS.

Table 1. Test scores and teachers' acquisition of HOTS questioning.

Range of score	F	%	Grade
0–19	0	0	E
20–39	0	0	D
40–59	26	43.3	C
60–79	24	40	B
80–100	10	16.7	A

M = 3.73, SD = .57.

Table 2. Mean ranks and sum of ranks for variables of specialization and marks in HOTS among the teachers.

	Specialization	N	Mean rank	Sum of ranks
Score in HOTS	Malay language	18	24.72	445.00
	Non-Malay language	42	32.98	1385.00
	Total	60		

Table 3. Significance of the Mann–Whitney U test[a].

	Score
Mann–Whitney U	274.000
Wilcoxon W	445.000
Z	−1.682
Asymp. Sig. (2-tailed)	0.093

[a]Grouping variable: specialization.

Table 4. Mean ranks and sum of ranks for variables of attending HOTS' courses and marks in HOTS among the teachers.

	Attending courses	N	Mean rank	Sum of ranks
Score in HOTS	Yes	36	29.83	1074.00
	No	24	31.50	756.00
	Total	60		

Table 5. Significance of the Mann–Whitney U-test[a].

	Score
Mann–Whitney U	408.000
Wilcoxon W	1074.000
Z	−0.363
Asymp. Sig. (2-tailed)	0.717

[a]Grouping variable: attending HOTS courses.

Table 6. Mean ranks for variables of teaching experiences and score of HOTS among the teachers.

	Teaching experience	N	Mean rank
Score in HOTS	<5 years	10	27.50
	6–10 years	12	32.17
	>10 years	38	30.76
	Total	60	

Table 7. Significance of Kruskal–Wallis test[a,b].

	Score
Chi-square	0.415
Df	2
Asymp. Sig.	0.813

[a]Kruskal–Wallis test.
[b]Grouping variable: teaching experience.

Analysis of the results obtained from the Mann–Whitney U-test showed the relationship between attending HOTS courses among Malay language teachers and their score of HOTS. It was found that there was no significant difference in the scores between the Malay language teachers who attended HOTS courses (Md = 3, n = 36) and those who did not attend (Md = 3, n = 24), $U = 408$, $Z = -0.36$, $p > 0.05$, $r = 0.05$. This showed that attending HOTS course was not related to the Malay language teachers' score of HOTS. Z value was not significant at $p < 0.05$, which indicated that the null hypothesis, "there is no significant differences between the level of acquisition HOTS questioning with teachers' attendance at the HOTS courses", was accepted. Therefore, the teachers who attended or did not attend HOTS course did not differ in their score in HOTS.

Nonparametric analysis, Kruskal–Wallis test, for three groups of teachers with different teaching experiences shows the following results: X^2 (2, N-60) = 0.415, $p < 0.05$, $r^2 = 0.006$. This showed that Malay language teachers did not differ significantly in terms of their teaching experiences. These findings indicated that although the teachers have different teaching experiences, they gain similar HOTS scores.

5 DISCUSSION AND INTERPRETATION

In this study, we found that the level of knowledge acquisition in HOTS among Malay language teachers was moderate, with a mean score of 3.73 and a standard deviation of 0.57. Findings clearly show that Malay language teachers already knew and understood the concept of HOTS. It is important to review that the teachers are updated, in line with the Ministry of Education, to develop quality teachers and education in the 21st century (Teacher Education Division, 2015).

The findings also showed that teachers were prepared to be evaluated and tested in terms of the implementation of the current HOTS questions in teaching the Malay language in the classroom. When teachers ask HOTS questions and encourage explanations, they help their students to develop important critical thinking skills directly. By modeling good questioning and encouraging students to ask questions of themselves, teachers can help students learn independently and improve their learning (Corley & Rauscher, 2013).

The Mann–Whitney test showed no significant difference between teachers with specialization in the Malay language as compared to the non-specialized teachers. Therefore, the null hypothesis, "there is no significant differences between the level of acquisition HOTS questioning with teachers' specialization", is accepted.

However, the data showed that the mean rank of teachers who are not specialized in the Malay language (MR 32.98) is higher than that of the specialized teachers (MR 24.72). This occurs because the teachers tend to be exam-oriented and focus solely on the content of the lesson. As a result of these circumstances, the specialized educators started to underestimate the value added in the teaching of Malay language that emphasize the implementation of HOTS questions and questioning compared to the non-specialized teachers who are more open to receive numerous reprimands and guidance from senior teachers in grooming their skills. However, this finding was contrary to the study conducted by Ahmad and Mahamod (2014). They found that the mean score of the Malay language-specialized teachers was higher than that of the non-specialized teachers, using the t-test result obtained in their study.

Similarly, the findings showed no significant differences between teachers who attended or never attended any HOTS courses. Therefore, the null hypothesis, "there is no significant differences between the level of acquisition HOTS questioning with teachers' attendance at the HOTS courses", is accepted.

However, the data showed that the mean score of teachers who did not attend the HOTS courses (MR 31.50) was higher than that of the teachers who attended (MR 29.83). The findings were parallel with the data obtained by Zamri Hasan and Mahamod (2015), which showed that the level of understanding of teachers who attended the courses and its impact on HOTS application were low (M 1.96, SD 0.21).

In other words, implementation of the HOTS courses that have been organized at various levels by the District Education Office or the State Education Department was ineffective. It also showed that such courses do not get the appropriate input or do not expose the teachers to the theory and application of HOTS questions and questioning application in teaching process.

National Research Council (2013) suggested that to add knowledge, skills, quality of work, performance, and quality of one's education, various forms of training and exercises are required. We believe that teachers should be made to undergo courses on a regular basis to strengthen their knowledge and skills to apply HOTS questions and questioning in classroom. This is supported by Robert (2016) who stated that effective intervention programs are needed to help teachers to blend HOTS in their teaching process skillfully.

These findings provided clues to the Division of Teacher Education in the MoE in order to organize courses or program on professional development of teachers more effectively. This is because the study showed no significant differences among the educators who attended the courses and those who did not attend it. Post-service trainings are found to be very important, where Teacher Education Division plays an important role for implementing instructional methods that contain the HOT questioning technique elements to educators. Teachers Education Division also need to find more effective methods of dissemination to the educators as HOTS is a demanding skills nowadays.

The Kruskal–Wallis test showed that the aspect of teaching experiences has no significant differences in the score in HOTS among the teachers. It also showed that experience is not the measure for teachers to practice HOTS questioning in classrooms. Therefore, the null hypothesis, "there is no significant differences between the level of acquisition HOTS questioning with teachers' teaching experiences", is accepted.

Experiences do not affect teachers' teaching HOTS because HOTS is a new element to be highlighted by MOE nowadays. All teachers equally face problems while practicing HOTS questioning in teaching the Malay language in schools. These findings are in line with the findings by David and Ambotang (2014), who found that no influence of the implementation of HOTS among the novice teachers in the Sabah state. As a new skill, HOTS can be mastered by all educators studying it. This statement is also contradicted by Low et al. (2014), who explained that culture of implementing HOTS among novice teachers in the teaching process showed moderate level of knowledge and understanding of HOTS in them (M 3.26 SD.63).

Overall, the Malay language test in this study was drafted based on HOTS items at the analyzing, evaluating, and creating levels. It is also noted that intellectual students will be produced through the effective teaching methods of the teachers and lead to thinking and learning skills of the students (Philips, 1996). These findings supported that good teachers can be trained through the confidence, skills, knowledge, adaptability, and willingness to disseminate thinking skills to students. Therefore, similar study can be extended to other countries due to the importance of question and questioning in teaching HOTS to students. The history of teaching by question and questioning can be traced back to Socrates, who proved the effectiveness of the teaching method.

6 CONCLUSION

In conclusion, the study indicated that most of the Malay language teachers' acquisition knowledge in HOTS questioning is at the average level. No differences were found between the levels of knowledge acquisition in HOTS questioning with social background among the Malay language teachers. Therefore, it is recommended that Ministry of Education should conduct effective HOTS courses in order to ensure success in the implementation of HOTS questioning technique in all schools because of its effectiveness to instill students' HOTS.

REFERENCES

Ahmad, A. & Mahamod, Z. 2014. Skills of Malay Language teachers in implementation of school-based assessment in secondary schools. In Zamri Mahamod et.al (Ed). *Proceedings of the 3rd Seminar on*

Post-graduate Education in Malay Language and Malay Literature (p. 95–110). Bangi: Faculty of Education, National University of Malaysia.

Anderson LW, Krathwohl DR. 2001. A taxonomy for learning, teaching, and assessing: a revision of Bloom's taxonomy of educational objectives. New York: Longmans.

Curriculum Development Division. 2012. I-Think Program Cultivate Thinking Skills [Malay language]. Putrajaya: Malaysian Ministry of Education.

Corley, M.A. & Rauscher, W.C. 2013. *Deeper Learning through Questioning*. American Institute for Research.

Chapple, L., & Curtis, A. 2000. Content-based instruction in Hong Kong: Student responses to film. *System, 28*, 419–433.

David, C. & Ambotang, A.S. 2014. Professionalism of novice teachers in knowledge management, preparedness of teaching and Higher Order Thinking Skills (HOTS) on the implementation of teaching in schools [Malay language]. Paper presented in the *National Seminar on Family Integrity, 2014*, December 11, 2014, Universiti Malaysia Sabah.

Low, S.F, Zainuddin, A., & Ruhani, R. 2014. Standard of Malaysian Teachers: Knowledge Level of Pre-service Teachers under the PISMP at IPG Kampus Ilmu Khas [Malay languge]. Downloaded April, 4, 2017 from http://ipgkik.com/v2/wp-content/uploads/2015/05/Artikel7_2014.pdf.

Mansor, N.R. 2009. *Questions and Questioning in Teaching and Lerning of Language* [Malay language]. Kuala Lumpur: Utusan Publications & Distributors Sdn. Bhd.

National Education Blueprint (2013–2025). 2012. Ministry of Education.

National Research Council. 2013. *Education for Life and Work: Developing transferable knowledge and skills in the 21st century.* New York: National Academy for Science.

Nik Harmi, N.N.F & Mahamod, Z. 2013. The effectiveness of the Thinking Maps (i-Think) method on achievement, attitude, willingness and enrollment of Form Four students [Malay language]. In Zamri Mahmod et.al (Ed). *Proceedings of the Second Regional Seminar for Postgraduate Pen-educated Malay Language and Malay Literature* (p. 423–440). Bangi: Faculty of Education, National University of Malaysia.

Philips, J. (1996). *Developing Critical and Creative Thinking In Children.* Shah Alam: Lingua Publications.

PISA 2015 Rresult in Focus. 2016. OECD.

Rajendran, N.S. 2001. Teaching Higher Order Thinking Skills: Readiness of Teachers in Teaching and Learning Process [Malay language]. Downloaded April, 1, 2017 from http://nsrajendran.tripod.com/Papers/PPK12001A.pdf.

Rajendran N.S. 2010. *Teaching & Acquiring Higher-Order Thinking Skills: Theory and Practice.* Tanjung Malim: University of Education Sultan Idris Publisher.

Radovic-Markovic, M. & Markovic, D. 2012. A New Model of Education: Development of Individuality through the Freedom of Learning. *Erudito, 1*(1), 97–114.

Robert, J. 2016. *Language Teacher Education.* London: Taylor & Francis Group.

Sarsar, M.N. 2008. Textbook Addiction Treatment: A Move towards Teacher and Student Autonomy. Downloaded April 3, 2017 from http://files.eric.ed.gov/fulltext/ED503393.pdf.

Supramani, S. 2006. Teachers' questioning: Catalyst Higher Order Thinking Skills among the students [Malay language]. *Jurnal Pendidikan, 26,* 225–246.

Tarone, E. 2005. Fossilization, social context, and language play. In Z. Han, & T. Odlin (Eds.), *Studies of Fossilization n Second Language Acquisition* (p. 157–172). Clevedon: Multilingual Matters.

Teachers Education Division. 2015. Pedagogy of Malaysian Language in 21 Century [Malay language]. Putrajaya: Malaysian Ministry of Education.

Tsui, L. 2002. Fostering critical thinking through effective pedagogy: Evidence from four institutional case studies. *Journal of Higher Education, 73*(6), 740–763.

Yee, M. H, Md Yunos, J., Hassan, R., Mohamad, M.M., Othman, W. & Tee, T.K. 2013. Evaluation of the integration manual quality of self-directed learning styles of Kolb and Marzano's higher order thinking skills [Malay language]. *Proceeding of the International Conference on Social Science Research 2013.* Penang: Malaysian Science University.

Zamri Hasan, N.H. & Mahamod, Z. 2015. Perceptions of Secondary School Malay language Teachers towards Higher Order Thinking Skills in Teaching and Learning [Malay language]. In Zamri Mahmod et.al (Ed). *Proceedings of the 4th Seminar for Postgraduate in Malay Language & Literature Education* (p. 25–44). Bangi: Facutly of Education, National University of Malaysia.

Using poems to increase phonological awareness among children

Chiu Tieng Lim & Fong Peng Chew
*Department of Languayge and Literacy Education, Faculty of Education,
University of Malaya, Malaysia*

ABSTRACT: This study was conducted to analyze the effects of using the phonology method and the traditional method in Malay poems on phonological awareness in children (n = 62) from government preschools. This quasi-experimental study was conducted to compare the aforementioned two methods. The conceptual framework applied in this study was based on the preparation of reading by Stewart (2004) and theory of cognitive development (Piaget, 1977). The experimental group (n = 30) was trained to read poems using the phonology method, whereas the control group (n = 32) was trained using the traditional method. Both groups underwent pretests and post-tests to compare the level of phonological awareness in recognizing the number of syllables in a word, rhyming syllable discrimination in words, producing rhyming syllable, and early and end of phonemic sound. The findings showed that the use of rhyming poems has enhanced the level of phonological awareness among preschool children after the treatment. By contrast, the traditional method showed no effect on phonological awareness. However, the Mann–Whitney U-test found no significant difference between the scores of males and females in the experimental group. Therefore, optimal use of poems was carried out by teachers in preschools to enhance phonological awareness because the genre of poem is identified as an effective way to increase the literacy of children.

Keywords: children, national primary schools, preschool, phonological awareness

1 INTRODUCTION

According to a report by UNESCO (2008) and Taguma et al. (2013), early childhood education plays an important role in the development of a child's mind and can reduce existing deficits in future. Studies conducted in the United States found that children who attended preschool were more committed to education and earned higher incomes in their future. Recognizing the importance of preschool education and the scope for Malaysia to increase the rate of enrollment, the preschool has become one of the main objectives under the national key result areas (NKRA) in education and National Education Blueprint (2013–2025). Malaysian government plans to develop the country as the educational center in Southeast Asia, mainly to increase its literacy rate.

2 STATEMENT OF PROBLEM

In Malaysia, reading problems among school students are always crucial for both parents and teachers. Ministry of Education found that the average dropout in national primary schools from 2006 to 2010 was about 0.3% for males and 0.2% for females. Low reading habit among school students in addition to the failure to master the reading, writing, and counting skills is identified as a factor contributing to the dropout problem (*Utusan Online*, April 16, 2011). This showed that using the Malay language as the medium of instruction is a problem in primary schools, which actually starts at the preschool level.

Mastery of language skills is closely related to the development of spoken language. Oral language development is a highly complex cognitive processes. Children produce spontaneous speech from 10 months to 2 years (Chang, 2010). However, they are unable to identify errors and corrections that need to be improved in speech. (Li, 2003).

Language development has led to the development of phonological awareness, which includes children who are sensitive to voice at preschool age (Yopp & Yopp, 2009). Phonological awareness is the ability to detect and manipulate the sounds of language without knowing its meaning (Lonigan, 2006). Phonological awareness has evolved from a two-dimensional form of simple to complex (Yopp & Yopp, 2009). Children aged 5 and 6 years are usually able to distinguish the characteristics of sounds (phonemes) and produce some sounds. This is the beginning of children to recognize songs, rhymes, and poems. Thus, phonological awareness is a key foundation for the success of the practice of reading and it is important to detect children who have trouble in reading (Ehri et al., 2001; Lonigan, 2003; Storch & Whitehurst, 2002).

The selection of reading materials is important in the process of reading. Quality reading materials such as poems and storybooks also have a positive effect on the level of phonological awareness. The beat and rhythm of the poem may improve short-term memory of children (Michalis et al., 2014). Preschool teachers can perform mechanical reading to draw attention to the sound of children reading. At the same time, children are encouraged to observe sound and language (Yopp & Yopp, 2009).

According to The Poetry Library (2003), poetry is the most appropriate art of literature to express any speech. Poems are favored by all levels of society, especially children. Children become interested in poetry from childhood. Children perceive teachers reading poem as their inspiration and role model. The practice of reading, writing, and listening to nursery rhymes is the most meaningful experience for children of various ages (KBYU Eleven, 2010).

Children with phonological problem are likely to face problems in the spoken language. They often make the mistake of replacing the sound in words such as "name" to "mama" and "blink" to drink". The origin of this problem is the delay in language phonology, aphasia, and central auditory processing disorder (CAPD). Some children acquire language later than usual, some children master the language although somewhat slower than others, and some children do not show changes at all and require early intervention. Therefore, early intervention is important to detect reading problems in the Malay language in order to solve the phonological problems at the early stage.

Many studies showed that females outperform males in reading and writing competencies. On the contrary, many researchers found no significant differences between the genders with the achievement of reading and writing in children. The lack of insight into this topic in Malaysia suggests a review of recent empirical studies on the topic.

Preschools in Malaysia emphasize reading and writing competency of children to prepare them with basic literacy skills before entering primary schools. This is the main objective of preschools as foundation to form literate young generation (Malay Language Standard Curriculum Document in Preschools, 2011).

In this context, the Malay language, as the medium of instruction in national preschools, is the phonic language, which can be learned and mastered easily using the phonic method (Yusoff et al. 2014). To increase phonological awareness of children, poems are the main literary material in the study.

Thus, this study aimed to answer the following research questions:

1. What is the level of phonological awareness of experimental group in pretest and post-test?
2. What is the effect of using poems to enhance the level of phonological awareness of children with different teaching methods?
3. Is there any significant difference between phonological awareness and gender after using phonic method?

3 CONCEPTUAL FRAMEWORK

Phonological awareness is one of the indicators of reading ability. Therefore, the conceptual framework applied in this study is based on the theory put forward by Stewart (2004), who suggested to read poems to children as their preparation of reading, as shown in Figure 1.

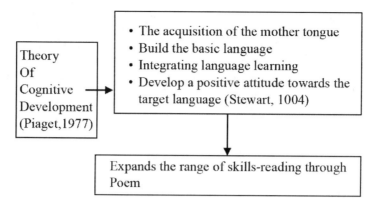

Figure 1. Conceptual framework.

Traditional poems interchangeable with modern poems are read mechanically to children to enhance their phonological awareness.

According to the theory of cognitive development (Piaget, 1977), children aged 5–6 years can identify and manipulate the sounds of vowels, syllables, words, and sentences but most likely still do not appreciate the true meaning behind the words and sentences.

3.1 *Research design*

This study used a quasi-experimental design and involved two groups, namely experimental group and control group. The respondents were selected based on the purposive sampling method. The experimental group received the phonological awareness intervention by using rhyming poems, whereas the control group used traditional teaching methods on the rhyming poems. Pretest in this study was based on the test that was developed by the researchers based on phonological awareness items. Both groups were tested separately, and post-test will be given after the children underwent the teaching session.

Training sessions were held for 2 months. The researchers developed the training modules based on transactional model of development, which states that meaning is constructed through reading and influenced by social, cultural, and environmental factors (Miller, 2000). This module allows the researchers to teach Malay language poems based on preschool program syllabus established by Ministry of Education (2010).

3.2 *Research instrument*

The test used in this study was based on the concept of phonological awareness proposed by Cisero and Royer (1995) who divided phonological awareness into three forms: (1) segment word into sounds; (2) divide each syllable of a word, and (3) divide into individual phonemes.

There were five sections in pre- and post-tests administered during the study, including:

a. Part 1: Recognizing syllables in words
b. Part 2: Discriminating rhyming sounds in syllable
c. Part 3: Producing rhyming syllables
d. Part 4: Early phonemic sound
e. Part 5: End of phonemic sound

Items used in the pre- and post-tests were the same and being administrated twice. Then, 2 months is sufficient to administer a pre-test and post-test holistically after the experiment was carried out.

Researchers modified the items in the Test of Phonological Awareness Skills guided by Criteria Reference Tests of Phonological Awareness by Robertson & Salter (1995) and Yopp-Singer Test of Phoneme Segmentation (Yopp, 1995). Content validity of the instruments was confirmed by a group of experts on the construction of each item in the test paper. They consisted of Madam N, who has more than 10 years experiences in teaching the Malay language in preschool, Madam A, who is Head of Malay language in National Primary schools and a Malay language lecturer at the university, and Dr. C.

Before the data collection, the test was checked for reliability. A pilot study was conducted among 30 pupils at a national primary school. Results of the analysis showed that the test has consistency, with a Cronbach's alpha value of 0.85. This indicated that the instrument was robust and can be implemented in the study.

3.3 Respondents of the study

For this study, 62 children were selected from two preschools in Kuala Lumpur, comprising 28 males and 34 females. Preschools A and B represented the experimental group (n = 30) and the control group (n = 32) by using rhyming poems with and without training and practice of phonological awareness, respectively.

In terms of ethnicity, Chinese were the majority, i.e., 42 children (67.7%), followed by Malays (12, 19.4%), Indians (6, 9.7%), and others (2, 3.2%). The age of the preschoolers ranged from 6 years 0 months to 6 years 11 months. There were 32 respondents in the control group and 30 children in the experimental group.

4 FINDINGS

Research Question 1: What is the level of phonological awareness of control and experimental group in pretest and post-test?

Each part of test carried 20 marks, with 5 parts carrying totally 100 marks. For instance, children who scored 20% meant that they can recognize syllables in words (Part I), 40% meant that they can discriminate rhyming sounds in syllable (Part II) and Part I, 60% meant that they can produce rhyming syllables (Part III), Part II and Part I, and so on.

Table 1 shows the overall pretest and post-test scores for the control group. Children with score ranges 0–20% and 80–100%, respectively, did not show any increase in performance after the post-test was conducted, followed by the reduction of 4 (12.6%) children at score range 21–40%. On the contrary, score ranges of 41–60% and 61–80% showed increase of 2 (6.3%) children. The frequency distribution of scores for the control group showed a decrease in the early stages of low range of score, followed by slow increase at moderate score range. This scenario presented that the traditional method of teaching poems in the Malay language class showed moderate level of phonological awareness among children.

Table 2 shows the overall pretest and post-test scores for the experimental group. On the basis of analysis of the frequencies that have been executed, available score ranges of 0–20% and 61–80%, respectively, decreased a total of 2 (6.7%) children after the post-test was conducted; score range of 21–40% presented a reduction of 1 (3.3%) child, whereas the

Table 1. Marks distribution of the control group for the pretest and post-test.

Range of score	Pretest F	%	Post-test F	%	Changes in F	Changes in %
0–20	5	15.6	5	15.6	0	0
21–40	9	28.1	5	15.6	−4	−12.6
41–60	4	12.5	6	18.8	+2	+6.3
61–80	5	15.6	7	21.9	+2	+6.3
81–100	9	28.1	9	28.1	0	0
Total	32	100	32	100	–	–

F = frequency, % = percentage.

Table 2. Marks distribution of the experimental group for the pretest and post-test.

Range of score	Pretest F	%	Post-test F	%	Changes in F	Changes in %
0–20	6	20.0	4	13.3	−2	−6.7
21–40	4	13.3	3	10.0	−1	−3.3
41–60	6	20.0	0	0	−6	−20.0
61–80	6	20.0	4	13.3	−2	−6.7
81–100	8	26.7	19	63.4	+11	+36.7
Total	30	100	30	100	–	–

Table 3. Comparison of the effect of using poem with gender using the T-test for the control group.

Poems	N	M	SD	t	p
Pretest	32	55.56	32.29	9.53	4.38
Post-test	32	61.88	31.57		

*Significant at $p < 0.05$.

score range of 41–60% showed decrease of 6 (20%) children. On the contrary, score range of 81–100% performed dramatic incensement with 11 (36.7%) children. The frequency distribution of scores for the experimental group showed a decrease in the early stages of low and moderate score ranges, followed by large increase especially at the maximum score range. This scenario illustrated that the use of the phonology method in teaching Malay language poems enhanced the level of phonological awareness among children significantly.

Research Question 2: What is the effect of using poems to enhance the level of phonological awareness of children with different teaching method?

Pair sample t-test analysis showed no significant difference in the overall score of the control group using the traditional method (chalk and talk) between the pretest (M = 55.56, SD 32.29) and post-test (M = 61.88, SD 31.57), t (32) = 9.53, p > 0.05. The mean score obtained clearly showed that the control group that learned poems using the traditional method did not show much progress. Thus, the null hypothesis (Ho1), "there is no significant difference of using traditional method to teach poem in increasing phonological awareness", is accepted.

Pair sample t-test analysis showed significant differences in the overall score of experimental group between the pretest (M = 55.93, *SD* 32.15) and post-test (M = 78.76, *SD* 32.98), $t(30) = 7.00, p < 0.01$. The value of the effect size for the overall score of the poem in the post-test was at 0.52, which showed a moderate effect size. These values clearly show that the use of a simple rhyme had a significant impact on phonological proficiency among preschool children. The mean score obtained clearly showed that the experimental group showed more progress after they went through the phonic training. Thus, the null hypothesis (Ho2), "there is no significant difference of using phonic method to teach poem in increasing phonological awareness", is rejected.

Question 3: Is there any significant difference between phonological awareness and gender after using phonic method?

Analysis of the results obtained from the Mann–Whitney U-test showed the relationship between the gender of children and their phonic awareness score in post-test. It was found that there was no significant difference in the score between the male pupils (Md = 3, n = 16) and female pupils (Md = 3, n = 14), $U = 100.50, Z = -5.30, p > 0.05, r = 0.97$. This showed that gender was not related to their score in post-test. The fact that the Z value was not significant at $p < 0.05$ indicated that the null hypothesis, "there is no significant difference between phonological awareness and gender after using phonic method", was accepted. Therefore, the male and female pupils did not differ in their phonic score in post-test.

Table 4. Comparison of the effect of using poem with gender using the T-test for the experimental group.

Poems	N	M	SD	t	p	n
Pretest	30	55.93	32.15	7.00	0.00*	0.52
Post-test	30	78.76	32.98			

*Significant at $p < 0.01$.

Table 5. Mean ranks and sum of ranks for score using the phonic teaching method with gender of the children.

	Gender	N	Mean rank	Sum of ranks
Post-test	Male	16	14.78	236.50
	Female	14	16.32	228.50
	Total	30		

Table 6. Significance of the Mann–Whitney U test[a,b].

	Score
Mann–Whitney U	100.500
Wilcoxon W	236.500
Z	−5.30
Asymp. Sig. (2-tailed)	0.596
Exact Sig. (2*(1-tailed sig))	0.637[b]

[a]Grouping variable: gender.
[b]Not corrected for ties.

5 DISCUSSION AND INTERPRETATION

Descriptive findings showed that the level of using poems among preschool children in the experimental and control groups can be detected through pretest and post-test of phonological awareness. On the basis of the results, it is proved that the phonic method influenced the process of teaching and learning in preschool in general and the Malay language in particular rapidly. By contrast, the traditional method of teaching poem was found to increase the level of phonological awareness slowly.

This scenario explained that the phonic method of teaching poems enables children who are at moderate levels to directly read the follow-up sensitivity to sound without understanding and knowing the meaning of words or sentences. Children can read spontaneously and smoothly without spelling out the syllables. Most children can manipulate sound after mastering five basic skills of phonological awareness, namely phonemic awareness, phonics, fluency, vocabulary, and comprehension to become good readers. Children are able to distinguish skills with segmenting the words to the sound, combining the sound of the word, and evaluate two words that have the same sound based on the scope and size of the components of the sound, depending on the component in the structure of words, emphasizing syllables or elements syllables as vowels, consonants, and rhyme. Thus, phonological awareness can build basic skills of language (Stewart, 2004).

This finding coincided with the study conducted by Yopp and Yopp (2009), who found that children can detect and manipulate the sound in speech without understanding the meaning of the language. According to Rober (2010), phonological awareness is the ability to identify and manipulate the language of words, syllables, phonemes, rhyme, intonation, and play with sound. Children are able to master the syllables, phonemes, rhyme, and so on when playing with the sound of the syllables in recognizing words, rang the sound of early and late sound, and produce rhyming word by word. The results of the studies conducted by Yopp and Yopp (2009) and Ambruster et al. (2010) found that similar sounds can be manipulated without knowing their meaning.

The above statement was supported by the theory of cognitive development (Piaget, 1977), according to which children can solve problems intuitively but cannot understand the concepts and ideas. Phonological awareness is actually only meant to help children in detecting sound externally to apply the sound to form a word, either easy or difficult. In this context, children can identify and manipulate the sounds of vowels, syllables, words, and sentences, but most likely still do not appreciate the true meaning behind the words and sentences. They simply read spontaneously by the mere mention of allowing children to master reading skills in a short period. Playing sound through syllables, words, and syntax in various forms not only amuse children, but can also enhance their reading capabilities. Therefore, it is true that the theory of cognitive development is indirectly practiced on children in this study.

Consequently, t-test analysis showed no significant differences of score in control group, but indicated significant differences of score in the experimental group between the pretest and post-test with moderate effect size. The findings of Jason et al. (2005) were equivalent to the results of this study that children learning poems using the phonic method have achieved progress significantly ($t(30) = 7.00$, $p < 0.05$, $n = 0.52$) compared to those learning using the traditional method ($t(32) = 9.53$, $p > 0.05$).

Analysis of the results obtained from the Mann–Whitney U-test showed that children who learn poems using the phonic method showed no significant difference in performance between male and female. At this stage, the child's cognitive structure has still not reached maturity as adults. They will follow their instincts as pioneered by Piaget's (1977) theory of cognitive development.

This statement was in line with the study of Siegel and Smythe (2005) and Chia and Kee (2013), who revealed no significant differences in children between the gender with regard to their ability to read and reading achievement. The study used three types of test kits to detect the level of the reading of 65 children, namely Word Recognition and Phonics Skills Test-Second Edition (Wraps-2) (Moseley, 2003), Comprehensive Receptive and Expressive Vocabulary Test-Second Edition (CREVT-2) (Wallace & Hammill, 2002), and Neale Analysis of Reading Ability-third Edition (NARA-III) (Neale, 1999).

However, the findings were contradictory to the results obtained by Below et al. (2010). Using a cross-sectional design and five dynamic indicators of basic early literacy skills measures, they tested for gender differences in reading skills for 1218 kindergartens through fifth-grade students. A series of two-way repeated measures analyses of variance with time of year (fall, winter, and spring) served as the within-subjects variable and gender serving as the between-subjects variable, which showed that females scored significantly higher than males on the four kindergarten measures; however, these differences were small.

As a result, teachers need to be more concerned with the selection of literary material in accordance with the children in order to foster their interest in reading. Selection of a variety of materials in accordance with the level of children plays an important role in stimulating the ability of reading and early literacy for children. The preschools should give full support to the efforts of teachers to provide reading material for children.

Disclosure of the holistic phonics techniques provides insight and awareness to preschool teachers that the text of the poem can be optimized as teaching material to achieve the goal of preschool education including the Malay language. Poem as a medium of attracting children to learn can lead them to track the correct reading skills.

6 CONCLUSION

In conclusion, we showed that the phonic method in teaching poem facilitates achievement of phonological awareness among preschool children rapidly, regardless of their gender. This means that the literary material should be considered for its role to assist children to master reading.

Thus, appropriate teachers should take the initiative and creativity to use the phonic method in teaching for developing and enhancing children's literacy. Future study may focus on the phonology awareness among children by using prose and poems to determine the best literary material.

REFERENCES

Ambruster, B.B., Lehr, F. & Osborn, J. 2010. *Put Reading First: Kindergarten through Grade 3* (3rd Ed.). National Institute for Literacy.
Aphasia definition. National Aphasia Association. Downloaded from https://www.aphasia.org/aphasia-definitions/.
Below, J.L., Skinner, C.H., Fearrington, J.Y. & Sorrel, C.A. 2010. Gender Differences in Early Literacy: Analysis of Kindergarten through Fifth-Grade Dynamic Indicators of Basic Early Literacy Skills Probes. *School Psychology Review, 39*(2), 240–257.
Chang, M.Y.S. 2011. Developmental pathways for first language acquisition of Mandarin nominal expressions: Comparing monolingual with simultaneous. *International Journal of Bilingualism, 14*(1), 11–35.
Chia, N.K.H. & Kee, N.K.N. (2013). Gender differences in the reading process of six-year-olds in Singapore. *Early Child Development and Care,* 183(10), 1432–1448.
Cisero, C.A., & Royer, J.M. (1995). The development and cross-language transfer of phonological awareness. Contemporary Educational Psychology, 20, 275–303.
Curriculum Development Division. 2011. Malay Language Standard Curriculum & Assessment Document in Preschools. Putrajaya: Ministry of Education, Malaysia.
Ehri, L.C., Nunes, S.R., Willows, D.M., Schuster, B.V., Yaghoub-Zadeh, Z., & Shanahan, T. (2001). Phonemic awareness instruction helps children learn to read: Evidence from the National Reading Panel's meta-analysis. Reading Research Quarterly, 36, 250–287.
Jason L. & Anthony, David J.F. (2005). Development of Phonological Awareness. Journal of Psychological Science, 14, 255–259.
Less rBeading contributor to dropout. Utusan Online, April 16, 2011.
Li, P. (2003). Language acquisition in a self-organizing neural network model. In: Quinlan P (ed) Connectionist Models of Development: Developmental Processes in Real and Artificial Neural Networks (p 115–149). New York: Psychology Press.
Lonigan, C.J. (2003). Development and promotion of emergent literacy skills in preschool children atrisk of reading difficulties. In B. Foorman (Ed.), Preventing and remediating reading difficulties: Bringing science to scale (pp. 23–50). Timonium, MD: York Press.
Lonigan, C.J. (2006). Development, assessment, and promotion of preliteracy skills. *Early Education and Development, 17*(1), 91–114.
Michalis, L., Kalliopi, T. & Elissavet, C. (2014). Effects of a Rhythm Development Intervention on the Phonological Awareness in Early Childhood. IPEDR, 78(10), 49–53.
Miller, W.H. (2000). Strategies for developing emergent literacy. Boston: McGraw Hill.
National Education Blueprint (2013–2025). 2012. Ministry of Education.
Piaget, J. (1972). The psychology of the child. New York: Basic Books.
Rhymers Are Readers: The Importance of Nursery Rhymes. KBYU Eleven.
Robert, P. 2010. *CSET California Subject Matter Examination for Teachers: Multiple Subjects* (3rd Ed.). California: Barron's.
Siegel, L., Smythe, I. 2005. Reflections on research on reading disability with special attention to gender issues. *Journal of Learning Disabilities, 38,* 473–477.
Stewart, M.R. (2004). Phonological awareness & bilingual preschoolers: should we teach it and if so, how? Early childhood education Journal 32(1): 31–37.
Taguma, M., Litjens, I. & Makowiecki, K. 2013. Quality Matters in Early Childhood Education and Care. OECD Publication.
The Poetry Library (2003). URL (consulted June 2004): http://www.poetrylibrary.org.uk/poetry.
Whitehurst, G.J. & Lonigan, C.J. (2002). Emergent literacy: Development from pre-readers to readers. In S.B. Neuman, & D.K. Dickinson (Eds.), Handbook of Early Literacy Development (pp. 11–29). New York: Guilford.
Yopp, H.K. & Yopp, R.H. (2009). Phological awareness is child's play. Beyond the Journal. Young Children on the Web. 1–9.

Web-based catering marketplace business plan: Cateringku.com

Budiman Susanto
Bina Nusantara University, Jakarta, Indonesia

Ford Lumban Gaol, S. Si & M. Kom
Department of Science, School of Information Systems, Bina Nusantara University, Jakarta, Indonesia

ABSTRACT: Food is one of fundamental human needs besides clothing and shelter. Humans need food whenever and wherever they live. Recently, the number of restaurants and catering services is growing rapidly. Because of that, individuals in the society can choose their favorite food more easily. Furthermore, the internet has changed the way humans conduct businesses. Today, there are many catering services which are managed through websites. However, the websites can only be used by the owners. Therefore, a marketplace for catering services needs to be created. This study discusses the business plan for making the catering marketplace. The results of this study indicate that the catering marketplace can be run and produce revenue based on the calculation of ROI and IRR.

Keywords: catering, marketplace, website, ROI, IRR

1 INTRODUCTION

Food is one of basic human needs besides clothing and shelter. Humans need food every day whenever and wherever they live. Nowadays, the number of restaurants and catering services is growing rapidly. Customers can choose what kind of food they want to have. Every caterer also tries to provide something new to their customers.

The rapid development of technology has resulted in the growth of internet users [1]. In 2016, the total number of internet users reached 132.7 million people. This number increases every year.

The internet has been used for various purposes. One of the applications is to make a marketplace. [2] Marketplace is a type of ecommerce that connects those that provide services to those who need the services. Through the marketplace, sellers and buyers can easily do transactions, creating efficient market.

Taking into account such an aspect, a web application for solving the problem will be created. [3] The web application can be accessed through web browsers with connection; it will be developed using a programming language such as HTML, or JavaScript. This system later will provide features which can be used by consumers to search or use catering services.

There are several critical success factors in this business which should be considered in the development of the marketplace. [4] Research conducted in 2000 identified six critical success factors for the development of e-business strategy: creating a consumer-centric strategy, embracing outsourcing to improve business performance; acting like a new entrant, using information management to differentiate products, and being part of an e-business community. This research used mini-case studies to illustrate the application in real businesses. The results of this research indicate that almost all businesses will become e-business. Those which do not change will be invaded by their competitors. Further, those accepting the challenge will benefit from the implementation of the 6 CSF.

As an online business, this business also needs strategies to start. [5] A study that was conducted with McCarthy's four marketing mix model and Porter's Five competitive forces model identified strategies for Internet companies that respond to the competitive forces and thereby achieve competitive advantages. The study provides significant insights into the development and implementation of e-business strategies that contribute to increased profit.

When developing the system, this study also created a business plan

2 METHODOLOGY

The system that will be created is a web application. The web application used for the marketplace has some features, such as price bidding, product offering, price dealing, price demand, and product demand. The design steps started from the idea determination, opportunity analysis, market analysis, product description, business model, prototype development, development planning, risk analysis, financial planning and creating a business plan.

2.1 Idea determination and opportunity analysis

This study is aimed at solving issues in catering industries. Based on surveys and observations, there were several issues in this industry, such as:

a. There is no space that unites sellers and buyers of catering food.
b. Sellers only got buyers from people that they know.
c. The cost when sellers make an online catering system is expensive.

2.2 Market analysis

After the idea determination and opportunities analysis, market analysis was conducted. SWOT analysis, Five Force Porters analysis, and market segmentation were conducted in the following steps.

a. SWOT analysis
There were some companies used as samples in this step, such: Berry Kitchen as an online catering service that provides daily catering, snacks and bento, and Klik-eat as an online service delivery. There are many brands that join Klik-eat and Gorry Gourmet as the pioneers of healthy catering.

b. Five Force Porters Analysis
This analysis was conducted to identify the competitive environment of the catering business and to find out how the competition involving external parties will affect the business of catering marketplace. There are several analysis performed in this step:

- Competitive Rivalry
- Threat of new entry
- Threat of substitution
- Buyer Power
- Supplier Power

c. Market Segmentation
The market of catering marketplace was segmented in this step. The segmentation was done based on:

- Geography
- Demography
- Psychographic
- Behavior

2.3 Defining product description

This step was done to provide detailed descriptions of the product and service that will be offered to consumers. The product is an online marketplace website for catering that can be used by traditional caterers, restaurants, and people who want to have their own catering business and also people who want to look for daily catering services for food needs.

2.4 Business model

In this step, a business model was created for catering marketplace. The business model refers to the business model canvas which covers 9 elements: segmentation, value proposition, key partner, key resource, key activity, cost structure, customer relationship, channel, and revenue stream. This business model is aimed at identifying the relation of opportunities and value proposition in this business. The business model canvas in the catering marketplace is divided into 2 parts:

- Buyer business model canvas
- Seller business model canvas

Below are the business model canvas designed for the business.

2.5 Prototype development

In this step, the prototype was built using waterfall method. The results of this step are user interface design and process flow of business.

2.6 Development planning

This step followed the prototype development step. In this step, the backend of the system was built.

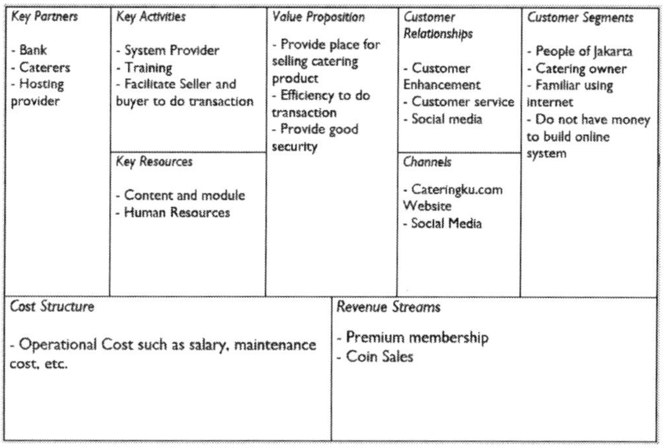

Key Partners	Key Activities	Value Proposition	Customer Relationships	Customer Segments
- Bank - Caterers - Hosting provider	- System Provider - Training - Facilitate Seller and buyer to do transaction	- Provide place for selling catering product - Efficiency to do transaction - Provide good security	- Customer Enhancement - Customer service - Social media	- People of Jakarta - Catering owner - Familiar using internet - Do not have money to build online system
	Key Resources - Content and module - Human Resources		Channels - Cateringku.com Website - Social Media	

Cost Structure	Revenue Streams
- Operational Cost such as salary, maintenance cost, etc.	- Premium membership - Coin Sales

Figure 1. Business model canvas for seller.

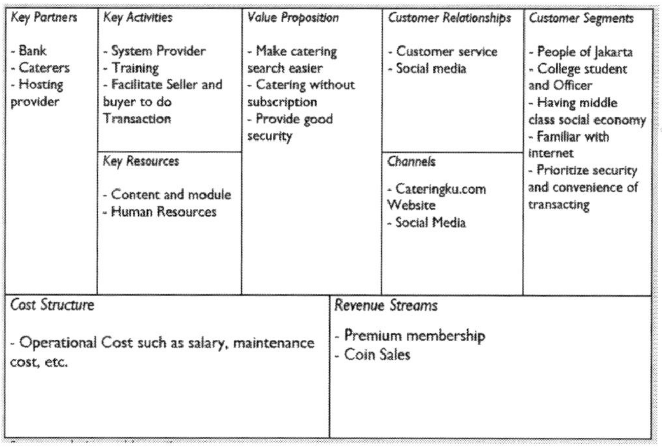

Key Partners	Key Activities	Value Proposition	Customer Relationships	Customer Segments
- Bank - Caterers - Hosting provider	- System Provider - Training - Facilitate Seller and buyer to do Transaction	- Make catering search easier - Catering without subscription - Provide good security	- Customer service - Social media	- People of Jakarta - College student and Officer - Having middle class social economy - Familiar with internet - Prioritize security and convenience of transacting
	Key Resources - Content and module - Human Resources		Channels - Cateringku.com Website - Social Media	

Cost Structure	Revenue Streams
- Operational Cost such as salary, maintenance cost, etc.	- Premium membership - Coin Sales

Figure 2. Business model canvas for buyer.

2.7 Risk analysis

This stage analyzed the potential risks that can cause problems to the business, as well as measures that can be used to prevent the occurrence and mitigate the risks.

2.8 Financial planning

At this stage, financial planning is made of business marketplace catering this. Financial planning includes projected revenue, expenses, cash flow, and financial analysis for estimated return of investment using Payback Period and IRR.

2.9 Business plan

In this step, a complete and systematic business plan was prepared based on the result of the previous step.

3 SYSTEM DESIGN

The prototype for the web application was created using waterfall method. For the programming language, this website used HTML and JavaScript to build the front end system, and Java to build the backend system. For the business flow, the system is split into 2 parts:

- Business process for the catering owner
- Business process for the consumer

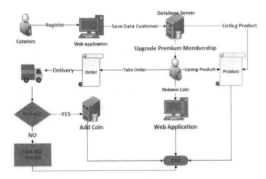

Figure 3. Business process catering service owner.

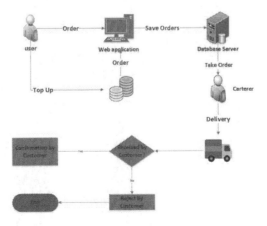

Figure 4. Business process catering consumer.

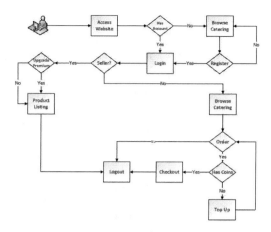

Figure 5. Web application navigation in Cateringku.com.

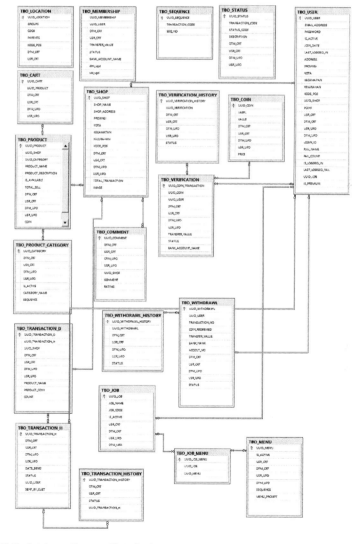

Figure 6. UML database diagram Cateringku.com.

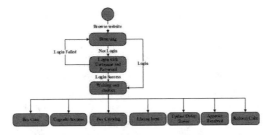

Figure 7. State transition diagram of Cateringku.com.

Figure 8. Homepage Cateringku.com website.

Figure 9. Homepage Cateringku.com website (2).

Figure 10. Catering shop list page.

Business process for the caterer starts when a user or consumer performs the registration. The caterer can do transaction such as: buying the premium membership, listing item, delivery order, and coin redeem.

The catering consumer can buy coins, and verify when catering is received.

The database diagram is as follows:

The state transition diagram of cateringku.com is as follows:

Below is the user interface design of Cateringku.com.

Figure 11. Catering shop menu list and order page.

4 BUSINESS PLAN

The business plan of the web consists of executive summary, industry, consumer and competitor analysis, marketing plan, operational plan, development plan, team, risk analysis, financial planning and analysis.

4.1 Executive summary

Cateringku.com is a marketplace website designed specifically for the sale and purchase of daily catering. Cateringku.com was built specifically to facilitate customers to order food.

The business model that runs is different from the common online catering services. Through Cateringku.com, the caterers can register their catering products in coins. Coins are a transaction tool used on Cateringku.com. Consumers will be able to buy the food they want by doing top up coins first on Cateringku.com. This cannot be done on other online catering websites. The reason for the use of coins is that transactions conducted can run safely. This is because the process of purchasing and redeem coins is done directly by Cateringku.com. In addition, the use of coin provides efficiency because the users only need to make top-ups over a number of coins and then they can use the coins to do transactions whenever desired. Products and services offered to customers include:

a. sales of premium account for caterers;
b. sales of coins for transactions on Cateringku.com;
c. business training and features on Cateringku.com.

4.2 Industry, consumer and competitor analysis

The competitive condition in the catering marketplace, as well as customer and competitor analysis can be explained as follows:

a. Industry analysis
Food industry is one of the promising business sectors. This is because humans need food. Besides that, the number of menu choices has also attracted customers.

Nowadays, there are many catering businesses managed in traditional ways. Catering business managed using a modern way are not common. The number of catering businesses in Jakarta is 3106 catering, which contains traditional catering and modern catering businesses. To obtain data from the industry, this study also conducted a survey that involved 72 people in Jakarta. The survey results are as follows:

– There was 70.8% of people who have used catering
– There was 61.4% of people who have difficulties in finding catering
– There was 81.7% of people that can be helped with the web application.

A web application is a technology that allows a catering to be managed in a modern way. A web application requires a browser and internet to run. The reason of the use of web application on Cateringku.com was because the web application can run on all devices. According to APJII, the number of internet users in Indonesia in 2016 reached 132.7 million people and it increases every year. According to GDP, in 2014 Indonesia was ranked 6th in terms of the number of Internet users in the world.

Nowadays, there are many catering businesses managed in a modern way. Online catering website is used as promotional and marketing media and will assist catering businesses to provide better services to the customer. The marketplace is built to give competitive advantages to traditional catering to promote their catering in inexpensive ways.

b. Customer analysis

In general, this catering marketplace can be used by everyone. However, the specific target customers of catering marketplace are individuals who have a catering business and restaurants, and customers who want to search/order daily catering, particularly office workers and students. The profiles of the target customers are presented in the table below. The customer profiles were obtained from the survey and are based on the segmentation criteria proposed by Kotler and Keller (2012).

c. Competitor analysis

Competition at this business can be described in 5 Force Porters:

– Competitive Rivalry

In the catering marketplace business, there are already several competitors who are well established and have a good reputation. Competitors in the catering marketplace are companies that offer services similar to those offered by the catering business. The main competitor of the catering marketplace is Berry Kitchen, which is an online catering service that provides daily catering, snacks, and bento.

The indirect competitors are Klik-eat which is a delivery service for food, and Gorry Gourmet which is a provider of healthy food, Go-food, and other traditional catering.

Each competitor has its own advantages and is in the area of Jakarta, such as Berry Kitchen which uses system points on the purchase as well as different menus every day. It has sold 1000 lunchboxes every day. Meanwhile, Klik-eat and Go-Food have more than 15,000 restaurants in their applications. In order to compete with its competitors, Cateringku.com should be able to attract traditional caterers to join the website. Also, the mobile application should be built so that it can reach consumers more easily. The mobile app of Cateringku.com was not available when Cateringku.com started operating; it will be developed as an additional product and service in the following year.

Table 1. Customer segmentation: Buyer.

Geographic	Located in DKI Jakarta
Demographic	– Students and Officers
	– Having middle to upper class economy
Psychographic	– Familiar with the internet
	– Prioritizing security and convenience when doing transactions
Behavior	Do not have time to prepare food

Table 2. Customer segmentation: Seller.

Geographic	Located in DKI Jakarta
Demographic	Having restaurants or catering businesses
Psychographic	– Familiar with the internet
	– Prioritizing security and convenience when doing transactions
Behavior	Do not have money to build a system

– Threat of new entry

The likelihood of a new competitor's entry can be difficult. This is because the development of website and business needs a lot of investments. The investment required is either in the form of capital for building the application and human resources to run the company's business activities. In the future, there may be new businesses similar to Cateringku.com. These newcomers can come from Indonesia as well as overseas, such as EZCater which is based in Boston and has served 17 million people and has about 50,000 catering businesses incorporated. EzCater will probably enter the Indonesian market. This is because EzCater own a business that has served almost all states in the USA so it is possible that the company expands its business to reach other countries. The business models offered by entrants can be different from or similar to those offered by Cateringku. However, to enter the competition in the marketplace, newcomers must provide substantial capital used for website development and business promotion.

– Threat of substitution

The replacement product of this marketplace catering business is online catering and food service delivery that has been running. Replacement products have their own uniqueness through the products offered to consumers. The threat of replacement products is highly dependent on price competition and products offered to consumers. If other online catering businesses have much lower prices and better products from the marketplace, then customers will switch to online catering services.

– Buyer Power

In this business, the power of the buyers' offerings has a strong effect. This is due to the many competitors in the food industry. However, the marketplace features that are used can be its strength because the marketplace built can be used by home catering businesses and restaurants. Therefore, if Cateringku.com can attract consumers and maintain their loyalty, it can minimize the strength of customer powers.

– Supplier Power

From the strength of supplier supply, suppliers have an important role that will affect the smoothness of the company's activities.

4.3 *Marketing plan*

Marketing plan in this business can be described as follows:

a. Marketing mix 7p
   – Product

   Cateringku.com offers a web application platform for catering marketplace that can be used for product listings and product purchases. In carrying out its business, Cateringku.com will offer the following services:
   - Free coins
   - Premium membership
   - Coins

   – Price

   The prices of services provided by Cateringku.com are as follows:
   - Account Services
   - Coin

   – Place

   The distribution channel used by Cateringku.com is Cateringku.com website that can be accessed by customers anywhere and anytime. The entire process that exists on Cateringku.com business, such as account upgrades and coin purchases, will be made through Cateringku.com website.

   – Promotion

   Promotions on Cateringku.com can be explained by AISAS model:
   - Awareness

   To increase awareness, Cateringku.com used ads and promotions through the internet, such as social media and advertisements in certain forums to make customers more familiar to the products.

- Interest

  To increase customer interest, a friendly user interface will be designed, to provide convenient experience to users using Cateringku.com products.
- Search

  For search, Cateringku.com will use SEO (Search Engine Optimization) to assist customers in making decisions.
- Action

  For action, the direct interaction will be done to sellers and buyers to give a good experience to customers. Examples of direct interaction are to maintain SLA for all transactions that exist on Cateringku.com and provide a good response to customer complaints.
- Share

  Once buyers make a purchase, they will be asked for feedback to identify the buyers' shopping experience and give a feature to share their experience to others so that people will increasingly know Cateringku.com.

– People

Human resources are an important factor in this business. Therefore, for the development and operation of this business, people who have experience in their field will be recruited through the process of recruitment and job expo.

– Process

The process of purchasing products and services both account and coin will be done through the website.

– Physical Evidence

The effects that can be caused by *Cateringku.com* are:
- providing ease of shopping for catering customers;
- Providing wider distribution access that can provide more revenue to caterers;
- improving shopping security for catering customers.

b. Differentiation

The differentiation made by Cateringku.com is to make traditional catering businesses can be managed in a modern way with the help of technology. Cateringku.com makes it easy for caterers to have a digital store at a free cost so that catering service owners can do promotion and receive orders. This can provide effectiveness and efficiency for catering service owners to reach catering users to increase the income of the catering service owners. In addition, catering customers can also search catering shops using Cateringku.com website. This cannot be done by traditional catering and other online catering websites.

c. Segmenting, Targeting and Positioning
  – Segmenting

  In general, this catering marketplace can be used by everyone. However, the specific target customers are as follows:

Table 3. Account price at Cateringku.com.

Product or Service	Price
Free account	Free
Premium membership	IDR 150,000/month

Table 4. Coin price at Cateringku.com.

Coin	Price (IDR)
5	25,000.00
10	50,000.00
25	125,000.00
50	250,000.00

- Targeting

  Humans need food every day, wherever and whenever they live. The target of Cateringku.com is to provide facilities for sellers and buyers to do transactions so that transactions and searches become easier to do.
- Positioning

  Catering is one of alternative food services. Cateringku.com, as the first catering marketplace in Indonesia, will position itself as a market leader in Catering field. The slogan of Cateringku.com is "Cateringku for all" which is made to position itself as a market leader in the field of the catering industry.

4.4 Operational plan

The operational activities of Cateringku.com were to help caterers to run their business, as well as do network monitoring and business training. To support operations, Cateringku.com rent an office in West Jakarta area. In this office, all operational activities of Cateringku.com will be implemented. The training process at Cateringku.com will also be held at the office by collaborating with entrepreneurial workshops in Jakarta.

4.5 Development plan

In the development plan, the strategy of web application development and timeline of web application work will be described. Below is the system information model used in Cateringku.com.

The system used is the same system; the same database was used for the admin, the seller, and the buyer. The distinguishing features are the menus and pages that exist on each subsystem that can be accessed by the users.

Web application development will be done by Cateringku.com development team. The development of web application will use waterfall method. In the initial development phase, the development team will create a prototype system to analyze the business process on the application to test whether the system runs as expected. After evaluation, the team will build a web application that can facilitate customer transactions. Features available in the first year are registration, listing of food, purchases, premium membership and coin conversion to IDR. With this feature, Cateringku.com is expected to be able to assist customers in running their business activities. In the next year, the system will be developed by utilizing the technology map where customers and owners of catering services will be able to see the tracking delivery

Table 5. Customer segmentation: Buyer.

Geographic	Located in DKI Jakarta
Demographic	– Students and Officers
	– Having middle to upper class economy
Psychographic	– Familiar with the internet
	– Prioritizing security and convenience when transacting
Behavior	Do not have time to prepare food

Table 6. Customer segmentation: Seller.

Geographic	Located in DKI Jakarta
Demographic	Having restaurants or catering businesses
Psychographic	– Familiar with the internet
	– Prioritizing security and convenience when transacting
Behavior	Do not have money to build the system

of catering products. The development of the mobile application will also be done to reach the community so that ordering and receiving orders can be done anytime and anywhere.

4.6 Team

Human resources who have certain skills and qualifications are needed to run the business. There were some positions needed for this business, including:

- Director
- Project Manager
- Web developer
- Accounting and Finance
- Customer Service
- HRD

In addition to organizational structure planning, planning is also required with regard to the number of human resources. The HR planning for the first 2 years is as follows.

After human resource planning, the planning of employee payroll system was done by setting salary increase of 10% per year. Below is the planning on employee salary:

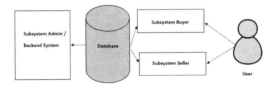

Figure 12. System information model.

Table 7. HR planning Year 1.

| Posisi | Tahun ke –1 ||||||||||||
	Jan	Feb	Mar	Apr	May	Jun	Jul	Aug	Sep	Oct	Nov	Dec
Direktur	1	1	1	1	1	1	1	1	1	1	1	1
Project Manager	1	1	1	1	1	1	1	1	1	1	1	1
Web Developer	2	2	2	2	2	2	2	2	2	2	2	2
IT Support dan QA	1	1	1	1	1	1	1	1	1	1	1	1
Accounting dan Finance	1	1	1	1	1	1	1	1	1	1	1	1
Customer Service	0	0	0	0	0	0	0	0	0	0	0	0
HRD	1	1	1	1	1	1	1	1	1	1	1	1

Table 8. HR planning Year –2.

| Posisi | Tahun ke –2 ||||||||||||
	Jan	Feb	Mar	Apr	May	Jun	Jul	Aug	Sep	Oct	Nov	Dec
Direktur	1	1	1	1	1	1	1	1	1	1	1	1
Project Manager	1	1	1	1	1	1	1	1	1	1	1	1
Web Developer	3	3	3	3	3	3	3	3	3	3	3	3
IT Support dan QA	1	1	1	1	1	1	1	1	1	1	1	1
Accounting dan Finance	1	1	1	1	1	1	1	1	1	1	1	1
Customer Service	2	2	2	2	2	2	2	2	2	2	2	2
HRD	1	1	1	1	1	1	1	1	1	1	1	1

Table 9. Salary planning.

Position	Year – 1	Year – 2
Director	IDR 6.500.000	IDR. 7.150.000
Project Manager	IDR 5.500.000	IDR. 6.050.000
Web Developer	IDR 4.500.000	IDR. 4.950.000
IT Support and QA	IDR 3.500.000	IDR. 3.850.000
Accounting & Finance	IDR 3.500.000	IDR. 3.850.000
Customer Service	–	IDR. 3.500.000
HRD	IDR 3.500.000	IDR. 3.850.000

Table 10. Initial capital.

Company establishment cost	Rp. 8.000.000
Office rent cost per year	Rp. 55.000.000
Furniture	Rp 10.000.000
8 laptops	Rp. 56.000.000
Server	Rp. 24.000.000
Operational cost	Rp. 500.000.000
Total	Rp. 653.000.000

Table 11. Optimistic income statement.

Income	Year 1	Tahun 2	Tahun 3	Tahun 4	Tahun 5
Account Premium Selling	–	244,800,000.00	367,200,000.00	489,600,000.00	612,000,000.00
Coins selling	–	1,200,000,000.00	1,800,000,000.00	2,100,000,000.00	2,400,000,000.00
Total income	–	1,444,800,000.00	2,167,200,000.00	2,589,600,000.00	3,012,000,000.00
Expense					
Salary	365,500,000.00	559,200,000.00	615,120,000.00	676,632,000.00	744,295,200.00
Promotion and advertising	50,000,000.00	60,000,000.00	65,000,000.00	70,000,000.00	75,000,000.00
Depreciation of fixed assets	18,000,000.00	18,000,000.00	18,000,000.00	18,000,000.00	18,000,000.00
Domain & interest cost	10,000,000.00	10,000,000.00	10,000,000.00	10,000,000.00	10,000,000.00
Rent cost	55,000,000.00	55,000,000.00	55,000,000.00	55,000,000.00	55,000,000.00
Utility cost	20,000,000.00	20,000,000.00	20,000,000.00	20,000,000.00	20,000,000.00
Legal	8,000,000.00	–	–	–	–
Training	–	20,000,000.00	22,000,000.00	25,000,000.00	30,000,000.00
Tax final	–	14,448,000.00	21,672,000.00	25,896,000.00	30,120,000.00
Income after tax	(526,500,000.00)	688,152,000.00	1,340,408,000.00	1,689,072,000.00	2,029,584,800.00

4.7 Risk analysis

Every business that runs must have different risks. In this case, the risks that will be faced by Cateringku.com are as follows:

- The existence of a gap in the security system that makes hackers able to enter the system. Solution: taking a high awareness of security in making the web application that is built by testing the gap that may be exploited by hackers and doing prevention.
- There are people who still do not understand the internet; so, they prefer the conventional ways to run their business.

Table 12. Pessimistic income statement.

Income	Year 1	Tahun 2	Tahun 3	Tahun 4	Tahun 5
Account Premium Selling	–	122,400,000.00	244,800,000.00	365,400,000.00	487,800,000.00
Coins selling	–	120,000,000.00	180,000,000.00	240,000,000.00	600,000,000.00
Total income	–	242,400,000.00	424,800,000.00	605,400,000.00	1,087,800,000.00
Expense					
Salary	365,500,000.00	559,200,000.00	615,120,000.00	676,632,000.00	744,295,200.00
Promotion and advertising	50,000,000.00	60,000,000.00	65,000,000.00	70,000,000.00	75,000,000.00
Depreciation of fixed assets	18,000,000.00	18,000,000.00	18,000,000.00	18,000,000.00	18,000,000.00
Domain & interest cost	10,000,000.00	10,000,000.00	10,000,000.00	10,000,000.00	10,000,000.00
Rent cost	55,000,000.00	55,000,000.00	55,000,000.00	55,000,000.00	55,000,000.00
Utility cost	20,000,000.00	20,000,000.00	20,000,000.00	20,000,000.00	20,000,000.00
Legal	8,000,000.00	–	–	–	–
Training	–	20,000,000.00	22,000,000.00	25,000,000.00	30,000,000.00
Tax final	–	2,424,000.00	4,248,000.00	6,054,000.00	10,878,000.00
Income after tax	(526,500,000.00)	(502,224,000.00)	(384,568,000.00)	(275,286,000.00)	124,626,800.00

Table 13. IRR assumptions.

Capital	653,000,000.00
Year 1	48,500,000.00
Year 2	724,052,000.00
Year 3	2,059,060,000.00
Year 4	3,736,532,000.00
Year 5	57,848,316,800.00
IRR	118%

Solution: providing guidance and technology training for the community.
– Inadequate human resources in the field and high employee turnover rates.
Solution: providing training for employees, especially programmers before entering the development project. In addition, knowledge management is needed so that existing information and knowledge can be maintained when there is employee turnover.
– The low income of catering customers and customer interest in the web application created.
Solution: doing promotion with a strategy so that customers become interested to try the website.

4.8 *Financial planning and analysis*

The industry target of the marketplace is catering service. From the survey involving 72 people, it can be concluded that:

1. 70.8% people have used the catering services;
2. 61.4% people find it hard to find the catering service;
3. 81.7% people will be helped by the web app that will be created.

The financial planning assumption of the project cost is describe in the following table:

On income statement, projection is done from both optimistic and pessimistic. The result is described in the following table:

Based on ROI Calculation, ROI in first and second year is:

$$ROI\ year-1 = \frac{-533,100,000}{653,000,000} \times 100\%$$

$$ROI\ year-1 = -82\%$$

$$ROI\ year-2 = \frac{235,852,000}{653,000,000} \times 100\%$$

$$ROI\ year-2 = 36\%$$

ROI in the first year is negative because there are no income in the first year. In the second year, the company already got income and obtained ROI of 36%.

The value of 36% indicates that the return value obtained from the total investment IDR 653,000,000 is 36%.

The payback period for this business is around November on the second year, where the cash flow is IDR 654,456,000.

The Internal Rate of Return from this business is around 118% that can be seen in this following table.

5 SUMMARY

The catering marketplace provides great business opportunities. This can be seen from the payback period of the catering business, which is 2 years 11 month, ROI 36% in second year, and the return investment is about 118% in the fifth year.

REFERENCES

[1] APJII, (2016). Data Pengguna Internet di Indonesia 2016. Retrieved Feb 20, 2010, from http://goukm.id/data-pengguna-internet-di-indonesia-2016/.
[2] Wertz, B., & Kingyens, A.T. (2015). A Guide to Marketplaces. Retrieved from http://versionone.vc.
[3] Al-Fedaghi, S. (2011). Developing web applications. *International Journal of Software Engineering and Its Applications*, 5(2), 57–68.
[4] Viehland, D. (2000). Critical success factors for developing an e-business strategy. *Proceedings of the 5th Americas Conference on Information Systems*, 1–7.
[5] Shin, N. (1999). Strategies for competitive advantage in electronic commerce, 164–171.
[6] Garrett, S., & Skevington, P. (2013). An introduction to electronic commerce. *BT Technology Journal*, 2(4), 190–193.
[7] FME. (2013). *SWOT Analysis: Strategy Skills. Free-Management-EBooks*. http://www.free-management-ebooks.com/dldebk-pdf/fme-pestle-analysis.pdf.
[8] Kotler, P., & Keller, K.L. (2012). Marketing management. Upper Saddle River, N.J: Prentice Hall.
[9] Alfiani, Permatasari, Wijaya (2014). Molla & Licker Model Untuk Analisis Critical Success Factor Website E-Commerce, Studi Kasus Bhineka.com.

Practice of mobile guide and learning system by GPS positioning technology

Kai-Yi Chin & Hsin-Ya Hsiao
Department of Digital Humanities, Aletheia University, Taiwan

Zeng-Wei Hong
Department of Computer Science, Faculty of Information and Communication Technology, Universiti Tunku Abdul Rahman, Malaysia

ABSTRACT: People absorb new knowledge by visiting museums and it becomes a kind of leisure activity with learning significance. However, in traditional museum guide, regarding education or training of guides, it required great amount of manpower cost. Therefore, many large-scale museums began introducing new guide methods to improve the traditional one. In recent years, with the progress of technology, many mobile devices are employed in different educational situations. This study develops a museum guide system upon Global Positioning System (GPS). Based on handheld mobile device and mobile communication technology as well as teaching material of history of Tamsui Mackay, it allows learners to independently experience learning guide in different locations. They can thus freely engage in learning guide and profoundly absorb the historic relics and history related to Tamsui Mackay culture.

1 INTRODUCTION

Most museums in early times adopted traditional human guides. Through their help, people could learn the museum's characteristics on the spot, and the guides enhanced visitors' absorption of the background, concept, content, and knowledge related to the exhibits. However, there is a disadvantage to traditional guides. For instance, museums have to rely on manpower. Education or training of on-site guides as well as any interpreters requires a great amount of manpower cost. In addition, there should be a certain number of visitors in order to apply for a guide. For visitors, it is also relatively inconvenient. Under a group guide, the guide might have to explain things several times in order to allow each visitor to obtain complete information. This increases the guides' working hours and work demand. Hence, traditional guides tend to be restricted by the format, resulting in a waste of time and manpower and restrictions to the labor force (Wang, 2014; Chang, 2015).

Due to the previous limitations, many large-scale museums began introducing new types of guides to improve upon the traditional one. For instance, Chao and Lai (2008) treated Taiwan's National Palace Museum as an example and adopted PDA and RFID to allow learners to acquire knowledge of paintings in the Sung Dynasty. Visitors could freely obtain guide information on the exhibits. The National Museum of History has employed multiple interactive systems like a "multi-modal multi-media retrieval system", and visitors must borrow a PDA from the museum. The system includes 3 kinds of retrieval functions: voice, key words of the document, and similar pictures. The visitors input or speak the key words of the historic relics, and they can search for related words and pictures by a wireless full text index (Hong, 2005). This system and device not only effectively lower the amount of explanations for group guides, but also emphasize the maintenance of quietness and comfort in the exhibition. However, the guide tends to be restricted by the device provided by the museum, and it is not based on the visitors' personal mobile devices. This situation produces an extra charge, and thus is difficult to extend into different application domains.. (Wang, Chiang, & Hod, 2012).

With the development of technology recently, many mobile devices can be applied in different educational and guide situations. With a high degree of portability and computer-like ability, mobile devices are widely employed to experience on-site guides (Huanga, Rauchb & Liawc, 2010). For instance, Wang (2014) integrated Augmented Reality and a mobile device to replace the traditional voice guide system. With the assistance of a handheld mobile device, it allows learners to scan the Quick Response Code (QR code) and immediately download multimedia teaching materials that are presented on learners' mobile phones by Augmented Reality. Learners can freely receive guide learning in the exhibition according to their preferences and needs. Chu (2014) treated Tainan Confucian Temple as a learning environment and coordinated PDA and the convenience of a wireless network to develop a complete, rapid, and multi-media cloud mobile guide system. The system enhances the playfulness of the learning process for learners who are experiencing the beauty of ancient temples. It strengthens the functions of education and tourism and increases the number of tourists to Tainan Confucian Temple. Zhang, Sung, Hou, and Chang (2014) adopted Augmented Reality to develop a digital learning system in order to help learners enhance their learning outcome and interest in astronomical observations and to break through current instructional limitations.

However, previous studies mostly restricted mobile guide applications to a single place. When guide content is dispersed over several locations, it is more difficult to guide learners to move between different places. Once learners leave the area, they cannot use the mobile guide system again and it restricts the usage rate of the system. For instance, Hwang and Chang (2011) conducted a learning activity study by integrating Personal Digital Assistant (PDA) and Radio Frequency Identification (RFID) and practiced the learning of history of a temple. Hwang, Wu, Chen, and Tu (2016) combined Augmented Reality and smartphone sensor to construct a virtual butterfly ecological learning system on campus. Through a mobile device and in a butterfly park on campus, learners acquired knowledge on the growth of butterflies and observed their relationship with natural enemies at different phases. Huang, Lin, and Cheng (2010) also proposed a mobile plant learning system to allow learners to integrate a mobile device and RFID, employing it in a botanical garden. However, this kind of system was restricted to one place instead of multiple locations. Once learners leave the designated places, they cannot obtain further learning data by mobile device.

This study therefore develops a museum guide and learning system based upon the Global Positioning System (GPS) and adopts handheld mobile device and mobile communication technology as well as teaching materials of the history of Tamsui Mackay to allow learners to independently acquire the advantages of a learning guide in different learning locations. By the positioning service function of GPS, learners can reach different destinations according to e-map instructions. Once they arrive at their designation, they click the targets on the e-map, and they can immediately and completely obtain learning information and content of teaching materials. Learners can freely practice the learning guide and profoundly absorb the historic relics and history of Tamsui Mackay culture.

The goal of this study is to both reduce the workload of employees or guides who work at museums and to provide a simple tour-guiding system to support visitors for learning about historic relics and culture. To test the real-life applicability of our system, six employees who work at Oxford College, the Taiwanese institution founded by Dr. Mackay, were invited to experience the proposed system. After the actual operation, questionnaires and interviews were adopted to investigate the employees' satisfaction with the museum guide and learning system in the initial stage. By presenting historic context in a mobile way, it aims to stimulate learners' interest in historic sites and gives them visual enjoyment. We also hope that the proposed system can improve the traditional tour guide, reduce labor cost, strengthen learning intention, and be widely adopted as a tour guide at historic sites.

2 LITERATURE REVIEW

2.1 *Application of GPS technology*

With the rapid progress of mobile devices, applications using GPS technology are very prevalent. Marshall, Fenger, & Moioli (2011) increased reliability and indoor positioning by augmenting GPS receiver data with mobile network cell attributes. Enge and Misra (1999) stated that GPS is widely seen as the most important gift from the Department of Defense to the civil world, perhaps with the exception of the Internet. Many e-products have been actively engaged in the planning and integration of navigation functions of GPS, such as e-map company, automobile navigation system, mobile device, PDA, etc. (Wang, 2000). Thus, technology is being combined with learning issues (Ashbrook & Starner, 2002).

Chou and Chan (2012) integrated GPS with e-map function and adopted Augmented Reality, Google Earth, and Google Maps to provide a campus guide system for users. Although GPS is a simple positioning tool, its operational interface is easy to comprehend. This system was used to integrate the learning content of different fields, and thus students could approach the knowledge of new technology through games and apply the acquired skills to daily life. Wang (2013) introduced the author's experience using the Online Positioning User Service provided by the National Geodetic Survey (NGS) and the Automatic Precise Positioning Service provided by the Jet Propulsion Laboratory (JPL) in teaching two undergraduate courses. Lee, Tewolde, & Kwon (2014) designed in-vehicle device works using Global Positioning System (GPS) and Global System for Mobile communication/General Packet Radio Service (GSM/GPRS) technology.

Together with a tablet computer, GPS technology, and Augmented Reality, Chiang, Yang and Hwang (2014) proposed a plant learning system of situational learning that was employed by students in southern Taiwan in reality. With their own tablet computers and teacher's instruction, students were naturally involved in learning situations to reinforce learning interest and efficiency. Giraldo, Arango, Cruz, and Bernal (2016) integrated GPS with a smartphone to propose a campus guide system. Learners could visit the campus with their own smartphones and absorb the information related to buildings on campus.

Based on the above, GPS applications have become prevalent in the educational field. Learners can obtain related information by a combination of GPS and other media. With the assistance of GPS in problem solving, they can acquire related learning skills and information. Upon coordination with a handheld mobile device, instruction can extend to outdoor places from the classroom, and learning is no longer limited by space. Knowledge and skills acquired can be applied in daily life (Chou, 2012).

Although a GPS-based system is not new, its potential in the educational domain has been less explored, especially for historical instruction. Thus, this study developed the museum guide and learning system, which guides learners to visit various historical buildings through the use of GPS technology. We hope that the proposed system can facilitate learner knowledge acquisition during museum visit activities.

2.2 *Museum guide*

The advancement of modern information technology has led to more diverse changes at museums. From traditional exhibitions, they have gradually developed various presentations and content. In early times, with the limitation of space, a great amount of description could not be included in the introduction beside the real objects. In addition, visitors could not obtain richer knowledge from simple descriptions. However, with the innovation of digital technology and devices, museums are applying digital technology to guides in order to provide visitors with different experiences (Serio, Ibáñez, & Kloos, 2013).

Many scholars have developed various types of digital guide systems. For instance, Chu (2014) combined PDA with a wireless network as a mobile guide system for Tainan Confucian Temple. In that system, a mobile phone was the browsing tool. It helped lower operating and learning thresholds, and visitors could immediately acquire information and pictures and also select their sequence of guide. Chao and Lai (2008) treated the National Palace Museum as an example and designed a guide game system by PDA and RFID to access data of historic relics in the museum. Learners are able to obtain knowledge of paintings in the Sung Dynasty. It added the value of a historic guide in the museum; in addition, when using the digital guide service, visitors could freely and naturally acquire guide information of exhibits and thus concentrate and enjoy the digital learning of historic relics.

This study hence combines museum guide and GPS positioning service function and adopts a smart mobile device to present an e-map with GPS function. According to the guidance of the e-map, learners can experience a learning guide for different locations. Without the limitation of time and space, they can freely accomplish all learning content and profoundly recognize the development of historic relics of Tamsui Mackay culture.

3 SYSTEM ARCHITECTURE

Figure 1 is a framework of the GPS mobile guide system developed by this study. The system includes 5 modules. The first is the main menu module, and there are 3 buttons for calling the front page, map, and point collection; and the learning module provides multimedia learning materials. The purpose is to allow learners to profoundly absorb the historic relics of Tamsui Mackay culture. It assists learners through pictures and words; the point collection module records the learning process by point collection to allow learners to recognize the places they have or have not visited. By this measure, it enhances learners' impression on learning locations.

The GPS e-map module guides learners to learning places by GPS technology and e-map. Learners can thus accomplish learning at their own speed without the limitation of time and space. An online video module is placed on the page of the learning module. When learners finish reading the learning information, they can press the button to connect to a guide video. The video also enhances learners' impression on learning content.

Figure 2 is the learning flowchart of the GPS mobile guide system. First, when learners enter the APP, the first image they see is the front page, including 8 thematic pages. When learners press the button at the bottom left of the front page, the main menu shows up on the screen. Learners can press the second button on the main menu to call up the GPS e-map module; they can now reach different learning locations according to the e-map instruction. The characters on the screen walk with the learners and move at the same time to guide them.

Once learners arrive at their destinations, they can click on the sign on the e-map and the corresponding learning information will be shown on the screen. The learning information mainly is used to illustrate the relationship between the museum and the history of Mackay. Learners can thus profoundly appreciate the development of Tamsui Mackay culture. Once

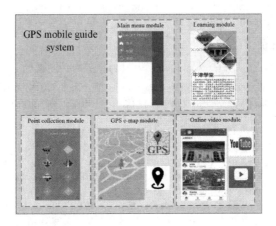

Figure 1. Framework of GPS mobile guide system.

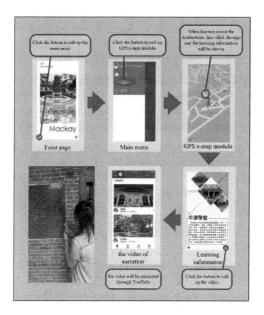

Figure 2. Learning flowchart of the GPS mobile guide system.

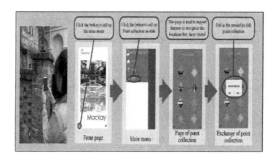

Figure 3. Point collection flowchart of the GPS mobile guide system.

learners finish reading the information, they can open the related video of narration through the use of YouTube. The learning objectives can now make a greater impression on visitors.

Figure 3 is a point collection flowchart of the GPS mobile guide system. The first image is the front page. By clicking the bottom left button, learners can enter the main menu to press the button of point collection and open the screen; the function allows learners to control their learning process and realize the learning locations they have visited or have not visited. Hence, they have clear learning objectives and directions. Once learners collect all points, they can go to Aletheia University Oxford College and show the image of point collection to the personnel in order to receive post cards. The goal is to reward learners who finish the guides of all locations.

4 EXPERIMENT RESULTS AND DISCUSSION

Our museum guide and learning system uses GPS to present an e-map instruction, so as to attract learners, meet their demand for visual enjoyment, and enhance their learning efficiency. In this study, six employees who work at Oxford College, were invited to operate the proposed system, and questionnaires and interviews were conducted to investigate their satisfaction with and acceptance of the system. The questionnaire was composed of eight questions that

Table 1. Results of the questionnaire survey on satisfaction.

Questions	Mean ± SD
1. Is it convenient to operate this tour-guiding system?	4.1 ± 0.52
2. Are you satisfied with the stability of this tour-guiding system?	3.2 ± 0.45
3. Has this tour-guiding system deepened your understanding of historic buildings?	4.0 ± 0.73
4. Do you think this tour-guiding system is fresh and attractive?	4.1 ± 0.62
5. Do you think this tour-guiding system can help visitors to learn related knowledge about historical buildings?	3.4 ± 0.55
6. Has this tour-guiding system left a deep impression of learning content on you?	3.8 ± 0.45
7. Has the e-map presentation of this tour-guiding system increased your intention to visit historical buildings?	3.6 ± 0.55
8. Are you satisfied with this tour-guiding system as a whole?	4.0 ± 0.78

must be answered using a 5-point Likert scale, with responses ranked from 1 (strongly disagree) to 5 (strongly agree). Table 1 shows the experimental results of the investigation.

According to the results of the experiment, the mean scores between 3 and 4 denote that most of the participants were satisfied with the system. The highest mean score corresponds to Question 1 (M = 4.1) and Question 4 (M = 4.1), indicating that most of the participants felt satisfied when operating our proposed system, and they also felt fresh and attractive toward this system. However, the score of Question 2 was the lowest. Because of the instability of the wireless network, the participants suggested that the stability of the system could be enhanced.

The participants were interviewed with the following questions: 1) What is the difference between this tour-guiding system and an actual tour? 2) What troubles do you have in the use of this tour-guiding system? 3) What impresses you about this tour-guiding system? 4) What is your suggestion for this tour-guiding system if you have more time to use it to learn?

For Question 1, most participants agreed that there was no time limit on the use of the system and convenient access could reduce the labor cost of the traditional tour guide. In terms of Question 2, some participants believed that the unstable network resulted in a slow reaction of the system, which is a problem to be solved. For Question 3, some participants agreed that the system deepened their understanding of the history of buildings. For Question 4, the participants expressed the hope of adopting multimedia materials and content to get more details, which will be our focus in further studies. One participant also expressed that he wanted to see the virtual 3D images of the interior or external structures of historical buildings, so as to get more details about buildings. Thus, in our follow-up study, we need to think about how to use Augmented Reality technology to extend the capability of our proposed system.

5 CONCLUSION AND FUTURE STUDIES

Different from the traditional guide system, the GPS mobile guide system proposed herein is based on historic relics and history of Tamsui Mackay culture. It collects all data related to historic relics of Tamsui Mackay as the main content of teaching materials. Therefore, the data are dispersed in several historic relics' locations instead of one location. Learners can see or acquire learning targets after walking for a while.

In order to effectively guide learners in learning, this study integrates GPS technology and e-map and adopts the presentation of a smartphone to efficiently guide visitors to learning locations. With instructions similar to Google maps, learners can walk with the characters on the map and adjust the directions at any time. The system thus provides a learning environment where they can control their own directions. In addition, the GPS mobile guide system proposed herein is based on point collection to allow learners to recognize learning locations and obtain points once they finish the learning activities in the places. Hence, they control the learning process and search for learning targets on the e-map and position their learning locations. Through this GPS mobile guide system, this study allows learners to further absorb the correlation between historic relics of Tamsui and Mackay culture, profoundly acquire Mackay culture and acquire rich knowledge.

As to the planning of future works, this study will first promote the usage rate of the GPS mobile guide system with the museum. Not only can students in school use the system, but tourists can also approach the Mackay culture by this system in order to promote tourism. In addition, this study expects the English translation to enhance diversity of teaching material in this system, and it will provide different versions to reinforce the utility of learning. Finally, since the development of the system has been accomplished, but not been used by students and tourists, the study's experiment has not yet been fulfilled. This aspect will be included in the planning of future works.

ACKNOWLEDGMENT

This work was supported by the Ministry of Science and Technology of Taiwan under grant number MOST 105-2221-E-156-007 and MOST 105-2632-M-156-001.

REFERENCES

Ashbrook, D. & Starner, T. (2002). Learning significant locations and predicting user movement with GPS. *Proceedings. Sixth International Symposium,* 101–108.

Chang, P. (2015). QR-based on Digital Guide System and Research User's Learning Effect. Aletheia University. Taipei, Taiwan.

Chao, J. & Lai, T.S. (2008, June). The mobile digital learning space of Taiwan national palace museum. *In EdMedia: World Conference on Educational Media and Technology,* 1, 5153–5161.

Chiang, T.H., Yang, S.J. & Hwang, G.J. (2014). Students' online interactive patterns in augmented reality-based inquiry activities. *Computers & Education,* 78, 97–108.

Chou, T.L., & Chan, L.J. (2012). Augmented reality smartphone environment orientation application: A case study of the Fu-Jen University mobile campus touring system. *Procedia-Social and Behavioral Sciences,* 46, 410–416.

Chu, H.C. (2014). Potential Negative Effects of Mobile Learning on Students' Learning Achievement and Cognitive Load-A Format Assessment Perspective. *Educational Technology & Society,* 17(1), 332–344.

Enge, P. & Misra, P. (1999). Special issue on global positioning system. *Proceedings of the IEEE,* 87(1), 3–15.

Giraldo, F.D., Arango, E., Cruz, C.D. & Bernal, C.C. (2016, September). Application of augmented reality and usability approaches for the implementation of an interactive tour applied at the University of Quindio. *2016 IEEE 11th Colombian Congress on Computing (11CCC),* 1–8, Popayán.

Hong, K.S. (2005). *A Digital Guiding System – Implementation and Research Using RFID and Wireless LAN Technology.* National Chung Cheng University. Chiayi, Taiwan.

Huang, Y.M., Lin, Y.T. & Cheng, S.C. (2010). Effectiveness of a mobile plant learning system in a science curriculum in Taiwanese elementary education, *Computers & Education,* 54(1), 47–58.

Huanga, H.M., Rauchb, U. & Liawc, S.S. (2010). Investigating learners' attitudes toward virtual reality learning environments: based on a constructivist approach. *Computers & Education,* 55(3), 1171–1182.

Hwang, G.J., & Chang, H.F. (2011). A formative assessment-based mobile learning approach to improving the learning attitudes and achievements of students. *Computers & Education,* 56(4), 1023–1031.

Hwang, G.J., Wu, P.H., Chen, C.C. & Tu, N.T. (2016). Effects of an augmented reality-based educational game on students' learning achievements and attitudes in real-world observations. *Interactive Learning Environments,* 24(8), 1895–1906.

Lee, S., Tewolde, G. & Kwon, J. (2014). Design and implementation of vehicle tracking system using GPS/GSM/GPRS technology and smartphone application. *IEEE World Forum on* pp. 353–358.

Marshall, C., Fenger, C. & Moiolo, S. (2011). Hybrid positioning and CellLocate. *White Paper Published by u-blox AG.* Retrieved from http://www.u-blox.com/sites/default/files/products/documents/CellLocate_WhitePaper_%28GSM-X-11001%29.pdf.

Serio, Á.D., Ibáñez, M.B. & Kloos, C.D. (2013). Impact of an augmented reality system on students' motivation for a visual art course. *Computers & Education*, 68, 586–596.

Wang, C.S., Chiang, D.J. & Ho, Y.Y. (2012). 3D Augmented Reality Mobile Navigation System Supporting Indoor Positioning Function. *IEEE International Conference on Computational Intelligence and Cybernetics*, pp. 64–68.

Wang, G. (2013). Teaching high-accuracy global positioning system to undergraduates using online processing services. *Journal of Geoscience Education*, 61(2), 202–212.

Wang, L.Y. (2014). *Combining Augmented Reality and 2D Barcode on a Mobile Guided System—A Case Study on Museum Guiding.* Feng Chia University, Taichung, Taiwan.

Wang, Y.H. (2000). GPS technology and development. *CTIMES*, 50, 108–116.

Zhang, J., Sung, Y.T., Hou, H.T. & Chang, K.E. (2014). The development and evaluation of an augmented realitybased armillary sphere for astronomical observation instruction. *Computers & Education,* 73, 178–188.

Issues and Trends in Interdisciplinary Behavior and Social Science – Lumban Gaol et al. (Eds)
© 2018 Taylor & Francis Group, London, ISBN 978-1-138-55380-4

Mathematical models for the spread of rumors: A review

Meksianis Z. Ndii
Department of Mathematics, University of Nusa Cendana, Kupang-NTT, Indonesia

Ema Carnia & Asep K. Supriatna
Department of Mathematics, Padjadjaran University, Jatinangor, Jawa Barat, Indonesia

ABSTRACT: Studies on how a rumor spreads have attracted increasing attention because rumors can shape public opinion and affect beliefs, thereby changing individuals' attitude toward social, economic, and political aspects. Rumor has negative impact on society and hence its spread needs to be clearly understood. However, understanding the spread of rumors in the society is challenging as many factors are involved, and their effects have to be determined. The use of mathematical models for analyzing the dynamics of the rumor spread is common, and a number of mathematical models have been developed to examine its spreading dynamics. Therefore, a systematic review about the available models is required. In this study, we review mathematical models for rumor spread in order to summarize several significant results and identify important aspects that require further investigation to advance our understanding of the dynamics of rumor spread.

1 INTRODUCTION

Rumor is a form of social communication and can shape the public opinion and affects the beliefs of individuals, there by leading to the changes of individuals' attitude toward economic, political, and social aspects (Kawachi, 2008, Misra, 2012). Rumor is unconfirmed truth and has negative impact on society, although an intervention can be implemented to minimize its negative effects (Huo *et al.*, 2011). Therefore, understanding the spread of rumor is significantly important to obtain scientific information and better strategies in reducing its negative impacts.

Dynamics of rumor spread can be investigated using mathematical models. A number of mathematical models have been developed to examine the dynamics of rumor spread and to assess the important factors contributing to it (Chierichetti *et al.*, 2011, Han *et al.*, 2015, Han *et al.*, 2014, Huo *et al.*, 2016, Huo et al., 2011, Kawachi, 2008, Ma *et al.*, 2016, Tian *et al.*, 2015, Wang *et al.*, 2014, Zhao *et al.*, 2013a, Zhao *et al.*, 2013b, Zhao *et al.*, 2012, Zhao *et al.*, 2013c, Zhao *et al.*, 2013d, Rabajante *et al.*, 2011, Piqueira, 2010). This study has contributed significantly to the body of knowledge of the dynamics of rumor spread. Therefore, a comprehensive review is needed to provide a critical summary of the existing research and to point out aspects that can be further investigated to advance our understanding of the dynamics of the rumor spread. Furthermore, to our knowledge, very few review papers on mathematical models for the spread of rumors are available.

2 METHODS

2.1 *Search strategy*

We perform a literature search in Scopus database using the keywords *rumor spreading*. We then limit our search using the keyword *network* to classify the work using network model.

We make sure that the list of publications uses mathematical approach to analyze the spread of rumor. We limit the time period from 1990 to 2016.

2.2 Paper selection

We select articles from journals and conference proceedings and then read abstract and the introduction of the paper to identify papers that utilize mathematical models for analyzing the spread of rumors. We then classify the work according to their purpose: understanding the dynamics of rumors spread with and without intervention. We obtained 620 articles, and only few review papers on mathematical models for the spread of rumors are available. We also added other papers about rumors that of interest were not obtained from the literature search in the Scopus database.

2.3 Model classification

We classify the mathematical models into the following two groups, network and non-network models (see, for example, (Funk *et al.*, 2013, Ganesh, 2015)). We define the non-network model as that formulated without consideration of the effects of network properties or topologies on the dynamics of rumor spread. A network model is the one that implements the concept of network and is formulated to examine the effects of network properties on the dynamics of rumor spread.

3 RESULTS

3.1 General overview

The aim of this study is to find how rumor spread has greatly attracted attention. Figure 1 shows an exponential growth in the number of publications on rumor spread in particular since 2009. The number of published articles has increased from only 2 articles in 1990 to 98 articles in 2016. By fitting data for the last 14 years to a simple exponential function, we found that it is likely that the number of publications would increase to around 200 articles per year by 2020. Furthermore, there are only very few review papers on the mathematical model for the spread of rumors.

Figure 2 shows that the research on rumor spreads has been the topic of interest in various fields, including Computer Science, Mathematics, Physics, Engineering, and Social Science.

The majority of mathematical models for the spread of rumors are inspired by compartmental-based epidemic models, where the population is divided into different compartments

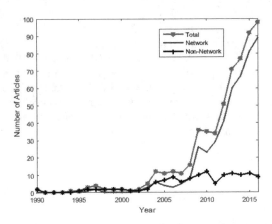

Figure 1. Number of published articles on rumor spreading between 1990 and 2016. Source: data from Scopus 1990–2015.

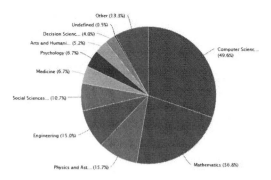

Figure 2. Classification of papers on rumor spreading based on subject areas.

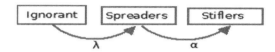

Figure 3. Schematic representation of rumor spreading models where the population is divided into ignorant, spreaders, and stiflers. The arrows represent the movement of individuals between compartments.

depending on their status. The population is generally divided into three groups: ignorant (individuals who do not know that rumor), spreaders (individuals who know and spread the rumor), and stiflers (individuals who know the rumor but do not spread it). The schematic representation of this model is shown in Figure 3.

Figure 1 shows that the ignorant individuals become spreaders at a rate λ and then become stiflers at a rate α. The model is then given by the following system of differential equations:

$$dX/dt = -\lambda XY/N,$$
$$dY/dt = \lambda XY/N - \alpha Y(Y+Z)/N, \qquad (1)$$
$$dZ/dt = \alpha Y(Y+Z)/N.$$

This is a deterministic version of the Daley–Kendall rumor spreading model proposed by Arias and Chaves (Thompson *et al.*, 2003). This model assumes homogeneous mixed population and does not take into account heterogeneity in the population. This is similar to SIR epidemic model except the movement from spreader (infected) to stifler (recovered) compartment. The concept of rumor spreading modeling is similar to that of epidemic modeling (Jin *et al.*, 2013, Ndii *et al.*, 2016b, Ndii *et al.*, 2016a, Supriatna *et al.*, 2012). In the epidemic model, the infected individuals leave the infected compartment linearly at rate α. On the other hand, in the SIR rumor spreading model, there is a nonlinear movement from the spreader to stifler compartment (see Equation 1): the spreaders (Y) move to stifler compartment (Z) after interacting with another spreader (Y) and stifler (Z) $\alpha Y(Y+Z)/N$. This model can be extended by adding other assumptions depending on the purpose of modeling (for example, see (Zhao et al., 2013a, Huo et al., 2011, Huo et al., 2016)). For example, this model has been extended by dividing the stifler compartment into two groups: individuals who hear the rumor and decide to spread it or not (Wang et al., 2014). Furthermore, a mathematical model for rumor spread has been formulated by including the forgetting rate variable (Zhao et al., 2013d).

3.2 *Non-network models*

In this section, we review the non-network mathematical models for the spread of rumors. A number of mathematical models for the spread of rumors have been developed. We observe that there are two major aspects for research on rumor spreading using non-network

Table 1. Summary of the work using the non-network models and their findings.

Title/Author	Study purpose	Findings
The Impact of Authorities' Media and Rumor Dissemination on the Evolution of Emergency (Zhao et al., 2012)	They explore the interplay mechanism among authorities' media, rumor dissemination, and the evolution of emergency. They propose a new interplay model and perform stability analysis by the Routh stability criterion	This research found the criterion for stability under which the deterioration of emergency can be effectively prevented. Media with high audience but low credibility will spread the panic. It is recommended that authorities should choose proper media to release the truth to public according to the seriousness of the situation and the rumor spreading rate.
Analysis of the Impact of Education Rate on the Rumor Spreading Mechanism (Afassinou, 2014)	They extend the SIR rumor spreading model by including the forgetting mechanism and the education rate of the population. This paper studied the effects of education in rumor spreading mechanism.	This research found that education affects the rumor final size. If there are more educated individuals in the population, the rumor final size has been reduced.
Dynamical Interplay between the Dissemination of Scientific Knowledge and Rumor Spreading in Emergency (Huo et al., 2016)	They develop a deterministic model where the population is divided into ignorant, spreaders, knowledgeable, and unknowledgeable.	The dynamics are affected by variations of the force of rumor spreading rates and the rates of scientific knowledge dissemination. Mathematical analysis showed that when the scientific knowledge dissemination rate reaches a critical value, the strictly positive interior equilibria will undergo Hopf bifurcation. The rumor is no more when the rate of rumor spreading is lower. All classes/groups coexist when the rates of scientific knowledge and scientific dissemination rates are in appropriate level.
On the Behavior of a Rumor Process with Random Stifling (Lebensztayn et al., 2011)	They generalize the Maki–Thompson rumor spreading model by assuming that each spreader ceases to propagate the rumor right after being involved in a random number of stifling experiences. They also consider the process with a general initial configuration and establish the asymptotic behavior (and its fluctuation) of the ultimate proportion of ignorants as the population size grows to infinity.	They prove that the ultimate proportion of ignorants converges in probability to an asymptotic value as the population size tends to infinity.
An Analysis of Rumor Spreading Model with Contra Productive Intervention (Aldila et al., 2017)	They develop a deterministic mathematical model by dividing the population into a general, semi-fanatical, fanatical, aware, and repented subpopulations. This model is then used to investigate the effects of mass campaign about the dangers of the rumor and the arrest of fanatical people on the spread of rumors.	They found that the model has three steady states: a general population only, aware and repented subpopulation, and a coexistence equilibrium. Furthermore, they found that mass campaign strategy has a potential to increase fanatical populations and hence should be implemented wisely.

models: understanding the transmission dynamics of rumors and the effects of intervention on the dynamics of rumor spread. For the first aspect, the research focuses on investigating the effects of factors/mechanism such as forgetting mechanism, emergency situations, and

Table 2. Summary of the work using network models and their findings.

Titles/Authors	Study purpose	Findings
Randomized Rumor Spreading in Poorly Connected Small-World Networks (Mehrabian et al., 2016)	They study the push–pull protocol on random k-trees, which are small world and have large clustering coefficient. The procedure of the push–pull protocol works is the following: A node knows a rumor and wants to spread it to all nodes in a network quickly. Every round, every informed node sends the rumor to a random neighbor, and every uninformed node contacts a random neighbor and gets the rumor from him if he knows it.	They found that the push–pull protocol can be efficient in poorly connected network
Total Flooding Time and Rumor Propagation on Graphs (Camargo et al., 2017)	They develop the model of a rumor propagation in a rumor time where each site has different initial information and investigate the number of conversation before the entire graphs know the all information.	They found that the ratio between the expected propagation time for all information and the corresponding time for a single piece of information is 3/2.
Why Rumors Spread so Quickly in Social Networks (Doer et al., 2012)	They develop a network model and simulate a simple information spreading process in different network topologies. They also analyze the information spreading in the mathematically defined preferential attachment network topology, which is a common model for real-world networks.	They found that news spreads much faster in the existing network topologies. They prove that a sublogarithmic time suffices to spread news to all nodes of the network. They also found that nodes with few neighbors are crucial for the fast dissemination.
Rumor Spreading Model with Noise Interference in ComplexSocial Networks (Zhu et al., 2017)	They modify the SIR model and explore rumor diffusion on complex social network. They consider variation of connectivity and assume it as noise.	They found that the variation of degree distribution and dynamics of rumor diffusion are related closely to noise intensity. They found a rumor diffusion threshold (in statistical average meaning) in a homogenous network. The variation in the connectivity speeds rumor diffusion and expands diffusion size, which is more significant in BA network in comparison to WS network.

(*Continued*)

Table 2. (*Continued*)

Titles/Authors	Study purpose	Findings
A Process of Rumour Scotching on Finite Populations (de Arruda et al., 2015).	They develop a competition-like model where spreaders try to transmit rumors and stiflers are trying to scotch it. They then study the influence of transmission/scotching rates and initial conditions on the qualitative behavior of the process. They also perform Monte Carlo simulations of the stochastic model on ER random graphs and BA scale-free networks.	They found that the choice of hubs as spreaders or stiflers does not affect the final size of ignorants.
SIR Rumor Spreading Model Considering the Effect of Difference in Nodes Identification Capabilities (Ya-Qi et al.)	They develop an SIR rumor spreading model to study the effect of different network node capabilities, which is the ability of individuals to identify rumors.	They found that the critical threshold increases, but the final size of rumors decreases when taking into account the node's identification capabilities. Furthermore, the differences in the node identification capabilities extend the rumor propagation time to reach a steady state and reduce the number of nodes that accept the rumor. Furthermore, the rumor transmission rate on the inhomogeneous networks is relatively larger than that in homogeneous networks.
Hawkes Processes for Continuous Time Sequence Classification: An Application to Rumour Stance Classification in Twitter (Lukasik et al., 2016)	They use Hawkes processes for classifying sequences of temporal textual data, which exploit temporal and textual information.	They found that the temporal information and textual content provide valuable information for the model to perform well.
Clustering Determines the Dynamics of Complex Contagions in Multiplex Networks (Zhuang et al., 2017).	They perform a mathematical analysis of generalized complex contagions in a class of clustered multiplex networks, aiming to understand the spread of influence or any other spreading process.	They found that the probability and the size of having global cascades decrease when the clustering increases. However, this changes with the average degree. Furthermore, there is a possibility for getting inaccurate conclusion about contagion dynamics when the link types and aggregating network layers are ignored.

individuals' decisions on the spread of the rumors and on the dynamics of rumor spread. For the latter aspect, this study focuses on the effects of intervention such as education, scientific knowledge, media, and government actions. A summary of the work on non-network models is given in Table 1.

Although the works on non-network model provide scientific information on the dynamics of rumor spread, they ignore the effects of network properties on rumor spreading dynamics.

Network properties such as clustering, cliques, and strength of node connection will significantly influence the dynamics of rumor spread. Furthermore, these properties represent the real structure of the population or society, which make the model more realistic.

3.3 *Network models*

The majority of work use network models (for example, see (Huo et al., 2016, Chierichetti et al., 2011, Onaga *et al.*, 2016)). This is an application of graph theory in mathematics. The network model is arguably the most representative approach for modeling the spread of rumors. This is because the properties of the network can represent the structure of the population.

Research shows that network structures affect the dynamics of rumor spread. For example, the variation in connectivity can accelerate rumor diffusion (Zhu *et al.*, 2017). Ignoring link types can lead to misleading conclusion about contagion dynamics, particularly when there is a high correlation of degrees between layers (Zhuang *et al.*, 2017). In comparison to the non-network model, the network models provide flexibility in terms of the effects of social structure in the dynamics of rumor spreading. A summary of work on network model is given in Table 2.

3 DISCUSSION

The advancement of communication technology results in the faster spread of information, rumors, and news. This can shape people opinions, behavior, and decision toward economic, political, and social aspects. This would be a problem when unconfirmed information or rumor has been widely spread in the society. Therefore, it is important to understand the dynamics of rumor spreads and the strategies to minimize its impact. A number of mathematical models have been formulated to understand the dynamics of rumor spread. The model can be categorized into non-network and network models. In general, research showed several important aspects that can reduce the spread of rumors, which are education and scientific knowledge, government roles, and individual decision (Huo et al., 2016, Afassinou, 2014, Aldila et al., 2017, de Arruda et al., 2015). Furthermore, the social structure of the population is also an important aspect, which can accelerate or decelerate the rumor spreading (Zhuang et al., 2017, Ya-Qi et al.).

4 CONCLUSION

Mathematical model is a useful tool for analyzing the dynamics of the spread of rumors. We can use non-network and network models depending on the purpose of the research. The models have been developed to investigate the better strategies to minimize the negative impact of rumors or understanding the effects of network properties on the dynamics of the spread of rumors.

The network model may provide flexibility for analyzing the dynamics of rumor spread because the population structure can be represented in network. An important property of the network theory, where its effects on the dynamics of the spread of rumor has not been fully investigated, is clique. The clique is a subset of vertices of undirected graph resulting in a complete subgraph. In social science, a clique can be defined as a group of individuals who share similar interest. This means that there are more than one clique in the society and their interaction may affect dynamics of idea, news, and rumor spread. Further research needs to be conducted to investigate the effects of clique in the dynamics spread of rumors. This may determine the time for rumors to reach all individuals in the society and the effects of clique in reducing the rumor spreading rate. This is particularly important when the number of nodes or individuals in each clique is different. The effects of clique on the dynamics of rumor and epidemic spread are currently under investigation.

ACKNOWLEDGMENT

This study was funded by the Directorate of Research and Higher Education through Penelitian Pascadoktor 2017 (grant no: 222/UN.15.10/LT/2017).

REFERENCES

Afassinou, K. 2014. Analysis of the impact of education rate on the rumor spreading mechanism. *Physica A: Statistical Mechanics and its Applications,* 414, 43–52.
Aldila, D., Paramartha, H. P., et al. 2017. An Analysis Of Rumor Spreading Model With Contra Productive Intervention. *International Journal of Pure and Applied Mathematics,* 112 519–530.
Camargo, D. & Popov, S. 2017. Total Flooding Time and Rumor Propagation on Graphs. *Journal of Statistical Physics,* 166, 1558–1571.
Chierichetti, F., Lattanzi, S., et al. 2011. Rumor spreading in social networks. *Theoretical Computer Science,* 412, 2602–2610.
De Arruda, G. F., Lebensztayn, E., et al. 2015. A process of rumour scotching on finite populations. *Royal Society Open Science,* 2.
Doer, B., Fouz, M., et al. 2012. Why Rumors Spread So Quickly in Social Networks. *Communications of the ACM,* 55, 70–75.
Funk, S. & Jansen, V. a. A. 2013. The Talk of the Town: Modelling the Spread of Information and Changes in Behaviour. *In:* MANFREDI, P. & D'ONOFRIO, A. (eds.) *Modeling the Interplay Between Human Behavior and the Spread of Infectious Diseases.* New York, NY: Springer New York.
Ganesh, A. J. 2015. Rumour spreading on graphs. *In:* BRISTOL, U. (ed.).
Han, Q., Wen, H., et al. 2015. Rumor Spreading and Security Monitoring in Complex Networks. *In:* THAI, M. T., NGUYEN, N. P. & SHEN, H. (eds.) *Computational Social Networks: 4th International Conference, CSoNet 2015, Beijing, China, August 4–6, 2015, Proceedings.* Cham: Springer International Publishing.
Han, S., Zhuang, F., et al. 2014. Energy model for rumor propagation on social networks. *Physica A: Statistical Mechanics and its Applications,* 394, 99–109.
Huo, L.-A., Huang, P., et al. 2011. An interplay model for authorities' actions and rumor spreading in emergency event. *Physica A: Statistical Mechanics and its Applications,* 390, 3267–3274.
Huo, L. A. & Song, N. 2016. Dynamical interplay between the dissemination of scientific knowledge and rumor spreading in emergency. *Physica A: Statistical Mechanics and its Applications,* 461, 73–84.
Jin, F., Dougherty, E., et al. 2013. Epidemiological Modeling of News and Rumors on Twitter. *The 7th SNA-KDD Workshop.*
Kawachi, K. 2008. Deterministic models for rumor transmission. *Nonlinear Analysis: Real World Applications,* 9, 1989–2028.
Lebensztayn, E., Machado, F. P., et al. 2011. On the behaviour of a rumour process with random stifling. *Environmental Modelling & Software,* 26, 517–522.
Lukasik, M., Srijith, P. K., et al. 2016. Hawkes processes for continuous time sequence classification: an application to rumour stance classification in twitter. *The 54th Meeting of the Association for Computational Linguistics.* Association for Computer Linguistics.
Ma, J., Li, D., et al. 2016. Rumor spreading in online social networks by considering the bipolar social reinforcement. *Physica A: Statistical Mechanics and its Applications,* 447, 108–115.
Mehrabian, A. & Pourmiri, A. 2016. Randomized rumor spreading in poorly connected small-world networks. *Random Structures & Algorithms,* 49, 185–208.
Misra, A. K. 2012. A simple mathematical model for the spread of two political parties. *Nonlinear Analysis: Modelling and Control,* 13, 343–354.
Ndii, M. Z., Allingham, D., et al. 2016a. The effect of Wolbachia on dengue dynamics in the presence of two serotypes of dengue: symmetric and asymmetric epidemiological characteristics. *Epidemiology and Infection,* 144, 2874–2882.
Ndii, M. Z., Allingham, D., et al. 2016b. The effect of Wolbachia on dengue outbreaks when dengue is repeatedly introduced. *Theoretical Population Biology,* 111, 9–15.
Onaga, T. & Shinomoto, S. 2016. Emergence of event cascades in inhomogeneous networks. *Scientific Reports,* 6, 33321.
Piqueira, J. S. C. 2010. Rumor Propagation Model: An Equilibrium Study. *Mathematical Problems in Engineering,* 2010.
Rabajante, J. F. & Umali, R. E. D. C. 2011. A Mathematical Model of Rumor Propagation for Disaster Management. *Journal of Nature Studies,* 10, 61–70.

Supriatna, A. K. & Anggriani, N. 2012. System Dynamics Model of Wolbachia Infection in Dengue Transmission. *Procedia Engineering,* 50, 12–18.

Thompson, K., Estrada, R. C., et al. 2003. A Deterministic Approach to the Spread of Rumors. *In:* LABORATORY, L. A. N. (ed.) *Working Paper.*

Tian, R.-Y., Zhang, X.-F., et al. 2015. SSIC model: A multi-layer model for intervention of online rumors spreading. *Physica A: Statistical Mechanics and its Applications,* 427, 181–191.

Wang, J., Zhao, L., et al. 2014. SIRaRu rumor spreading model in complex networks. *Physica A: Statistical Mechanics and its Applications,* 398, 43–55.

Ya-Qi, W. & Jing, W. SIR rumor spreading model considering the effect of difference in nodes' identification capabilities. *International Journal of Modern Physics C,* 0, 1750060.

Zhao, L., Cui, H., et al. 2013a. SIR rumor spreading model in the new media age. *Physica A: Statistical Mechanics and its Applications,* 392, 995–1003.

Zhao, L., Qiu, X., et al. 2013b. Rumor spreading model considering forgetting and remembering mechanisms in inhomogeneous networks. *Physica A: Statistical Mechanics and its Applications,* 392, 987–994.

Zhao, L., Wang, Q., et al. 2012. The impact of authorities' media and rumor dissemination on the evolution of emergency. *Physica A: Statistical Mechanics and its Applications,* 391, 3978–3987.

Zhao, L., Wang, X., et al. 2013c. A model for the spread of rumors in Barrat–Barthelemy–Vespignani (BBV) networks. *Physica A: Statistical Mechanics and its Applications,* 392, 5542–5551.

Zhao, L., Xie, W., et al. 2013d. A rumor spreading model with variable forgetting rate. *Physica A: Statistical Mechanics and its Applications,* 392, 6146–6154.

Zhu, L. & Wang, Y. 2017. Rumor spreading model with noise interference in complex social networks. *Physica A: Statistical Mechanics and its Applications,* 469, 750–760.

Zhuang, Y., Arenas, A., et al. 2017. Clustering determines the dynamics of complex contagions in multiplex networks. *Physical Review E* 95, 012312.

Relationship between emotional intelligence and job performance among early childhood professionals: A preliminary study

Nurul Ghanieah, Fonny Hutagalung, Noor Aishah Rosli & Chin Hai Leng
Department of Educational Psychology and Counseling, Faculty of Education, University of Malaya, Malaysia

ABSTRACT: Emotional intelligence is important for teacher effectiveness in the field of early childhood education. It is defined as the ability to process emotional information efficiently and accurately, and by recognizing our own feelings and emotions and those of others to manage the relationships with children. In order to become creative and effective educators of young children, they must consider their teaching competence beyond their academic qualification, professional skills, and intellectual competencies. The implications of the emotional competence level of teachers can be manifested in the form of performance of students. The objective of this study was to identify the relationship between emotional intelligence and job performance among early childhood professionals. The findings of this study will contribute to the education field, especially to the fields of early childhood education such as teaching, policy-making, and parenting.

Keywords: Emotional intelligence, job performance, early professionals

1 INTRODUCTION

The National Association for the Education of Young Children (NAEYC) stated that early childhood education is the main years where children from zero to eight years old having their early-years lesson in life to help children identify their ability, recognize their skills, and improve their development. The most important thing part of preschool professionals is acquiring emotional intelligence. Emotional intelligence (EQ) has its own practical value in the workplace. Teachers play an important role in school setting. Irene Becker (2015) stated that EQ is the skill of understanding and managing our own emotions and others'. A high level of EQ will lead to effective job performance.

In order to become creative and effective educators of young children, teachers must consider their teaching competence beyond their academic qualification, professional skills, and intellectual competencies. The implications of the emotional competence level of teachers can be manifested in the form of performance of students (Brackett & Katulak, 2006). Inability of a teacher to create a classroom environment that supports the fast-paced and quick learning among students can inhibit the performance of the pupils, while the teacher performance is explained through their quality of interaction with students. With this background, researchers have suggested inclusion of teacher training that specifically aims at the development of their level of emotional competence (Hawkey, 2006).

Professionals dealing with the problem of stress not only have a negative impact on workplace, but also present a negative behavior within the children environment. Consequently, their performance as educators declines, eventually resulting in poor skills or performance of the children. Being a part of the educational sector as an instructor entails high level of professional stress for the teachers, its impact on the classroom behavior is also evident. Teachers who are exposed to high level of work-related stress have been found to carry a greater risk of developing negative attitude about their jobs and its related responsibilities (Klassen, 2010).

Ismail and Idris (2009) analyzed the personality of Malaysian teachers by studying the linkages between their displayed emotions and handling of students within the classroom. It has been found that increased use of negative emotions within the classroom has inhibited the learning of students, thus suggesting that negativity in teacher's personality highly influenced their performance as educator. This research supported Pianta (1999) that the psychosocial and academic development of students is facilitated through the establishment of warm interpersonal relationships within the classroom with the teachers. As early childhood professionals, their responsibility is to emphasize the growth and understanding of young children. Lack of quality and awareness of being competent will lead to levels of children performance. It can rise to problematic behavior and inappropriate approaches in handling them.

On the basis of the statistics of cases of physical abuse against children obtained from the Division of Investigation of Sexual, Abuse and Children (D11), criminal investigation by Police Department of Malaysia indicates that caregivers are those most involved in crimes, with 439 cases in child care centers. This shows the importance of persons or the community responsible for educating, nurturing, and teaching children to be better human beings. There were a total of 1372 cases of abuse motif, hot-temper, and impatience. These statistical data have prompted the research of the relationship between emotional intelligence level of early childhood professionals and their job performance. Emotional intelligence is positively genuine in achieving good performance among early childhood education professionals that can help enhance their positive output to the profession.

On the basis of this explanation, the purpose of this study was to determine the relationship between emotional intelligence and job performance among early childhood professionals.

1.1 *Research questions*

The aim of this study was to examine the relationships between emotional intelligence and job performance among early childhood professionals. The overall guiding questions for this study were:

i. What is the relationship between self-awareness and job performance among early childhood professionals?
ii. What is the relationship between self-management and job performance among early childhood professionals?
iii. What is the relationship between self-motivation and job performance among early childhood professionals?
iv. What is the relationship between empathy and job performance among early childhood professionals?
v. What is the relationship between social skills and job performance among early childhood professionals?

1.2 *Literature review*

Studies on enthusiastic knowledge have increased after the study of Daniel Goleman, in which he exhibited his thoughts regarding the elements of the wonder, investigating the way of passionate knowledge and its suggestions for a person in individual and occupation-related circles of his presence. One of the main patrons in the EI hypothesis is Goleman (1996) who proposed that subjective insight normally spoke to by intelligence quotient (IQ) does not adequately clarify the abilities that should have been beneficial inside expert and social circles of life. The apparent absence of studies situating feelings as a part of wise conduct has likewise activated examination in such manner. The utilization of insight cannot bolster people in managing work environment issues, for example, hierarchical clash, reacting to change, and being creative.

According to Murphy (1989), human personality is impacted by three key factors, namely mental procedures, feelings, and the level of inspiration of the person. Contemplates have concentrated on investigating the linkages between the level of insight of an individual and the execution at the working environment. However, analysts have likewise begun scrutinizing genuineness of insight remainder (IQ) as the sole element that can be used to foresee or comprehend the execution of representatives. Goleman (1998) expressed that IQ does not adequately clarify the reasons why people with a higher level of specialized aptitudes cannot perform well in their jobs that require initiative position. Again, a representative with normally specialized abilities is ready to exhibit a more noteworthy level of ability to take care of such assignments. One of the conceivable reasons in such manner is that the level of execution is an impression of the level of enthusiastic skill controlled by the workers. The higher the level of the enthusiastic skill, the more prominent is the likelihood of adequately associate with the different representatives, along these lines bringing about better execution (Goleman, 1998; Carmeli and Josman, 2006).

Various studies have situated passionate insight (EQ) as a point of convergence of their investigative procedures, meaning to test the linkages of EQ with the yield of a person as the part of an association (Lam and Kirby, 2002; Rapisarda, 2002; Sy, Tram and O'hara, 2006; O'Boyle et al., 2011). Instructive organizations are esteemed as the key wellspring of creating scholastic fitness among the understudies. Instructors should guarantee that the data given to them are appropriately comprehended. Another key obligation connected with the educators is to design the educational modules, as well as to guarantee finish of the syllabus within the predetermined time span of the understudies. Notwithstanding this, supporting classroom training, overseeing association with understudies, taking adequate care of cooperation with guardians are also necessary obligations, which should be satisfied by the educators. It is likewise essential to note that a teacher's part is significantly more than just being a wellspring of scholarly learning for the understudies.

The part of instructors turns out to be more imperative as they encourage the mental development of the understudies as well. There is a plausibility that educators can recognize their own enthusiastic state and additionally the passionate condition of their researchers, in this way creating a more profound comprehension of why their understudies and they have a tendency to act in a specific way. Then again, absence of mindfulness about the passionate condition of self or others can offer insight into a dangerous translation of the conduct and additionally reception of unseemly ways to deal with handle them. Passionate knowledge is seen going up against a focal position in creating great execution among instructors in schools, schools, and colleges (Jennings and Greenberg, 2009). Therefore, to turn into powerful instructors, they should consider their showing capability past the scholarly fitness, proficient abilities, and scholastic capability. The ramifications of the level of enthusiastic fitness of the educators can be shown in the type of execution of understudies (Brackett and Katulak, 2006). Powerlessness of an instructor to make a classroom domain that supports the quick-paced and fast learning among understudies can repress the execution of the process, while the educator execution is clarified through the nature of cooperation, which has been set up with the understudies. In view of this foundation, the analysts have proposed consideration of preparing for educators particularly aiming at the advancement of their level of enthusiastic fitness (Hawkey, 2006).

1.3 *Definition of concept*

Emotional intelligence (EQ): Refers to how to manage self in feeling or emotions to maintain or motivate self or in our relationships (Goleman D., 1998).

Job performance: Refers to behavior of each employee toward his/her work quality and quantity (Jex, 2002).

Early childhood professionals: Refers to an individual or community who serves children aged from zero to eight years old and be responsible for their learning development towardreal-life experience (NAEYC).

2 RESEARCH METHODOLOGY

2.1 *Research design*

In this study, we employed quantitative approach; therefore, questionnaire survey acts as the best method in collecting the data regarding attitude, orientation, and preferences (Robinson, 2007). Quantitative approach helps offering review information on numerous characteristics,

Table 1. Seven dimensions of the USMEQ-i.

No	Domain	Question number
1	Emotional control	4, 7, 10, 11, 12, 25, 32, 38 & 44
2	Emotional maturity	14, 23, 30, 33, 34, 37, 42 & 43
3	Emotional conscientiousness	5, 9, 17, 20 & 26
4	Emotional awareness	22, 28, 29, 40 & 41
5	Emotional commitment	15, 16, 36 & 45
6	Emotional fortitude	1, 3, 31 & 46
7	Emotional expression	2, 8, 19 & 35
H	Faking index	6, 13, 18, 21, 24, 27 & 39

which become more useful for testing (Hair et al, 2007). Research design of this study tends to be correlational and descriptive as the researcher is interested in delineating the important variables associated with the problem (Sekaran & Bougie, 2010). Data obtained were analyzed using the Statistical Package for Social Science (SPSS) software.

2.2 Instrument

2.2.1 USM Emotional Quotient Inventory (USMEQ-i)

In this study, USM Emotional Quotient Inventory (USMEQ-i) was adopted and used as the main instrument to assess the emotional intelligence levels of the early childhood professional in collecting the required information, which was developed and adapted on the basis of the variables stipulated in the framework from the earlier study. With the aim to test the research objectives stated in Chapter 1, the designed questionnaire comprises 46 questions, consists of seven dimensions developed using the mixed-model theoretical approach of EQ.

It was developed by Yusoff, Abdul Rahim AF & Esa AR (2010) grouped into seven domains namely emotional control, emotional maturity, emotional conscientiousness, emotional awareness, emotional commitment, emotional fortitude, and emotional expression. A faking index domain was designed and included in the USMEQ-i to measure the tendency of respondents to overrate themselves. It consists of 39 statements representing the emotional intelligence dimensions and 7 statements representing the faking index domain. Each statement was rated under five categories of responses, namely "not like me", "a bit like me", "quite like me", "a lot like me", and "totally like me", to indicate how close the statement describes respondents' behavior. The seven dimensions are shown in Table 1.

3 VALIDITY AND RELIABILITY

The USMEQ-i initially consisted of 50 items representing the seven emotional intelligence dimensions. After validation procedures, 6 items out of the original 50 items remained. The final validation process was performed on 469 students who came from all over Malaysia; the sample was multiethnic, multi-religion, and multi-cultural. The validation procedure found that the USMEQ-i has good psychometric properties; it is a valid and reliable instrument that can be used to identify students' EQ. Factor analysis showed that all the items are well distributed according to the seven EQ dimensions. Reliability analysis showed that the USMEQ-i has a high internal consistency; Cronbach's alpha coefficient value was 0.96, which is more than the acceptable cutoff point of 0.7 (Downing S.M, 2004). The Cronbach's alpha coefficient values for each domain are shown in Table 2.

4 CONCLUSION

The findings of this study will be useful to the government of Malaysia, school administrators, child care institutions, and communities for developing their emotional intelligence in

Table 2. Cronbach's alpha value for each personality domain.

EQ domain	Cronbach's alpha value
Emotional control	0.899
Emotional maturity	0.816
Emotional conscientiousness	0.827
Emotional awareness	0.789
Emotional commitment	0.773
Emotional fortitude	0.656
Emotional expression	0.603
Faking index	0.827

order to reach their goals toward the development of young children. This study is in accordance with the aspirations of Malaysia Education Blueprint (2013–2025) to fulfill the educational needs of people, to enable the country to produce a generation that is knowledgeable, competent, capable, and competitive globally.

REFERENCES

Bracket, M.A., and Katulak, N.A. 2006. Emotional Intelligence in the Classroom: Skill-based Training for Teachers and Students. *Applying Emotional Intelligence: A Practitioner's Guide,* 1–27.

Carmeli, A., and Josman, Z.E. 2006. The Relationship among Emotional Intelligence Task Performance, and Organizational Citizenship Behaviors. *Human Performance*, 19(4), 403–419.

Downing, S.M. 2004. Reliability: on the Reproducibility of Assessment Data, Medical Education. 38, 1006–1012.

Goleman D. Emotional Intelligence. New York: Bantam 1995.

Goleman, D. 1998. The Emotionally Intelligent Workplace: An EI-Based Theory of Performance (Chapter Three). Cherniss C. & Goleman D. (Ed). Consortium for Research on Emotional Intelligence in Organizations.

Goleman, D. 1998. Working with emotional intelligence. New York: Bantam Goleman, D. 2005. Emotional intelligence U.S.A: Bentam book.

Hair, Jr., J. F., Money, A. H., Samouel, P. & Page, M. 2007. Research Methods for Business. Chichester: John Willey & Sons Ltd.

Hawkey, K. 2006. Emotional Intelligence and Mentoring in Pre-Service Teacher Education: A Literature Review. *Mentoring and Tutoring,* 14(2), 137–147.

Idris, R. 2003. Profil personaliti guru-guru sekolah. Unpublished Master's Thesis. Bangi, Selangor: Universiti Kebangsaan Malaysia.

Irene Becker http://justcoachit.com

Ishak, N. and Idris, K.N. 2009. The effects of classroom communication on students' academic performance at the International Islamic University Malaysia (IIUM). Unitar e-journal, 5(1), 37.

Jennings, P.A, and Greenberg, M.T. 2009. The Prosocial Classroom: Teacher Social and Emotional Competence in Relation to Student and Classroom Outcomes. *Review of Educational Research,* 79(1), 491–525.

Klassen, R.M. 2010. Teacher Stress: The Mediating Role of Collective Efficacy Beliefs. *The Journal of Educational Research*, 103(5), 342–350.

Lam, L.T., and Kirby, S.L. 2002. Is Emotional Intelligence an Advantage? An Exploration of the Impact of Emotional and General Intelligence on Individual Performance. *The Journal of Social Psychology,* 142(1), 133–143.

Murphy, K.R. 1989. Is the Relationship between Cognitive Ability and Job Performance Stable Over Time?. *Human Performance*, 2(3), 183–200.

Pianta, R.C. 1999. Enhancing Relationships between Children and Teachers. Washington, DC: American Psychological Association.

Robinson, D. T., Brown, D. G., Parker, D. C., Schreinemachers, P., Janssen, M. A., Huigen, M. and Berger, T. 2007. Comparison of Empirical Methods for Building Agent-Based Models in Land Use Science. *Journal of Land Use Science*, 2(1), 31–55.

Sekaran, U., and Bougie, R. 2010. Research Methods for Business: A Skill Building Approach (6th ed.). West Sussex, UK: John Wilet & Sons Ltd.

Yusoff MSB, Rahim AFA, Esa AR. The USM emotional quotient inventory (USMEQ-i) manual.

Russian partnership banking integration in Islamic financial system

N.F. Yalalova & A.A. Daryakin
Kazan Federal University, Kazan, Russian Federation

ABSTRACT: The article studies the perspectives of partnership banking and finance development in Russia, and challenges of Russian banking system integration in current Islamic financial system. The paper covers the results of marketing research about perception of Islamic banking in the Republic of Tatarstan, as one of potential pilot regions for partnership banking introduction in Russia. The article briefly describes existing experience of Russian financial institutions acting on Sharia principles. Furthermore, the authors analyze possible application of Islamic financial products within current legal environment and provide recommendations related to further development of the industry.

1 INTRODUCTION

1.1 *Moderate recovery of the global economy*

According to the United Nations report (2017), in 2016 the world economy expanded by just 2.2 per cent. Among the main reasons for weakening economic activity are low level of global investment, shrinking of world trade volume, slowed down productivity growth and high levels of debt.

Despite continuing economic sanctions and low oil prices Russian economy is also overcoming the recession started in 2014. Gross Domestic Product decreased 0.2 per cent in 2016 with positive forecast in 2017 (Bank of Russia, 2017).

Islamic finance, regarded as one of the most dynamic industries in the world with average annual growth around 15.4 per cent (IFSB, 2016), promoting trade, fair economic relations, higher social responsibility of businesses, just wealth distribution and prohibiting interest, uncertainty and gambling, could become an intelligent and innovative solution for the world economy stagnation.

1.2 *World Islamic financial market*

Islamic financial services industry has shown some slowdown during the last two years caused by currency depreciation in Asia and MENA region excluding GCC. Islamic Financial Services Board stated that Islamic financial assets amounted to USD 1.89 trillion in 2016, including Islamic banking at USD 1.5 trillion, outstanding Sukuk at USD 318.5 billion with Jordan and Togo as newcomers, Takaful (by contributions) at USD 25.1 billion and Islamic funds at USD 56.1 billion (IFSB, 2017).

The biggest Islamic banking market share belongs to Iran—about 33% of global Islamic banking, followed by Saudi Arabia and Malaysia (about 21% and 9% of global Islamic banking accordingly) (IFSB, 2017). Other notable countries in order of share of global Islamic banking are UAE, Kuwait, Qatar, Turkey, Bangladesh and Bahrain.

Though Islamic financial services industry remains relatively small—less than 2 percent of global finance (IMF, 2017), there is a strong interest and growing acceptance of Islamic finance in non-Muslim countries. Islamic financial services are represented today in UK, Germany, France, Luxemburg, Hong Kong, South Africa, USA, Australia, Japan and many other countries.

The former Soviet countries having strong historical and economic relations with Russia, including but not limited to cooperation within Shanghai Cooperation Organization and Eurasian Economic Union, are also developing actively Islamic financial institutions. Republics of Kazakhstan and Kyrgyzstan have already adapted their legislation to the needs of Islamic financial industry. The governments of Azerbaijan, Uzbekistan and Tajikistan have also announced initiatives facilitating the introduction of Islamic financial products and services. As a result, a number of financial institutions in the region meet the needs of Muslim population and have access to Islamic financing from the Middle East.

2 MARKET POTENTIAL

2.1 *Muslim population of Russia*

As per to the census (perepis2002.ru) conducted in 2002 the number of traditionally Muslim nations in Russia amounted to 14.5 million people—around 10 per cent of the overall population of Russia. Most of these ethnic Muslims reside in 7 regions of Russia including Republic of Tatarstan. Not all ethnic Muslims practice Islam: religious identity of people was lost during Soviet times, when people were forced to be atheists. According to the research conducted in 2012 (sreda.org), only 42% of ethnic Muslims were following religious principles in their everyday life.

2.2 *Awareness and acceptance of Islamic finance*

In 2015 there were two surveys about awareness and interest toward Islamic financial products in Russia.

One of them covered 46 regions and included 1600 people (NAFI, 2015). The results of the survey found out that only 1 per cent of interviewed has ever rejected traditional banking guided by religion. 89 per cent of interview participants have never heard anything about Islamic finance and only 1 per cent are well-informed about Sharia-compliant banking. 12 per cent are ready to become clients of Islamic banks, while 69 per cent will likely deny the possibility.

Another survey was conducted in the Republic of Tatarstan—the region with half ethnic Muslim population. The level of population awareness about Sharia-based financial system is quite high. 20 per cent of 1000 interviewed have heard about Islamic banking, and more than half mentioned mass media as a source (CASD, 2015). 36 per cent would like to know more about Islamic banks and more than 40 per cent of participants are interested in becoming Islamic bank clients. Around 1 per cent mistakenly believe that Islamic financial products are free of charge and almost 4 per cent believe that Islamic banking is only for Muslims.

Nearly 20 per cent of respondents would be attracted to Sharia-compliant products and services. Almost 15 per cent of interviewed are interested in trying Islamic banking products even if they are more expensive than traditional. Among factors that could attract potential clients are lower financing rate and higher deposit interest rate were mentioned by 60 per cent of interviewed. Excluding alcohol, tobacco, weapon and gambling industries from bank's investment portfolio was stated by 42 per cent of interviewed as one of motivators to use Islamic banking products.

In general citizens have positive expectations about future of the industry: nearly 56 per cent of respondents believe that Islamic banking has good perspectives in Tatarstan.

Thus, around 6 million people in Russia are potential clients of Islamic banks and at least 1,4 million do not invest their money in traditional banks because of religious principles. Regions with traditionally Muslim population demonstrate higher interest and readiness to accept new financial products and services based on Sharia principles.

3 ISLAMIC FINANCE APPLICATION IN RUSSIA

3.1 *The development of academic thought*

The possible implementation of Islamic financial services and products within Russian legislation is highlighted in a number of studies undertaken by Russian scholars and academicians.

Thus, Bekkin (2009) gives a brief overview of Islamic financial services development and application in former Soviet Union with a deeper attention to Takaful. Trunin and coauthors (2009) review the historical development of Islamic finance industry in the world and study different ways of state regulation of the industry. Chokaev (2015) examines possible ways of Islamic finance application in Russia and possible economic effects following the industry development. Gabbasova (2016) investigates the economic and legislative background for Islamic finance development in Russia and highlights the possible solutions for appropriate changes to financial and banking legislation. The possibilities of sukuk application are examined in works of Pakhomov (2009), Dadalko & Masyuk (2012) and Gafurova (2011).

The aim of this paper is to provide recommendations for further development of the industry in Russia taking into consideration the current level of awareness and existing legislative environment.

3.2 Local initiatives

The foundation for modern relations between Russia and Islamic world was laid in 2005, when Russia has got an observer status in Organization of Islamic Cooperation (oic-oci.org).

Started in 2007 with signing of a Memorandum of Understanding, a fruitful cooperation of Islamic Development Bank and Tatarstan Republic has resulted in a number of considerable milestones. One of them is KazanSummit (kazansummit.com) held annually in the Republic since 2009 and remaining the main and unique platform for cooperation between the Russian Federation and the Organization of Islamic Cooperation member states. The summit now has a federal level importance and brings annually high-level regulators and practitioners to share their experience, competence and set basis for future Islamic finance projects in Russia. As an example, Tatarstan International Investment Company (tiic.bz) and Eurasian Leasing company (euroasianlease.ru) were founded in 2010 and 2012 accordingly, both acting on Sharia principles, following the agreements signed on KazanSummit.

Among several private initiatives the one with relatively long and successful story is Financial House Amal (fdamal.ru), founded in 2010, and working in partnership with one of local banks. The bank's mission is to perform settlement and cash services and to keep clients' money on separated correspondent account. Amal offers auto-, home- and commodity financing and investment accounts to individuals and leasing and working capital financing to corporate clients. Around two thousand people have become Amal's clients since foundation. Organizational structure of Amal combines a group of companies. Cooperative society operating as a financial center of the group attracts investments from individuals and corporate clients and distributes them among other companies in the group. Leasing company within the structure offers ijarah and deferred payment contracts to big companies working with VAT, while trade company and retail company of the group perform trade operations with deferred payment with individuals, entrepreneurs and small businesses exempted from VAT.

Similar interaction scheme between a local bank and a group of financial companies providing Islamic banking products and services was applied at the Center of partnership banking (CPB), founded in 2016 and considered on federal level as a pilot project for introduction of Islamic banking products and services in Russia. The CPB didn't have any special rights or tax exemptions and offered limited range of products to the market. Unfortunately, the partner local bank has lost its license in 2017 and CPB had to stop operations too.

There are two similar companies in the republic of Dagestan, another Russian region with Muslim population: Lariba finance (lariba.ru) and Financial House Masraf.

Besides financial companies working with individuals and small business, there is also a success story of AK BARS Bank (akbars.ru), the largest local bank in Tatarstan. In 2011 and 2014 it has raised the amount of USD 60 million and 100 million respectively under its Syndicated Murabaha Islamic financing deal. The funds raised were allocated to finance a priority service and transport infrastructure development projects in the Republic of Tatarstan. The deals were structured with precious metals as underlying financial asset. First deal was awarded the title of "Deal of the Year 2011 in Europe" by the Islamic Finance News for the innovative character and the uniqueness of the transaction. The deal was the first of its type in the CIS, and opened the doors to the Islamic financial market for Russia.

Since 2015 there is also a takaful product "Halal invest" on Russian market, provided by well-known insurance company – Allianz Life (allianzlife.ru). The company offers 5-years long life insurance contract enabling insured clients to allocate their savings in Sharia-compliant foreign investment funds.

As a brief summary for the above we see that Islamic finance industry was developing in Russia due to private initiatives carried by local regional interested parties for more than 10 years. Though some of them hasn't found the expected interest in the market, local initiatives drove Islamic finance on higher federal level. Their impact has resulted in a number of success stories and possible ways for future industry development.

3.3 *Federal initiatives*

Most experts agree that one of the crucial factors for the development of Islamic finance industry in the country is political will. This is especially important in Russia, where the growth of the industry is restrained by legal environment.

For the effective development of Islamic finance industry several important steps were done on federal level. In 2014 the Central bank of Russia (CBR) has sent queries to financial institutions asking to indicate the main issues and concerns related to Islamic finance application in Russia. Islamic banking was finally recognized by CBR and called partnership banking (to avoid religious issues). Working groups for partnership banking were created within the Central bank of Russia, the Federation Council (upper house of parliament), the State Duma (lower house of parliament) and Vnesheconombank (government-owned development bank) in 2014–2015.

In March 2016 Central bank's working group approved the Road Map on Development of Partnership Banking and Related Financial Services in the Russian Federation for the Period of 2016 to 2017. The document promotes the relationships with foreign countries and institutes, implementation of pilot projects and programs in the industry. It aims to widen the application of instruments of partnership banking, to improve financial literacy and raise awareness of population about partnership banking products and services, and to increase financial inclusion. The road map also includes studying expediency of legal amendments necessary for the development of partnership banking in the Russian Federation.

In 2015–2016 eight bills were submitted to the State duma facilitating the development of partnership banking and financial services in Russia. Unfortunately, proposed amendments were rejected.

3.4 *Restraining factors*

Within Russian legal framework limited Islamic financial products and services could be provided. They include mudarabah and musharakah based products applied for equity financing and trust management, murabahah and ijarah based products for debt financing, leasing and deferred payment, takaful for life and commodity insurance.

As for the operation of a full-fledged Islamic bank in Russia there are several legal, administrative and financial restrictions, making application of Sharia-based financial products impossible or too expensive for banks and consumers. In 2014 the Association of regional banks of Russia and the Association of Russian banks has prepared letters in response to the Central bank's query, indicating those restrictions.

First of all, weak awareness of Russian population about Islamic finance and banking, its advantages and specific features, resulting in religious concerns (only for Muslims), misunderstanding, misbeliefs (terrorism financing), wrong expectations (free of charge, guaranteed income), etc. These factors would highly affect the industry development in the very beginning and could be solved only by joint long-term efforts of government, financial and religious authorities, educational institutions and mass media through constant provision of balanced, objective information about Shariah-based financial products and services, their benefits and features.

Another huge problem is absence of infrastructure, including Islamic interbank market and investment funds, takaful insurance companies, arbitrage. Within current environment existing Islamic financial institutions face liquidity problems, loose potential income, take higher risks and finally have to raise the cost of funding for consumers.

Lack of standards and legislation regulating Islamic finance activities, from one side restrains the development of the industry, as it is more risker, requires more efforts and resources to launch Islamic business, but from another side lead to financial innovations and development of practices based on the needs of market players. In 2016 AAOIFI standards were translated into Russian. Hopefully, this would harmonize the application of Islamic contracts in Russian practice.

Existing banking regulation doesn't allow banks to perform Islamic banking operations. Among the main nonconformances are Civil Code of Russia containing several articles to be amended about credit and deposit operations indicating interest as a mandatory condition in the contract; and the federal law No. 395-I "About banks and banking activity" mentioning chargeability as one of the principles of banking and forbidding production, trade (except financial assets) and insurance activities to banks.

Russian banks are not allowed to invest attracted funds in share capital of newly established companies. As a result Islamic bank would be limited in application of Musharaka contracts and sukuk.

Islamic banks' investment operations would cause higher risks comparing to credit operations and consequently would require higher reserves and reduction of bank's capital followed by liquidity problems (Igonina, 2015).

Administrative barriers occur while registering property rights in purchase-sale contracts. Such kind of operations need more time and financial resources. There is also no special software for accounting and banking operations and as a consequence higher operational risks.

Islamic murabahah specific mechanisms in most countries are regarded as financial operations, while in Russia VAT is chargeable in trade operations, even though this would be done for financing purposes in Islamic banks. In this way conventional banks don't pay VAT and can offer better financing conditions to their clients (Chokaev, 2015). Besides, income received from investment accounts is charged with higher taxes than income deriving from bank deposits creating even more tax burden to the clients of Islamic banks.

And finally, there are no specialists as the industry is too young. There are educational programs providing necessary education but number of students is not high as the employment perspectives are not clear yet. A newly established Islamic financial institution in Russia would need to educate its personnel in one of few universities providing Islamic finance courses in Russia, or independently—attracting specialists from the market.

3.5 *Further steps to be done*

Considering all the above the following measures could be recommended for further development of the industry.

First, to launch a pilot project on the base of a large federal bank with enough financial resources and manpower in a region with Muslim population, and to empower the bank to provide all existing Islamic financial products and services, enacted and protected by an act specifically created for that purpose. According to Gabbasova (2016), revocation of the license and bankruptcy of the financial institution was the most often reason for the termination of activities. Should we follow this practice, people would associate Islamic banking with suspicious from the Central bank's side activity or unprofitable, not trustworthy and risky business Performing the project in a huge bank will ease the introduction of new financial products in the region due to enough financial and time resources, good planning and management, will help to avoid mistakes and negative experience, to attract more clients as a reliable institution with a good-will, to overcome mistrust and misbeliefs related to new financial products based on Sharia-principles.

Experience of the pilot project will not influence the conventional system and secular banks. It will show the real demand and attitude of population, demonstrate weak areas to be under special control. Later the experience could be replicated in other regions if successful.

Second, to introduce a special license for partnership banks empowered to perform trade operations, invest in share capital, etc. This will segregate funds, protect consumers and control the industry, as this kind of activity is riskier and more familiar to developed countries with mature financial system, well-tried mechanisms of capital and risk-evaluation.

Third, to create supervisory and standard-setting body under the Central bank of Russia. This will create uniform platform for all existing institutions, protect the industry from fraud and abuse.

Fourth, to provide tax neutrality to the industry, so that both partnership and conventional banks will have equal conditions in the market.

Fifth, to develop non-banking Islamic financial institutions. They include insurance companies, investment funds, investment and asset management companies, leasing companies. This will create necessary infrastructure for the industry and meet the market demand till Russian banks will be able to perform Islamic banking.

3.6 *Economic benefits*

Economic effects from the introduction of Islamic finance in Russia could be divided into external and internal benefits. External benefits would include potential investments from OIC countries to halal industries and infrastructure projects. As a consequence of Islamic finance industry development new halal industries would grow and international trade with Muslim countries would also increase. Foreign players attracted by new Islamic banking market would create a healthy competition and heal Russian banking system.

As for internal benefits, they may include mobilization of internal funds (up to 20 million Muslims living in Russia) and meeting the demand for Islamic banking products and services in Russian market. Besides, the development of ethical, green and social types of financial instruments, replacement of debt-based instruments with investment- and partnership-based types of economic relations will recover and strengthen national financial system.

4 CONCLUSION

Modern Islamic finance industry has shown qualitative and quantitative growth within last fifteen years, widening the geography and introducing new financial instruments.

Current macroeconomic and geopolitical conditions create a new environment for Russian economy. One of the serious challenges became limited access to financial resources and external funds.

Prevailing conditions require new sources and ways of economic and financial growth enabling effective development of the area within current economic unions, diversification of risks through the development of alternative socially responsible types of financing and mature investment mechanisms application.

Financial system based on Sharia principles might become justified solution for current challenges as it would promote the development of social justice, growth of investments and as a consequence reduction of debt, increase equitable global trade, facilitate ethical financial products and support green technologies.

An attempt to evaluate the potential of Islamic financial products and services market in Russia was made. The market surveys has shown positive attitude of the market to a new financial system, based on Shariah, in the regions with high share of Muslim population. Nevertheless the level of awareness and understanding is relatively low even among practicing Muslims. A broad and long-term educational campaign should be carried to raise the financial grammar of the population and increase the level of acceptance of new products and services.

Without government support the industry can't grow. Existing limited supply of Islamic financial products and services will never meet the growing demand. Present legal environment doesn't allow Russian conventional banks to accommodate Islamic financial products, currently available in other financial institutions but to a limited extent.

Private local initiatives undertaken in regions within last 10–15 years have set a solid background for federal authorities to support the development of the industry. The government of the country has created working groups on different levels to reveal existing restraining factors and drive the industry further.

Successful cases of Muslim countries and non-Muslim secular states shows possible benefits the country will get from the developing of alternative financial system. Creating equal conditions for Islamic and conventional banking will increase healthy competition, attract Muslim capital from inside (Muslim population) and outside resources (external financing from rich oil countries).

Islamic financial system is more sustainable in financial crises as every contract has an underlying real asset, it is based on the principle of risk-sharing, connects financial sector with real economy, strives for financial inclusion and social well-being of all segments of population.

REFERENCES

Bank of Russia. Talking Trends, Macroeconomics and markets. *Research and Forecasting Department Bulletin*, No. 3 (15), April 2017. Retrieved from http://www.cbr.ru/eng/ec_research/wps/bulletin_17-03_e.pdf.

Bekkin, R. 2009. Islamic finance in the former Soviet Union, 179–209. Retrieved from http://www.bekkin.ru/downloads/rb1391505391.pdf.

Burnashev, N. 2016. Developing a sound Islamic banking system in Kazakhstan. *Kazakhstan Islamic finance*, 84–87. Retrieved from https://ceif.iba.edu.pk/pdf/ThomsonReuters-KazakhstanIslamicFinance2016ANewFrontierforIslamicFinance.pdf.

CASD (Center for analytical research and development) 2015. *The survey about perception of Islamic banking and financial products in the Republic of Tatarstan*. (Unpublished).

Chokaev, B. 2015. Islamic Finance: Opportunities for the Russian Economy. *Voprosy Ekonomiki, (6), 106–127*. (In Russian).

Dadalko S. & Masyuk T. 2012. Sukuk securities: concept and market development tendencies. *Bankovskiy vestnik, No.19, 38–43*.

Gabbasova, R. 2016. First steps towards Islamic Finance regulation in the Russian Federation. *Journal of King Abdulaziz University, Islamic Economics*, 29(1), 127–138. http://doi.org/10.4197/Islec.29-1.11.

Gafurova, G. 2011. Sukuk in the system of Islamic financing. *Vestnik AGTU, Ekonomika, No.2, 90–94*.

IFSB 2017. *Islamic Financial Services Industry: Stability Report 2017. Islamic Financial Services Board*. Retrieved from http://www.ifsb.org/docs/IFSB%20IFSI%20Stability%20Report%202017.pdf.

IFSB 2016. *Islamic Financial Services Industry: Stability Report 2016. Islamic Financial Services Board*. Retrieved from http://www.ifsb.org/docs/IFSI%20Stability%20Report%202016%20(final).pdf.

Igonina A., Vagizova V. & Batorshyna A. 2016. Liquidity management in Islamic banking industry. *Social Sciences and Interdisciplinary Behavior*—Proceedings of the 4th International Congress on Interdisciplinary Behavior and Social Science, ICIBSOS – 2016. – Vol., Is. – P.265–268.

International Monetary Fund 2017. Ensuring Financial Stability in Countries with Islamic Banking. *Policy papers*. Retrieved from http://www.imf.org/~/media/Files/Publications/PP/PP-Ensuring-Financial-Stability-in-Countries-with-Islamic-Banking.ashx.

International Monetary Fund. Middle East and Central Asia Dept. 2017. Multi-Country Report: Ensuring Financial Stability in Countries with Islamic Banking-Case Studies-Press Release; Staff Report. *IMF Country Report*, (17/145), 0–101. Retrieved from http://www.imf.org/~/media/Files/Publications/CR/2017/cr17145.ashx.

Khaki, G.N., & Malik, B.A. 2013. Islamic Banking and Finance in Post-Soviet Central Asia with Special Reference to Kazakhstan. *Journal of Islamic Banking and Finance*, 1(1), 11–22.

NAFI (National Agency for Financial Researches) 2015. *Clients of Islamic Bank—who they are?* (In Russian).

Pakhomov, S. 2009. Problems and possibilities of Islamic finance application. *Depozitarium, Zarubezhniy opyt, No.11(81), 24–27*.

Salikhova, A. & Rustemov, T. 2016. Challenges and perspectives of Islamic finance education in Kazakhstan. *Kazakhstan Islamic finance*, 62–65. Retrieved from https://ceif.iba.edu.pk/pdf/ThomsonReuters-KazakhstanIslamicFinance2016ANewFrontierforIslamicFinance.pdf.

TIDA 2017. *Programme of the 10th International economic summit "Russia—Islamic World: KAZAN-SUMMIT 2017."* Retrieved from https://kazansummit.com/programme/.

Trunin P., Kamenskikh M., Muftiahetdinova M. 2009. Islamic banking system: Present state and prospects for development. *Nauchnye Trudy IEPP, No. 122, 1–84*. (In Russian).

United Nations 2017. *World Economic Situation and Prospects 2017.* http://doi.org/10.1007/BF02929547.

Yandiev M. 2011. Actual problems of Islamic financial system foundation: Philosophy and practice. *Rynok tsennykh bumag, No. 9, 54–57*. (In Russian).

Zaher, T., & Hassan, M. 2001. A Comparative Literature Survey of Islamic Finance and Banking. *Financial Markets, Institutions & Instruments*, *10*(4), 155–199. http://doi.org/doi: 10.1111/1468-0416.00044.

Influencing factors of ethnic tolerance among multiethnic youths

Abdul Razaq Ahmad, Abdul Aziz Abdul Rahman & Mohd Mahzan Awang
Faculty of Educational Studies, National University of Malaysia, Malaysia

Fong Peng Chew
Faculty of Education, University of Malaya, Malaysia

ABSTRACT: This study aims at investigating the contributing factors of ethnic tolerance among multiethnic youths. The following four factors were identified in this study: socioenvironment, socioparticipation, knowledge, and practices of patriotism. Ethnic tolerance can be defined as ethnic relationships and the unity among myriad ethnic groups who practice a variety of cultures, religions, and lifestyles which are different from each other, and who can live together without feeling prejudiced at each other. Ethnic tolerance also refers to the concept of integration in the context of ethnic diversity in the multiethnic society in Malaysia (Syamsul Amri, 2012). This survey included 2600 multiethnic youth aged between 19 and 29 years in Malaysia, which comprises the Peninsular area (n = 1600), as well as Sabah and Sarawak (n = 1000). The peninsular zone was divided into four zones, namely north, south, east, and west. A set of questionnaires was used for data collection. A pilot study was carried out before data collection. Cronbach's alpha results indicate that the items for all constructs are valid and can be used (0.806 and 0.953). The results from this study revealed that all factors had a significant relationship toward racial tolerance. Detailed findings show that socioparticipation is the largest contributor to tolerance (34.4%), followed by knowledge, practices, and socioenvironment. Hence, all parties have to work together to organize activities that can enhance the friendship relationships between different races. Integrity, moral, and empathy should be used by every individual in learning and practicing tolerance. The importance of tolerance as a social value cannot be understated in the life of a harmonious society.

Keywords: socioenvironment, socioparticipation, knowledge, tolerance, patriotism, multiethnic youth

1 INTRODUCTION

Social harmony and unity can be achieved via many ways, but the most prominent factors are through education and social interactions. In a Malaysian context, the national education system plays important roles in promoting unity and harmony. History is one of the academic subjects that aims at promoting unity and social harmony. The 13th May 1969 tragedy was a tragic moment in Malaysia due to ethnic conflicts. The importance of unity became a lesson for political leaders, which leads to the introduction many programs to unite them. Fostering unity and appreciation of patriotism values begins in elementary school. The Malaysian government then realized the importance of patriotism among nations. Ministry of Education, Malaysia (2000) urged that all schools in Malaysia irrespective of the types to promote patriotism and ethnic tolerance via academic subjects as well as co-curricular activities. Academic subjects introduced were History, Moral Education, Geography, Islamic studies, and others. In addition, the government emphasizes the importance of integration activities among school students via various academic and nonacademic activities. However, despite being in education process for 13 years, the patriotism spirit among school leavers is loose and still at moderate level (Mohamed et al., 2011). This shows the importance of this study. On the contrary, we must also be aware that "unity" is the key variable in determining the stability of politics, society, and economy.

2 STATEMENT OF PROBLEM

2.1 *Social support, socioenvironment, and patriotism*

Social support is a vital factor for behavioral changes. The importance of the relationship between social support and patriotic behavior was rarely investigated in sociological studies. Earlier sociological studies conducted in Malaysia focused on describing interethnic interaction using philosophical paradigm. Studies on social support and socioenvironment are more likely to focus on educational outcomes rather than understanding the link with patriotism.

Zhang et al. (2015) reviewed the relationship between the quality of relationships in school, social support, and loneliness in adolescence. Using a sample of 1674 adolescents who were randomly selected from secondary schools, the study found that boys' loneliness is influenced by the characteristics of the opposite gender, students, teachers, as well as sexual relationships, while girls' loneliness is only affected by sexual relationship. In addition, social support mediated the relationship between sexual support and the relationship between students and teachers as well as loneliness. In addition, the quality of sexual relationships showed a strong association with the loneliness among both girls and boys. Finally, the quality of sexual relationships also showed the strongest association with boys' loneliness compared to the relationship with opposite sex as well as relationship between students and teachers. These findings are discussed to explain the possible mechanisms by which interpersonal relationships can influence loneliness. In future studies, the causal relationship and other factors influencing loneliness should be further reviewed.

Tian and Huebner (2015) conducted a longitudinal study for 6 weeks to study the model of intermediation medium that can explain the relationship between social support related to school (i.e., teachers' support and classmates' support) and subjective optimum of well-being at school among adolescents (N = 1316). The analysis confirms the hypothesis model that the efficiency of academic competence partially mediated the relationship between social support related to school and subjective well-being in school. The results suggest that both social context factors (e.g., school-related social support) and the factors within the system (e.g., academic competence and social acceptance) are crucial for optimal subjective well-being of teenagers in school.

2.2 *Socioparticipation and patriotism*

Past studies focusing on the link between socioparticipation and patriotic behavior among multiethnic youth have had limitations as many previous studies have tended to focus on the philosophical aspects of ethnic tolerance. Socioparticipation has been found to be a great platform to promote patriotism and ethnic tolerance. Indeed, the focus on socioparticipation and patriotism is essential as it would provide empirical evidence on the association of these two variables. Volunteerism teaches individuals to put away their egocentric nature aside (Mohd Mahzan, 2017). Patriotism in this study refers to a set of social behavior that portrays the feeling of loving to a country. This includes the use of national language in formal communication as well as proud to be a national citizen among multiethnic societies. Boekel et al. (2016) conducted a study on the effects of participation in school sports in academic and social functions. For many students, organized sports by school play an important role in students' academic and social experience. This study focuses on the effects of participation in school-organized sports toward academic achievement and students' perceptions toward support from family, teachers, and community, as well as school safety. Although the results of this correlation study cannot be interpreted as causal, the involvement in sports among high-school seniors can help improve the visible support from family, teachers, and community as well as school safety, which eventually will lead to better academic achievement in school.

2.3 *Sociosupport and socioparticipation*

Although the concepts of sociosupport and socioparticipation were interchangeably used in previous studies, both concepts have some clear distinctions. Sociosupport refers to how socioenvironment members, including peers, family members, and virtual friends, provide support and encouragement toward multiethnicity to be involved in social activities. Sociosupport provided by them could be in various forms, either directly or indirectly. On the contrary, socioparticipation refers to the involvement of youth in social activities, e.g., community-based activities, hobby-based activities, or educational based-activities. In this study, the focus of socioparticipation is the youth involvement in various activities that may have a direct impact on patriotism. This includes the youth involvement in civic and political activities.

2.4 *Knowledge and patriotic behavior*

Ong Puay Liu's (2009) observations on the comments toward the Article 153 regarding the privileges of Malays and native people in Malaysia revealed that there is no threatening action that can harm the integrity among racial tolerances when discussing this issue, which revealed the level of maturity among Malaysian in expressing their thoughts. According to Wong (2005), the lack of awareness among the youth on the challenges to be faced by them to conduct their responsibilities as citizens is the issue of fading nationalism values among the younger generation today. This is because they are the generation of post-independence and enjoy all the opportunities that have been available without the bitter toil nightmare that had to be experienced by previous generations in the struggle to achieve independence for the prosperity of today's generation. Saraswathy (2003) found that national media such as radios and television played important roles for knowledge distribution via drama and advertisement. Brown (2005) stated that the knowledge and introduction toward patriotism knowledge should be applied starting from the children so that they are sensitive to their origins. Patriotism is a manifestation of feelings, attitudes, values and worldview of the country or the state itself, and the feelings such as love, respect, or care. A country's idealism and patriotism can be interpreted in various ways. For example, singing the national anthem is one way to express the spirit of nationalism and patriotism. Another example is eager to protect and stand in order to protect the country from various types of threats, both internal and external. This is because, the threats can diminish the patriotic spirit among the community itself (Mohamed et al., 2011).

Therefore, this study aims at identifying the influencing factors for ethnic tolerance among multiethnic youth. Four factors identified to be examined were socioenvironment, socioparticipation, the knowledge on patriotism, and patriotism practice.

3 METHODOLOGY

This study used descriptive statistics (mean and standard deviation) and inferential statistics (regression) through 7-Likert scale questionnaire as an instrument. The 7-Likert scale was used to give respondents more choices in providing response to the questions. It also provides wide ranges of feedback by respondents. This kind of scale was commonly used in high level of academic survey. According to Rahman (2017), the use of 7-Likert scale is useful as it helps respondents to choose the best response. The sample was based on the sampling schedule (Krejie & Morgan 1970), involving 2600 youth of different ethnicities aged between 19 and 29 years, comprising Peninsula (n = 1600) consisting of a zone north of Penang (n = 400), east zone in Pahang (n = 400), the central zone in Selangor (n = 400), and the southern zone of Johor (n = 400) Sabah (n = 590) and Sarawak (n = 410). This study included multiethnic youth from Peninsula Malaysia because the majority of them are exposed to various social environments that may have implication toward their patriotic behavior. Indeed, there is a different situation of social cultural among multiethnic youth in Peninsula Malaysia and Borneo. This may be due to social historical of their life that might affect their attitude toward other ethnics. The sampling method used in this study was stratified as random sampling. This study contains 105 items consisting of five variables, namely socioenvironment influence (21 items), socioparticipation (19 items), ethnic tolerance (20), patriotism knowledge (22 items), and patriotic behavior (23 items), as well as demographic, consisting of 10 items. Social environment influence assessed in this research comprises the following elements: the influence of family environment; peer influence; school influence; community influence; and social media influence. Meanwhile, the social participation items include the involvement in community-based activities, neighborhood activities, nongovernmental activities, and family activities. The ethnic tolerance measured the extent to which the youths are able to compromise many things with other ethnic groups. The patriotism knowledge includes the youths comprehension toward the basic historical background of Malaya, Malaysian administrative, the country ideology, and philosophy (Rukun Negara). Patriotic behavior focused on overt behavior portrayed by youth in various social actions, including negotiation situations as well as daily conversation with others.

This questionnaire was developed by a number of studies related to the knowledge of patriotism and ethnic tolerance, which have been adapted from various empirical evidences. The results of the pilot study for the whole construct is a high level of reliability, which is between 0.806 and 0.953 and can be used in the actual study. With regard to gender, 1276 (49.1%) were men and 1324 were women (50.9%), while for the location, 1315 of the respondents came from urban areas and 1285 of the respondents live in rural areas. With regard to family income: MYR2000 (45.7%), MYR2001 to MYR3000 (24.2%), MYR3001 to MYR4000 (17.9%), and MYR4001 (12.2%). Several ethnic groups are involved in the study, namely Malay, Chinese, Indian, Sabah and Sarawak Malay, Sabah and Sarawak Chinese, and Sabah and Sarawak Bumiputera. Data were analyzed using SPSS version 22.0. The mean score interpretations are: 1.00–2.20 (low), 2.21–3.40 (moderately low), 3.41–4.60 (moderate), 4.61–5.80 (moderately high), and 5.81–7.00 (high).

4 FINDINGS

4.1 *The level of encouragement from socioenvironment, socioparticipation, knowledge, and patriotic behavior variables toward racial tolerance*

Descriptive analysis involving mean and standard deviation was conducted to analyze the overall level of influences from socioenvironment, socioparticipation, patriotism knowledge and practices, as well as tolerance. Figure 1 shows the overall mean of each variable in the study.

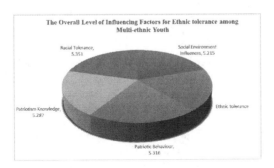

Figure 1. Levels of socioenvironment influence, socioparticipation, patriotism knowledge, patriotism practice, and ethnic tolerance variables.

Figure 1 shows the result of descriptive analysis of the levels of socioenvironment influence, socioparticipation, patriotism knowledge, and patriotism practices. Overall, all the aspects are recorded to be moderately high. In detail, ethnic tolerance variables show the highest scores (mean = 5.351), whereas socioparticipation shows the lowest (mean = 4.804). This shows that the level of socioparticipation among youth is still at an unsatisfactory level. This result is in agreement with the study of Rusimah Sugianto et al. (2009) who stated that many of the activities among youth whether Malays, Chinese, and Indians only involve their own ethnicity. This means that the influence of social environments on ethnic tolerance is still unsatisfactory. Supposedly, the multiethnic society in Malaysia should have strong spirit to encourage ethnic tolerance as it essential for promoting harmonious society.

This finding is supported by Konstantinova (2016), which revealed that they are more comfortable because of their similar culture, language, and requirements in carrying out activities. This study is also consistent with studies by Ghazali Shafie (2001), who found that residential segregation by race led to a lack of race relations in the community. This problem also occurs in the artistic and cultural aspects where there is lack of involvement from other ethnicities in cultural activities despite various attempts made by the government through various agencies.

4.2 *How socioenvironment, socioparticipation, patriotism knowledge, and patriotism practice promote ethnic tolerance?*

Before discussing the details, it is needed to clarify the distinctions between the variable of "knowledge and patriotism practice" and the variables of "socioparticipation and patriotism". The variable of "knowledge and patriotism practice" has two main aspects: (1) the understanding and comprehension of patriotism and (2) the patriotic behavior of multiethnic youth. The variables of "socioparticipation and patriotism" refer to the involvement of multiethnic youth in social activities that may promote their patriotism values.

A multiple regression analysis was conducted to identify the contribution of variables, namely socioenvironment influence, socioparticipation, patriotism knowledge, and patriotic behavior toward ethnic tolerance among Malaysian youths. Before the multiple regression analysis is carried out, the researchers first ensure and verify whether the score distribution of questionnaires is no MYRal and linear or otherwise. This is done by means of the residual scatter plot and normal regression plot that can be obtained from the subprogram "Linear Regression: Plots" contained in SPSS. On the basis of the plot of the distribution, the distribution of scores available study questionnaire is normal and linear.

In addition, before that, the researchers also determine the correlation between the independent variables and independent variables to determine whether there is multicollinearity. It was found that no independent variable in this study have multicollinearity. Therefore, researchers using stepwise multiple regression analysis as recommended by Hair *et al.* (1995).

Results of stepwise multiple regression analysis, involving four independent variables on the dependent variables, tolerance practices, revealed that there are four independent variables, which had significant correlations contribution ($p < 0.05$) toward the variance.

Multiple regression analysis *(stepwise)* shows that independent variables of patriotism knowledge, socioparticipation, and socioenvironment influence are predictors that have correlation and contributions (51.90%) that are significant ($p < 0.05$) toward patriotic behaviors.

Table 1. Analysis of variance.

Analysis of variance	Sum of square	dF	Mean square	F	Sig. value
Regression	989.93	4	247.48	430.92	0.00[a]
Error	916.02	1595	0.57		
Total	1905.95	1599			

Source: Analysis of variance was $F(4.159) = 247.483$ and the significant level was $p = 0.00$ ($p < 0.05$).

Table 2. Contribution of socioenvironment influences, socioparticipation, patriotism knowledge, and Patriotic behavior toward ethnic tolerance.

Variables	Unstandardized coefficients B	Std error	Standardized coefficients Beta	T	Sig.	R2	Contribution
Socioparticipation	0.258	0.019	0.293	13.72	0.05	0.34	34.4%
Patriotism knowledge	0.332	0.25	0.280	13.21	0.00	0.46	11.8%
Patriotism practice	0.264	0.26	0.215	10.05	0.00	0.50	4.3%
Socioenvironment influences	0.23	0.33	0.14	7.06	0.000	0.52	1.5%
Constant	−0.29	0.15		−1.95	0.05		

The main and highest predictor of tolerance and among Malaysian youth is socioparticipation ($\beta = 0.293$, t = 13,725, and p = 0.051), which contributes toward 34.40%. This shows that when socioparticipation score increased by one unit, the practice of tolerance will be increased by 0.293 units. This finding clearly shows that socioparticipation is a key factor that contributed 34.40% to the practice of tolerance among Malaysian youths. The second most important predictor, which accounted for 11.8% to the practice of tolerance among youth, is the patriotic knowledge ($\beta = 0.280$, t = 13,211, and p = 0.000). This shows that when patriotism knowledge score increased by one unit, the practice of tolerance increased by 0.280 units. This indicates that the knowledge of patriotism among the youth are also important factors contributing to the practice of their patriotism.

The third most important predictors, which accounted for 30.4% against the practice of patriotism among youth is the patriotic behavior ($\beta = 0215$, t = 10,052, and p = 0.000). This shows that the patriotic behavior among youth is also an imperative factor contributing to the tolerance practice. The fourth most important predictors accounted for 1.50% to the practice of patriotism among youth is socioenvironment influence ($\beta = 0.144$, t = 7,057, and p = 0.000). This shows that a socioenvironment influence among youth is also an important factor contributing to the practice of their tolerance.

It shows that the activities related to patriotism are also identified as an important factor that contributes to racial tolerance. Thus, continuous efforts must be made as it can change the tolerance behavior especially among youths of various ethnic groups in Malaysia. This finding supports the study conducted by Fatimah Daud (2004), which have seen the extent to which young people socialize and interact in a community college and also affects the way their relationship and interaction with the outside community structure that is more complex. The results of this research program involving joint activities and exchange of cultural programs have increased the level of language proficiency, inter-religious understanding, and mutual understanding and solidarity between the youth of different ethnic groups. Therefore, involvement in social activities can promote tolerance between races. Jamil Wong et al. (2004) also showed the strengthening of interethnic relationships among the youth through sports and cultural activities that have been organized. However, there are still negative perceptions and prejudices in their minds as a result of the values instilled from home.

5 SUGGESTION AND STUDY IMPLICATIONS

The study shows that socioparticipation in activities involving youth of different ethnicities have not reached a satisfactory level. The results also showed that the activity of the program involving various ethnic groups is a major contributor to racial tolerance. Therefore, the Ministry of Youth and the National Unity and Integration Department should conduct a joint program, regardless of the term. Furthermore, the programs must involve ethnic diversity held at organized city or rural areas because it can foster good relationships among the youths. They will learn directly about other ethnic cultures. This is because at the school, they are still being separated based on race, which give less opportunity to engage in activities, events, or programs.

The variable patriotism knowledge and practices has not reached a satisfactory level, and there is a significant gap among the youth from various aspects related to patriotism and ethnic tolerance in Malaysia, where Malay youth was proven to have higher patriotism and tolerance compared to Chinese and Indian. This happens because many organizations are only attended by the Malay community. Therefore, programs related to patriotism should be extended to Chinese and Indian communities as happened to the Malay community nowadays. It may also be caused by less social interaction because the Chinese youth tend to focus solely on economic field. Therefore, college or youth training centers related to patriotism should be established at the national level. The program should be coordinated with the requirement of today's youth using relevant approaches and can be implemented in either short term or long term.

In addition, the open-house program in the community for festivals in Malaysia has to be expanded not only at the national level, but also at the district, state, and local communities because it is proven in providing the opportunity for different ethnicities to interact. However, it was revealed that the youth participation in activities through non-formal education is less efficient in promoting ethnic tolerance as compared to the effect of formal schooling in shaping behavior toward ethnic relationships among the youth. Therefore, schools must play a more structured role by providing a variety of activities and curriculum, which are capable of improving the environment, especially to encourage patriotism behavior and racial tolerance.

6 CONCLUSION

Tolerance is the key to maintaining harmonious and peaceful social relationships, with respect to the differences among people, culture, lifestyle, and religion. It is both a good character and behavior that should be studied. Tolerance practices are not a forced acceptance but the willingness to compromise in order to achieve a goal. In addition, the concept of interethnic relationships should be corrected so that every citizen has the awareness and a sense of responsibility for maintaining the country's security by strengthening its unity. All of this have to be the core in ensuring a smooth development process in line with the aspiration of making Malaysia a unique model country by uniting diversity and building a nation which is united regardless of the differences in race, religion, culture, and language.

ACKNOWLEDGMENT

This study was based on the empirical research funded by the Malaysian Council of Former Elected Representatives (MUBARAK) and the Ministry of Finance Malaysia.

REFERENCES

Ahmad Ali Seman. 2011. Keberkesanan Modul Pengajaran dan Pembelajaran Sejarah Berteraskan Perspektif Kepelbagaian Budaya Terhadap Pembentukan Integrasi Nasional. Unpublished Ph.D Thesis Faculty of Education, Universiti Kebangsaan Malaysia.

Azwani Ismail. 2012. Kesan Model STAD terhadap patriotisme. sikap dan kemahiran komunikasi pelajar terhadap mata pelajaran sejarah. Tesis Doktor Falsafah. Universiti Kebangsaan Malaysia. Bangi.

Boekel, V.V., Bulut, O., Stanke, L., Zamora, J.R.P., Jang, Y., Kang, Y., & Nickodem, K. (2016). Effects of participation in school sports on academic and social functioning. Journal of Applied Developmental Psychology, (46) 31–40. http://dx.doi.org/10.1016/j.appdev.2016.05.002.

Cham, 1975, Class and Communal Conflict in Malaysia, Journal of Contemporary Asia, 5, 446.

Bourdieu, P. 1986. The forms of capital. In J. Richardson (Ed.) Handbook of Theory and Research for the Sociology of Education (New York, Greenwood), 241–258.

Bronfenbrenner, U. 2005. Making Human Beings Human: Bioecological Perspectives on Human Development. Thousand Oaks, CA: Sage Publications.

Graham Brown 2005. Making Ethnic Citizens: The Politics and Practice Of Education In Malaysia. Crise Working Paper. No. 23. University Of Oxford: Queen Elizabeth House.

Doganay, Y. 2013. The Impact of Cultural based Activities in foreign language teaching at upper-intermediate (B2) level. Education Journal, 2(4): 108–113.

Fatimah Daud. 2004 Polarisasi kaum di kalangan pelajar-pelajar universiti. Kuala Lumpur: Jabatan Antropologi & Sosiologi, Fakulti Sastera & Sains Sosial, Universiti Malaya.

Furnival, G.M. 1971. All possible regressions with less computation. Technometrics 13, 403–8.

Ghazali Shafie. 2001. Pembinaan Bangsa Malaysia. In Ismail Hussein et al (eds.). Tamadun Melayu Menyongsong Abad ke–21. (2nd Ed.) (p.174). Bangi: Penerbit Universiti Kebangsaan Malaysia.

Hair, J.F., Black, W.C., Babin, B.J., Anderson, R.E., & Tatham, R.L. 2006. Multivariate data analysis. 6th Ed.. New Jersey: Prentice Hall.

Hazri, J., Molly, N.N.L., Santhiram, R., Hairul Nizam, I., Nordin, A.R., Wah, L.L., Nair, S. 2004. Ethnic Interaction among Students in Secondary School in Malaysia: USM.

Kementerian Pendidikan Malaysia (KPM). 2000. Education Statistics of Malaysia. Kuala Lumpur: KPM.

Konstantinova, V.V., Kolomiets, D.L., Glizerina, N.D., Tihomirova, V.T., Kuznetsova, L.V., Golovina, N.N. 2016. Patriotic education of school children in a multiculture educational environment. Medwell Journals, The Social Sciences, 11(8): 1923–1928.

Krejcie, R.V., & Morgan, D.W. 1970. Determining sample size for research activities. Educational and Psychological Measurement, 30, 607–610.

Matulis, B.S. & Moyer, J.R. (2016). Beyond Inclusive Conservation: The Value of Pluralism, the Need for Agonism, and the Case for Social Instrumentalism. A Journal of the Society for Conservation Biology. 00(0), 1–9.

Mohd Mahzan Awang. Januari 25, 2017. Semai Patriotisme menerusi kesukarelawan. Utusan Online.

Mohamed, P.A.S., Sulaiman, S.H., Othman, M.F., Che Jumaat Yang, M.A. & Haron, H. 2011. Patriotism Dilemma among Malaysian Youth: Between Strategy And Reality. International Journal of Business and Social Science, 2 (16).

Ong Puay Liu. 2009. Identity Matters: Ethnic Perceptions and Concerns. In Lim Teck. Ghee and Alberto Gomes (Eds.). Multiethnic Malaysia: Past, Present and Future. Petaling Jaya: MiDAS & SIRD. Chapter 24: 463–481.

Najamuddin Bachora. 2014. Kecenderungan Interaksi Sosial Ke Arah Penghayatan Nilai Perpaduan Dalam Kalangan Guru Pelatih. Tesis Ph.D Belum Diterbitkan Fakulti Pendidikan Universiti Kebangsaan Malaysia.

Putnam, Robert D. 2001. Social capital. Measurement and consequences." Canadian Journal of Policy Research 2, 41–51.

Rusimah Sayuti, Mohd Ainuddin, Iskandar Lee Abdullah dan Salma Ishak. 2009. Kajian hubungan kaum dalam kalangan pelajar Sekolah Pembangunan Sosial, Sintok: Universiti Utara Malaysia (UUM).

Saraswathy Chinnasamy. 2003. Media TV RTM sebagai saluran pembangunan negara. Kuala Lumpur: UM.

Shamsul Amri Baharuddin. (Ed.). 2012. Modul Hubungan Etnik, Institut Kajian Etnik, Universiti Kebangsaan Malaysia: Bangi, Selangor.

Tian, T., Zhao, J., & Huebner, E.S. 2015. School-related social support and subjective well-being in school among adolescents: The role of self-system factors. Journal of Adolescence, 138–148. http://dx.doi.org/10.1016/j.adolescence.2015.09.003.

Tyler, T.R. 1990. Why People Obey the Law. London: Yale University Press.

Wong Shui Wah. 2005. Nasionalisme Malaysia Di Kalangan Pelajar Sekolah Menengah Kebangsaan dan Pelajar Sekolah Menengah Swasta Sekitar Johor Bahru. Universiti Teknologi Malaysia. Tesis Sarjana Muda.

Zhang, B., Gao, Q., Fokkema, M., Alterman, V., Liu, Q. 2015. Adolescent interpersonal relationships, social support and loneliness in high schools: Mediation effect and gender differences. Social Science Research 53, 104–117. http://dx.doi.org/10.1016/j.ssresearch.2015.05.003.

Analysis of innovation cluster formation in industrially developed countries

I.M. Ablaev
Kazan Federal University, Kazan, Russia

ABSTRACT: In this study, we present the analysis of the European and the American experience of cluster development to be adopted in the process of Russian cluster policy development. We summarize the current trends of cluster development, and highlight the key factors providing the favorable conditions for cluster formation and its further growth. The goal of this study is to single out the adaptive methods of cluster formation, which can be adopted to the Russian economic background and the cluster development policy. This study was carried out by using the expert assessment and statistical analysis methods.

1 INTRODUCTION

The growth of the world economy has given rise to a large number of successful cluster projects, and this wide experience can be adopted in the process of Russian cluster policy development. There are many examples indicating that the competitive advantages of certain product manufacturers and service providers in the global marketplace were achieved by means of clusters: the IT cluster in the Silicon Valley (USA), film production in Hollywood, giant automotive innovative clusters in United States, Germany, and France, and so on. The matter of great interest for Russia is the experience of industrially developed countries, such as the countries of Europe and the United States, nowadays remarkable for the considerable number of various cluster projects being successfully run in different sectors of the global economy, for the carefully worked out and approved methods of cluster development.

2 THEORY

The development level of European countries and the United States is mostly the same. Hence, it is possible to single out some general trends in the development of the economy, and particularly in cluster development. However, the formation process and the current state of economies have their peculiarities determined by the following factors:

– comparatively independent development of industrial sectors and
– different traditions and national features of economic management.

The innovation process has become a leading factor of economic growth in developed countries. For European countries, the use of cluster potentials is the main instrument for achieving the social and economic goals (European commission 2010).

In Europe, there is no lack of clusters, but there is a certain need of global scale clusters and the clusters in the high-technology sectors (SEC 2008–2018). In order to encourage the entry of European clusters to the world market, a special attention and support is given to the cross-border cooperation. The latest investigations reveal the retardation of European clusters in the sphere of IT (DG Enterprise and Industry Report, 2008), which actually could have been the strong driver of the total economic growth in whole Europe, and in particular, the use of technology could foster the interaction between the existing clusters, thereby enhancing their capacity.

Great efforts are made for the creation of interaction platforms for the members of innovation technology clusters intended for accumulation of all the necessary information about the clusters. One of the platforms working in this direction is the online network, The European Cluster Observatory, which provides unified access to the data about the clusters and cluster initiatives. There is a "cluster map" created on this basis suggesting the variety of instruments for analysis: it allows making standard and specific inquiries. Besides this platform, there are other ones, among which the research works of Harvard Business School can be singled out, although these studies are mainly casual, not continuous.

The next program—European Cluster Alliance—was created within the PRO INNO Europe program as means for the establishment of the dialogue between the national and regional authorities in order to promote efficient cluster policies, eliminate duplicating functions, and reduce the fragmentariness of cluster initiatives in Europe (DG Enterprise and Industry Report 2008).

The business line of European Cluster Excellence Initiative is the search for the optimum schemes of cluster management, the expected result of which is the improvement of the cluster management system.

Moreover, regular meetings of experts are organized to provide the EU Commission with questions about elaborating cluster development programs. Thus, the results of European Cluster Policy Group are EU Commission recommendations as for the priorities in further cluster development.

Besides the policies aimed at cluster development, there are other European programs which somehow affect it. Among these are Europa INNOVA, Regions of Knowledge, and Enterprise Europe Network. A large number of various programs on the regional and national levels, as well as on the level of EU Commission, often lead to the duplication of their functions and investment that causes now the necessity of a certain coordination mechanism (SEC 2008).

An important place in Europe is given to the biotechnological clusters. Biotechnologies, alongside with nanotechnologies, micro- and nanoelectronics, photonics, and engineering of "new" materials, are called "key enabling technologies", which in perspective may form the basis of EU economic evolution (SEC 2009).

Biotechnology is important for Europe for the following reasons (European commission 2008):

– as a required condition for achieving the food security in the region;
– as a response to the climate change in the region (mitigation of consequences and adaptation to them); and
– job-creation and maintenance of regional economic competitiveness.

There are a number of large biotechnological clusters functioning currently in the EU, such as Medicon Valley (Sweden, Denmark), BioValley (Switzerland), and BioCat (Spain), and the cluster of Picardie and Champagne-Ardenne regions (France). Besides, there are other examples of bioclusters, attractive for the business entities due to the potential profit on the technology onrush and for the local authorities who thereby may acquire an extra tool for the social and economic development of regions.

There are projects realized within the EU and aimed at the development of biotechnology and bioclusters. The most significant among them is European Council of Bioregions (CEBR), established in 2006. The main objective of CEBR is the competitive recovery of the biotechnological products that can be achieved through solving of the three core problems: the reduction of the fragmentariness of European companies and regions; the transformation of the competitiveness between the regions of Europe into cooperation; and the formation of the platform for the biotechnological initiatives in the EU in general (Rekord 2010).

At present, two projects are being run under the patronage of CEBT (Rekord 2010):

– BioCT within the "Regions of Knowledge" project, which is aimed at the elaboration of the unified action course of biotechnological production development;
– ABC Europe, which unites 14 European biotechnological clusters.

A separate purpose-made platform NetBioCluE (Networking activity for Biotechnology Clusters in Europe), which is a part of the Europe INNOVA project, was created in order to

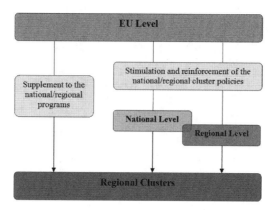

Figure 1. Role of the EU in developing regional clusters.

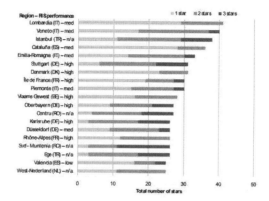

Figure 2. European regions by cluster portfolio strength.

encourage the cluster interaction in the spheres of biotechnology and healthcare. The reports for European authorities on the current situation and the recommendation provision concerning further development of bio-industries are performed by EuropaBio.

The research work by Skoch (2007) presents the analysis of 33 clusters in various economic sectors, located in different geographic regions of Europe, and on this basis, the author suggests the average profile of the European cluster. Most of the examined clusters have been recently originated ones: only 5 clusters appeared before 1940, 8 clusters were created in the period 1940–1970, while the rest 20 were formed after 1970. The recent origination is typical of not only the knowledge-intensive cluster, but also traditional clusters. Approximately 2/3 of the regional clusters unite less than 200 firms, and the employment in most of the clusters exceeds 2000 people.

Over the last decade, the number of employees has increased in the majority of the clusters, while 70% of the clusters showed increase in the quantity of firms (Skoch, 2007). The employment data contrast with the average sector indicators, which are notable for the poor growth and even decrease in production.

The regional cluster policy proves greater efficiency in comparison to the federal or national ones, because it considers the geographical peculiarities and the peculiarities of the business located in that region.

The observation of European cluster practice (DG Enterprise and Industry Report 2015) allows us to determine the following peculiar features:

– An insufficient attention is paid to the cross-national interaction. The cluster policy is reduced to the establishment of the cooperative interaction between the scientific organizations and the production; moreover, only a limited number of clusters are intent on the international competitiveness.

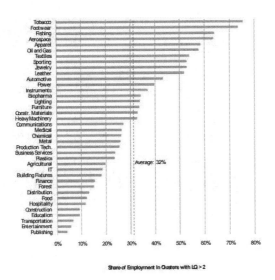

Figure 3. Geographic concentration of employment by cluster category.

- There is a tendency for creation of the new high-tech clusters, while the service, financial, and transportation and logistics clusters are poorly understood and left unappreciated.
- The all-European cluster policies are closely coordinated with the leading ones. The spread and share of cluster activity within branches are shown in Figure 4.
- A multitude of projects are aimed at innovation and cluster development, a steady proliferation of which leads to the incoordination and nonsystematic, fragmentary analysis of the previous experience.

The most important objectives of cluster activities in European countries within the period 2007–2016 are shown in Figure 4 and refer to the following spheres: human resources upgrading; business development; commercial cooperation; R&D and innovation, business environment improvement, and so on. In summary, cluster activity initiatives could increase the competitiveness level of the regional economy and accelerate the labor productivity.

Because the US economy contributes the largest solid economic complex, it takes precedence over the EU economy. Within the period 1990–2005, one-third of the world GDP growth belonged to the United States. In 2008, the US GDP per work hour was the highest in the world. The absence of grave commercial, investment, and migration barriers in the United States gives reasons for the high industrial intensity. According to the reports by The European Cluster Observatory, in 32 out of the 38 spheres under analysis, the territorial concentration in the United States is higher than that in Europe. Moreover, the United States and Europe differ in the entrepreneurial culture; thus, 61% of the Americans are interested in their own business, while in Europe, it is 45% (SEC 2008).

The last decades marked the shift of the economic structure toward the production of high-tech goods and services; a new knowledge-based economy has been created. The evolution of the economy became possible due to the high innovation potential; for example, the United States takes the leading place in the patent quantity per capita. Nevertheless, the United States takes additional measures on maintenance and strengthening of the competitiveness in all the active economic regions (see Figure 5).

The US economic policy is remarkable for the restricted interference into the cluster development process. The US authorities do not consider it necessary to scrutinize the commercial success of the businesses if it does not harm the US economy. Moreover, the United States, as distinct from Western Europe, is characterized by the high absorption of market and the strong diversification of the sectorial structure of the economy. In our opinion, it is the reason why the United States has adopted the so-called "upward" model of cluster development, which presupposes the business initiatives as the starting point, and the function of the state

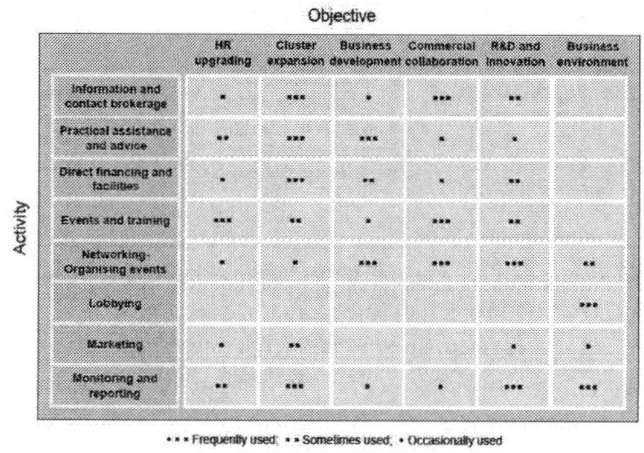

Figure 4. Objectives and activities of cluster-specific programs and initiatives.

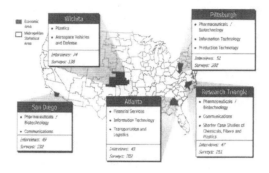

Figure 5. Clusters of innovation initiative regional surveys.

authorities consists in the development of the sci-tech partnership, in the tax credit extension and the provision of concessional taxation for the enterprises and firms involved into the state and the private R&D programs (Menshenina & Kapustina, 2008).

Notwithstanding the absence of the federal cluster policy, there is a multitude of regional programs. A good example of the effective cooperation between the R&D organizations, the business entities, and the authorities of the state of California is Silicon Valley. Silicon Valley, located close to Stanford University (San Francisco) and the University of California (Berkley), displays numerous companies producing computers, computer accessories, and software (Menshenina & Kapustina, 2008).

One of the most high-capacity clusters in the United States is the Michigan Automotive Cluster. Currently, 61 out of the 100 agencies representing the leading motorcar manufactures are located in Michigan. The cluster comprises 250 technology centers and 58 colleges and universities on the motorcar industry. At present, the Michigan Automotive Cluster can be considered a developed cluster with almost exhausted potential, and its further development is possible only via transformation.

The Michigan Automotive Cluster started forming in the 20th century with the emergence of the launching of the motorcar production by H. Ford, W. Durant, and R. Olds. At that time, Michigan was the leader in the production of wood and copper, the ironstone of Minnesota was available as well. The Great Lakes provided access to the dynamically developing states, to Chicago and New York, the railway connected Michigan to the south and the west of the country. Michigan in its turn was attractive due to the available high-skilled labor force and the low level of the underdeveloped trade unions. The presence of steel-producing companies and the production of locomotives and trains determined the availability of the workers familiar with metal working. Moreover, the accumulated income from the forest industry and the mineral industry, and the availability of the venture capital allowed investing in the motorcar industry (Porter & Ketelhohn 2009).

Originally, the Michigan Automotive Cluster specialized in the passenger car production was characterized by greater demand as compared to the lorry production. By 1910, 458,000 motorcars were released and this sector turned into one of the key industries of the country (Porter & Ketelhohn 2009).

Further development of the Michigan Automotive Cluster was conditioned by the large steady market: in 1956–1958, the American Government introduced the program on improvement of the highway service between the states; the growth of the suburbs caused the increase of the customers dependent on the transportation; the car acquires the status of an icon; and the Americans become eagerly anticipant of the new model releases. Until the 1970s, the American car manufacturers operated in almost noncompetitive conditions, and by the end of the 1970s, the car market grew up to 111 million items. However, the economic recovery of Europe and Japan called forth the car import, and the oil crisis of the 1980s highly influenced the customer preferences and caused the drastic demand for the economy cars (Porter & Ketelhohn 2009).

The increase in imports entailed the emergence of foreign motorcar plants in the United States and the retardation of American manufacturers became obvious. The market crash in 1987 resulted in serious difficulties that the American producers had to face.

3 CONCLUSION

From our viewpoint, the American and European practice of cluster formation and development has its strong points that may be adopted and used in the process of innovation cluster development in Russian regions, particularly the American "upward model" of cluster development, if combined with a new generation technology platform. A high level of cross-border cooperation is observed in European countries such as Germany, Denmark, and France.

Hence, it is considered efficient to take over the competitive mechanism conceived in American automotive clusters which also combined the method of product range expansion with the cross-investment into the clusters from related economic sectors. The implementation of these mechanisms in the Russian cluster development presupposes the introduction of the cross-shareholding practice within the companies belonging to the innovation cluster. Consequently, the cluster with respect to ownership is to constitute the solid holding in the form of financial and industrial groups.

REFERENCES

DG Enterprise and Industry Report. 2015. Innovation Clusters in Europe: A statistical analysis and overview of current policy support.
European commission. 2010. Europe 2020. A strategy for smart, sustainable and inclusive growth. Brussels.
European commission. 2012. Innovating for sustainable growth: a bioeconomy for Europe. Brussels.
Menshenina, I.G. & Kapustina, L.M.. 2008. Clustering in regional economy: monograph. Yekaterinburg: Publishing house of the Ural State University of Economics.
Official site of the Michigan economic development corporation. [ONLINE] Available at: http://www.michiganbusiness.org/grow/industries/#automotive [Accessed 11 May 17].
Porter, M. & Ketelhohn, N. 2009. Automotive Cluster in Michigan (USA). Cambridge, Massachusetts: Harvard University.
Rekord S.I. 2010. Industrial and innovative clusters development in Europe: evolution and contemporary discussion. Saint Petersburg: The Publishing House of Saint Petersburg State University of Economics and Finance.
SEC. 2008. The concept of clusters and cluster policies and their role for competitiveness and innovation: main statistical results and lessons learned. Commission Staff Working Document SEC. Luxembourg: Office for Official Publications of the European Communities.
SEC. 2009. Preparing for our future: developing a common strategy for key enabling technologies in the EU. *In Communication from the commission to the European Parliament, the Council, the European Economic and Social Committee and the Committee of the Regions.*
Skoch A. 2007. International experience of formation of clusters. *Kosmopolis* 2(16). [ONLINE] Available at: http://www.intelros.ru/index.php?newsid=352 [Accessed 11 May 17].

An approach to developing a programmer competency model in Russia

L.V. Naykhanova, N.B. Khaptakhaeva, N.N. Auysheeva & T.Y. Baklanova
East Siberia State University of Technology and Management, Ulan-Ude, Buryatia, Russia

ABSTRACT: In this article, we describe the approach to developing a competency model. This approach is based on professional standards. At present, professional standards and competence approach are being actively introduced in the organizations of Russia. Human resource is one of the most strategic resources for every type of organization. Therefore, it becomes important to offer a reasonable and intelligent system for them. For this purpose, we have to focus on providing correct information about the labor functions of employees in the organization. The proposed competency model consists of main and variable components. The main component contains knowledge, skills, and behaviors of a profession. A variable component contains knowledge, skills, and behaviors that reflect the activity of the specialist in the concrete organization. Such an approach allows us to create a universal and adaptive model.

1 INTRODUCTION

The history of the development of a competency-based approach in human resources management dates back to the 1970s, which was founded by D. McClelland, R. Boyatzis, and L. Spencer. In Russia, the competence approach has become an integral part of personnel management system for the last 10–15 years. Modern organizations believe that the staff as a strategic asset must have, above all, competence, which includes all necessary knowledge, skills, and a pattern of behavior for the effective implementation of work (Chulanova, 2013).

The major problem encountered by HR managers is the high labor intensity of the management, vast majority of tasks, functions, and processes, which should be managed. Human resources management system (HRMS) is to be established to support the effective process management.

Creating of HRMS requires formalizing all information. However, in many countries, HR management is traditionally based on job descriptions. Instructions are developed by employers in free form. Therefore, some of the information in the instructions is difficult to formalize. Information of professional standards is much more formalized.

It is only in recent years that the professional standards are being rapidly developed in Russia. These standards contain requirements to staff competencies such as knowledge, skills, and behaviors. These elements have a good characteristic, that is, they can be formalized. This will describe the organizational capacity of the employee to perform his/her duties, evaluate the capabilities of the employee, determine the direction of its development, and ultimately adjust the strategic goals of the organization with the capabilities of employees and their potential contribution to the development of the organization (Shemetov. 2010).

The purpose of our research is developing the integrative tool on the basis of the systematic, structural, functional, and institutional approaches to organization's HR management.

During the research, it is important to find an approach that will contribute to responding flexibly and quickly to changes in the external environment, and to allow easy adjustment of requirements to the employee (knowledge, skills, and behaviors) to ensure business continuity, competitiveness, and growth. In the modern world, there are regular changes in the external environment: changes in the development of business; changes in technologies, methods, and means of doing business; and changes in the professional standard.

The purpose of this work is to create a competency model that can be changed quickly when the external environment changes.

Thus, we report the first phase of this study describing an approach to developing a competency model. The competency model lays the foundation of HRSM (Shemetov, 2010).

2 DEVELOPMENT OF THE COMPETENCY MODEL

The competency model is based on the professional standard. Therefore, initially, we should analyze the Russian standard before developing the competency model. Professional standard for a programmer is used as an example.

2.1 *Basic data source*

Professional standard aims to state structured requirements for the content and quality of work in certain professional activity. Requirements define necessary knowledge and skills of the person in a given area of work (Bateman & Coles & Keating and Keevy, 2014). The need to develop and introduce professional standards is initiated by the presidential decree No. 597 dated 7 May 2012 "On measures to implement state social policy".

Professional standards are complex and identify necessary knowledge and skills of an employee for performing official duties. Professional standards are based on the layout of the professional standards, approved by the Ministry of Labour and Social Protection of Russia, and shall enter into force following approval by the Ministry. Information about standards can be found on the Website of The National Agency for Qualifications Development (http://www.nark-rspp.ru).

At present, in accordance with the European Commission's requirements for competence, all IT specialists in European Union countries are divided into six main groups: "Business Management", "Technology Management", "Design", "Development", "Service and operation", and "Support" (European ICT Professional Profiles). Each of these groups has three to five professional categories totaling 23 European ICT Professional Profiles (Lebedeva, 2016).

There is conformity of European ICT Professional Profiles and IT professions in Russian professional standards to the above six groups.

For example, group "Development" includes:

1. European ICT Professional Profiles:
   – Developer;
   – Digital Media Specialist;
   – Test Specialist.
2. IT-professions in Russian professional standards:
   – Developer;
   – Technical Writer;
   – Web & Multimedia Master;
   – Graphic Design and User Interface Design Specialist;
   – IT Test Specialist.

Both European ICT Professional Profiles and Russian professional standards have a profession "Developer".

The main terms and their definitions in professional standards (hereinafter standard) are as follows:

- Professional activity is a collection of generic job responsibilities having a similar characters, results, and working conditions.
- Generic job responsibilities are a set of coherent job responsibilities as a result of division of labor in specific production or business process.

- Job responsibilities is a task system under generic job responsibilities.
- Task is an interaction process between employee and subject of labor to achieve the target.

The proposed model has two components: main and variable (Naykhanova & Baklanova, 2016). The main component should contain the competence of standard, and the variable component must have competencies of a certain company.

2.2 *Developing the main component of the competency model*

Developing the main component is based on the structure of the standard:

– Career level has several generic job responsibilities.
– Position and job responsibilities correspond to generic job responsibilities.
– Job responsibilities are described by behaviors, necessary knowledge, and skills.

As seen from the description, the standard is organized hierarchically. According to this fact, the competency model should be presented as an oriented tree G.

The first level of tree G is composed of professions of IT specialists; the second level is a career level; the third level includes generic job responsibilities; the fourth level contains job responsibilities; and the fifth level is behaviors, necessary knowledge, and skills.

Definition "IT-specialist" is a root. Arcs with the node, containing kind of IT-specialist (professions), come out of the root.

Node-profession is expanded with using, for ex-ample, a professional programmer's standard [8]. This standard was approved by a Ministry of Labour and Social Protection No. 679n dated November 18, 2013. The range of values is in the region of [3–6]; therefore, four arcs come out of node-programmer (Fig. 1).

In turn, arcs with nodes, containing m generic job responsibilities, come out of every node-"career level" (Fig. 2).

Arcs with nodes, containing *k* job responsibilities and positions, come out of every node-"generic job responsibilities" at the third level (Fig. 3).

At the fourth level, arcs with nodes, containing behaviors, knowledge, and skills, come out of every node-"job responsibilities" (Fig. 4).

An employee must have certain skills and knowledge required to assume behaviors. It is, therefore, nodes-"behaviors", "knowledge", and "skills" that are expanded at the fifth level (Fig. 4).

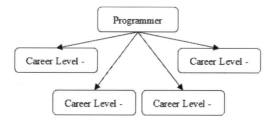

Figure 1. First level of the competency model.

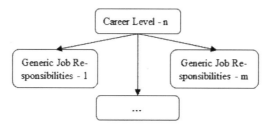

Figure 2. Second level of the competency model.

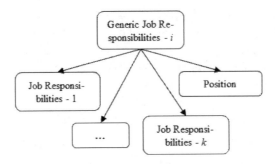

Figure 3. Third level of the competency model.

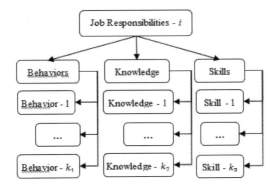

Figure 4. Fourth and fifth levels of the competency model.

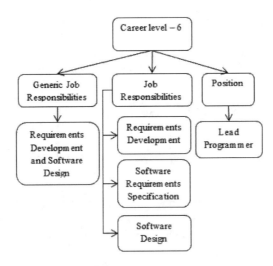

Figure 5. Structure of the sixth career level.

We consider the structure of the sixth career level (Fig. 5) by combining graphs from Figures 2 and 3.

Furthermore, the structure of the labor function "Requirements Development" is also considered (Fig. 6).

The structure shown in Figure 6 corresponds to the graph presented in Figure 4.

The constructed graph is the main component of programmer's competency model that corresponds to the Russian standard. However, every IT company has own direction and

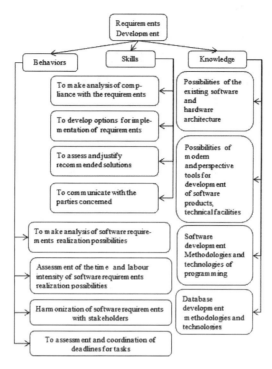

Figure 6. Structure of the job responsibilities "Requirements development".

uses a certain methodology, technology, and ways. It means that variable component must be added to the main component.

2.3 *Developing the variable component of the competency model*

The variable component is unique for every IT-company.

Terminal vertex (knowledge, skills, behaviors), shown in Figure 4, should be expanded to design the variable part.

We consider terminal vertex "Software development methodologies and programming technologies", presented in Figure 7 and coming out of node-"Knowledge", for a more understandable presentation.

In contrast to European and American professions, there is no profession called "Software Developer" in Russia. Job responsibilities in software development are related to programmers of the sixth career level in Russia.

Almost all existing software development methodologies (Waterfall, Agile, V-model, Iterative, RAD, Incremental Model, Spiral, etc.) are presented in the example shown in Figure 7. These methodologies are adopted in all countries. For the organization, nodes that identify only those methodologies that are used in this organization must be present. This also applies to technologies of programming.

Furthermore, all terminal vertices of the example, shown in Figure 7, must be expanded. The example of the expansion of the node, containing "V-Model" software development methodology, is shown in Figure 8. Decomposition of elements such as knowledge, skills, and behaviors lasts until the level satisfies the company requirements when the selection and certification of staff proceeds. Such an approach allows us to frame questions to an appraisal of staff automatically.

If the competency model is developed by a professional IT association, the company must leave only necessary nodes according to a direction of the organization and delete others.

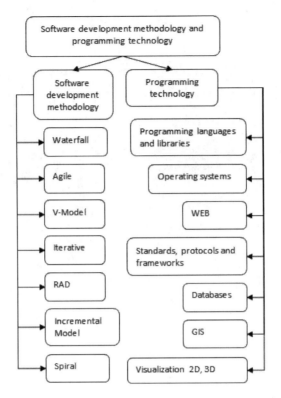

Figure 7. Example of the expansion of the node "Software development methodologies and technologies of programming".

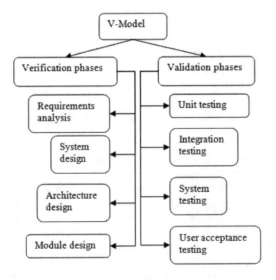

Figure 8. Example of the expansion of the node "V-Model".

3 IMPLEMENTATION OF THE COMPETENCY MODEL

The proposed model is implemented as ontology. Then, an expert system is developed as decision support system for human resource management. The competency model (ontology) is a core of a knowledge base of the expert system (EHRMS).

EHRMS is designed to support decision in personnel operations such as recruitment, job rotation, training, and evaluation. At present, the only assessments for staff were developed.

Facts and rules of the expert system are designed through the application of the ontology.

EHRMS improves traditional processes and decision-making processes (Chugh, 2014). For example, if an employee has a career level 4 and obtains successful assessments, then he/she can proceed for career level 5. If the result is successful, then there is an opportunity for promotion. The opposite result is an opportunity for demotion when an employee has a negative result.

4 CONCLUSION

In this paper, we considered a technique for developing a competency model, with characteristics of universality, adaptability, and scalability.

Universality is the ability of the technology to develop a competency model in any field of activity. Adaptability is the ability of the variable component to use configuration options directed at a particular company. The other positive property is the automatic generation of classes: that is, when developed in Protégé-Frames, ontology is exported to the expert system developed in CLIPS.

The approach outlined in this paper can be used in any country. To our knowledge, the competency model should be developed by a professional association. In this case, the company only has to configure it in their business processes.

REFERENCES

Bateman, A. & Coles, M. & Keating, J. & Keevy, J. 2014. Flying blind: Policy rationales for national qualifications frameworks and how they tend to evolve, *International Journal of Continuing Education and Lifelong Learning,* vol. 7, issue 1: 17–46.

Chugh, R 2014. Role of Human Resource Information Systems in an Educational Organisation, *Journal of Advanced Management Science*, vol. 2, no. 2: 149–153.

Chulanova O.L., Kibanov A.Ya., Mitrophanov E.A. 2014. The concept of competence-based approach in human resource management. Moscow: INFRA-M.

European ICT Professional Profiles CWA 16458. Available at: http://www.ecompetences.eu/ict-professional-profiles/.

Lebedeva T., Muravyov S. 2016. Professional standards in the IT-sector as a guide to action. *Professional Education in Russia and Abroad,* 4(24): 92–97.

Naykhanova, L.V. & Baklanova T.Y. 2016. Build competency model for it professional in the context of programmer's occupational standart.

Professional standart "Programmer". 2013. Available at: http://www.apkit.ru/committees/education/meetings/standarts.php (accessed 29.06.2017).

Shemetova N.K. 2010. Use of model of competences as a method of adoption of administrative decisions in the sphere of human resource management. *EGO: Economy. State. Society*, 1(1): 13–17.

Organizational behavior norms as a factor boosting the emotional stability of staff at housing and utility services enterprises

M.D. Mironova
Kazan Federal University, Kazan, Russia

R.M. Sirazetdinov & M.Y. Virtsev
Kazan State University of Architecture and Engineering, Kazan, Russia

ABSTRACT: This article examines the issues related to changes in qualification requirements for personnel employed at housing and utility services enterprises and improvements in the qualification upgrading system as part of the improvement of the management system used by enterprises in this field. The authors have proven that the shift in the competition focus among housing and utility services enterprises services from the price factor to the quality of goods and services can be achieved through the development of the skill level of personnel at enterprises and the development of the organizational quality philosophy.

Keywords: housing and utility services, organizational behavior, emotional stability, labor quality evaluation

1 INTRODUCTION

Economic reforms in the Russian economy that have been ongoing since the 1990s, the transition of the housing and utilities services sector to the market conditions resulted in profound changes in the consumer behavior of the new class of consumers of housing and utilities services that is homeowners. The development of the homeowner institution rendered the requirements to the quality of housing and utilities services and the activities of management companies in general significantly more rigorous. All this prompts the management of enterprises engaged in the housing and utilities services to devise and implement new development strategies based on innovative principles and aimed at expanding the range and improving the quality of services and minimizing the risks of accidents and disasters. Thus, the need to introduce systemic changes affecting organizational structures, resource and human resource subsystems has become of importance.

The effectiveness of economic reforms in the service sector is largely determined by the strategic focus of the management system, allowing enterprises to adapt to the changing conditions in the process of their functioning and development. The common and the most important property of all services is their direct and inseparable connection with human: services are produced by a human and to meet human's needs. The society is dependent on the personal qualities of workers, managers and engineering and technical workers in the service sector, on their professional competence, responsibility, and decency. The price of knowledge, skills and professional competencies of the employee's responsible behavior, depending also on the level of his intelligence, emotional maturity and mental state, increases many times.

We consider the model of management of the housing and communal sector based on the provisions of the social and cultural management concept (SCCM), which is based on the value of self-organization and self-control of specialists and managers, on the one hand, homeowners on the other hand. In the process of adaptation of housing and utilities services enterprises to new market relations, new strategies of innovative management take shape, a specific type of organizational behavior which is characterized by a dynamic equilibrium between the processes occurring in the environment and the system. These processes reflect the ability of the control system to adjust in a timely manner, to react to the dynamic changing conditions of the external environment through innovative management actions flexibly and adequately. It should be noted that the use of foreign management experience in the training of managers in the service sector will be successful only if national and cultural aspects of the organizational behavior of Russians are taken into account.

It is known that a common and important feature of all services is their direct link with the human: services are produced by the human to meet human needs. In the production of services, including housing and utilities, the consumer is influenced by different personal qualities, in particular communication skills of the entire staff of the company engaged in direct interaction with customers. The effectiveness of this interaction will directly depend on professional competence, responsibility and integrity of the staff. In studies of J. Coleman(1986), P. Bourdieu (1988) it was shown that not only formal institutions (laws, regulations) play an important role in the development of the services market, but also the specificity of the social and cultural potential of the industry.

Thus, the success of an enterprise operating on the market of housing and communal services will be directly determined by the level of development of labor resources which are a combination of experience, knowledge, qualifications, communicative abilities, skills of conflict management and decision-making. In our opinion, an important factor in the success of the service sector is the development of emotional intellect, whose value was proved by D. Goleman (1995).

The responsible behavior of staff engaged in the services sector depends on their qualification and intelligence, emotional maturity and mental state Factors of professional stress as a source of professional burnout are considered in the work of E. Berezentseva (Berezentseva E., 2014).

The conditions for the formation of the emotional stability of personnel as conditions for improving the quality of labor in the service sector were investigated in the work of Kruglov E. (Kruglov EA, 2013). Influence of the norms of organizational behavior on the regulation of the emotional stability of the organization's personnel is shown in the work of Mironova MD. (Mironova MD, 2015).

2 RESEARCH METHODOLOGY

During the research, such research methods were used as the study and generalization of the experience of personnel management in the service sector, the method of expert assessments, the analysis of the performance of workers in the housing and communal services sector. The concept of research uses the fundamental principles of management theory, the concept of social and cultural management, the theory of innovative development of economic systems in conditions of increasing complexity and dynamism of the external environment, the theory of emotional intelligence, on the basis of which the management system of the service enterprise is improved.

3 RESULTS

The development of key skills of specialists working in the housing and utilities services field is a most urgent task of innovation management. In fact, any enterprise has an unlimited number of existing and potentially necessary skills and knowledge so that their creation and acquisition may be quite chaotic in nature. For a company striving to be competitive, the development of key knowledge and skills due to the need of adaptation to the external environment must consciously be controlled by its managers. To this end, they must at least monitor the changes happening in the external environment, set forth goals and objectives and make sure they are achieved.

That being said, one can distinguish three main groups of key professional skills required for the success in the activities of an employee or a chief of any level at an enterprise engaged in the services sector:

- technical skills, i.e. the ability to apply the acquired knowledge in practice;
- communicative skills;
- a conceptual skill based upon the ability to analyze complex situations and define problems, the ability to find alternative solutions and choose optimal solutions from among them.

There is also the problem of ethno-cultural peculiarities of the organizational behavior of Russian citizens. The process of them selecting their corresponding forms and methods of decision-making, effective management and communications requires a serious research. Nevertheless, some general conclusions regarding the manifestation of Russian ethno-cultural values in the control system and organization have already been made. Thus, the work of R. Lewis divides several hundred of national and regional cultures into three groups. The author believes that the Russians belong to a multicultural group which determines the specifics of their relationship within the organization and in business negotiations. R. Lewis also highlighted the leadership characteristics of the Russians and their attitude to time as a resource (Lewis,1999).

Daniel Goleman, a well-known researcher of professional skills, factors and conditions of their formation, believes that workers in the sector of services attach particular importance to emotional intelligence based on five constituent elements: self-awareness, motivation, self-regulation, empathy and the art of building and developing interpersonal relationships. It is on these components that the ability to serve customers well rests (Goleman, 1995).

At the same time, it is the significant emotional load associated with the predominance of subject-subject relations in the sector of services that creates a phenomenon negative for the organization known as emotional burnout of staff. The modern understanding of employees' behavior in organizations shows that the chronic stress generates an emotional burnout of employees. The main adverse consequence of this phenomenon is his or her emotional exhaustion.

The results of the survey conducted among 173 employees of the three largest management companies in Kazan operating in the field of housing and utilities services showed that the main sources of stress were work overload caused by a large number of service requests, difficulties in establishing interaction between employees and customers, lack of effective interaction within the organization both horizontally (between employees) and vertically (between managers and employees), as well as the threat of penalties, lack of time for recuperation after work and the threat of losing one's job (see figure).

The discussion resulted in learning that the main factor of emotional burnout among employees within the housing and utilities organizations examined was uncertainty which finds expression in the absence of standards and regulations when performing one's duties. Also, interruptions in the provision of utility services due to technological failures and wear of municipal infrastructure transform into a negative emotional load for employees working directly with customer complaints. Spontaneous attempts to overcome emotional stress/tension on the part of employees at management companies reduce the time spent on working contacts with colleagues and clients, chronic fatigue, increased emotional instability,

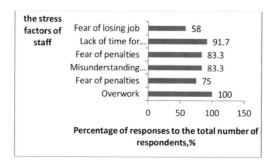

Figure 1. The sources of stress for staff in the sector of housing and utilities services, %.

conflict, reduced motivation, etc. An increasing staff turnover can be observed with elevated levels of absenteeism, an increase in the number of conflict situations which also negatively affects the performance indicators.

Thus, there is a contradiction between the goals of the organization and the limited opportunities to achieve them, associated in particular with the peculiarities of the manifestation of the emotional and volitional and intellectual aspects of one's personality. This contradiction can be solved by investing into the staff within an organization. Courses of professional training and retraining are highlighted among the areas to invest into. Then it is expenses associated with the treatment of employees and preventive measures, the construction of physical and health centers, preschool institutions and other social investments. Certain costs are associated with the development of the corporate culture.

Speaking of investments on the part of a company in training employees, Becker singled out a special training and a general one. Special training imparts knowledge and skills of interest only to the company where they are acquired (e.g., familiarizing beginners with the structure and internal regulations at the enterprise). During the general training, the worker acquires knowledge and skills that can find application in many other companies. Becker demonstrated that the general training is indirectly paid for by employees themselves when they accept a lower wage during the training period when seeking to improve their skills and through this derive an income from it. On the contrary, a special training is paid for by companies and they also derive an income from it as otherwise losses would be borne by employees when being fired at the initiative of the companies they worked for (Lewis, 1999).

In the study by Y. Korchagin, the company's investment in human capital is divided into investment in education and investment in knowledge (Korchagin, 2005). The author considers investment of companies in personnel training as an investment in education, while financial investments in professional knowledge for work in the field of science and innovation are regarded as investments in knowledge.

The effectiveness of investments into human capital is still an important issue. Investments into human capital are justified in cases when a company tends to increase its efficiency and when the contribution of a certain employee to that increase can be seen. Professor V.V. Lukashevich believes the investments into human capital result in enhanced labor productivity. According to his theory, the correlation between these indicators is expressed with the following formula:

$$A_i = (B - B_n) \times C/K, \qquad (1)$$

where A_i indicates the effectiveness of investments into human capital at the i-stage; B_n indicates the performance of an employee before the training; B indicates the performance of an employee after the training; C indicates product unit price; K indicates the investments into human capital (Lukashevich, 2002).

The disadvantage of this model is that it is not able to reflect the total growth of the employee's performance in the service sector. In our opinion, taking into account the emotional components in the service sector activity the evaluation of investments into human capital will have a two-component structure $\{A_i, A_j\}$, where A_i indicates the effectiveness of investments into human capital (1), and A_j is an indicator of the emotional and volitional aspects dynamics of employees before and after the training.

In order to upgrade skills of employees at management companies (Kazan, Tatarstan), courses were arranged on the premises of the Kazan State University of Architecture and Engineering where classes on overcoming stress situations and reducing conflict when working with customers were given.

To overcome stress factors, special attention is paid to the development of inner positive thinking among employees. For this we propose that staff seek to identify and consolidate the strong and positive sides of their personality. Next, employees were asked to take a psychological training with the aid of a professional psychologist who helped master some self-regulation techniques that shaped a positive attitude and helped to boost resistance to stress.

As shown by our study, methods aimed at boosting resistance to stress among staff employed by a housing and utilities services organization are implemented in the most efficient manner simultaneously with a training aimed at shaping organizational culture norms which define the attitude of employees to the work they do. Such norms include the following: priorities in achieving personal goals set by an employee and those pursued by the organization; order; discipline; protection from unnecessary information and relieving stress through time set aside for leisure. Order means a clear definition of objectives, functions, duties of each employee. This said, the formation of a system of values within an organization interiorized by each and every employee will be of great importance: sport as the norm of life, positive mood, a striving towards personal and professional growth and team spirit.

In our view, methods aimed at boosting resistance to stress among staff employed by a housing and utilities services organization are implemented in the most efficient manner simultaneously with a training aimed at shaping organizational culture norms which define the attitude of employees to the work they do. Such norms include the following: priorities in achieving personal goals set by an employee and those pursued by the organization; order; discipline; protection from unnecessary information and relieving stress through time set aside for leisure. This said, the formation of a system of values within an organization interiorized by each and every employee will be of great importance: sport as the norm of life, positive mood, a striving towards personal and professional growth and team spirit.

In our opinion, the issue of integration of an organization's values into the evaluation criteria of the quality of work performed by employees in housing and utilities services remains an acute issue. The staff labor efficiency ratio calculated using increasing and decreasing values of the corresponding ratios is generally known.

The increasing labor efficiency ratios include:

- an increase in the volume of services due to the growth of labor productivity;
- achieving an upsurge in productivity through the improvement and personal development of the workforce;
- the introduction of innovative technologies and management techniques.

The decreasing labor efficiency values include:

- poor-quality performance of professional duties;
- not meeting the deadlines set for the execution of assignments;
- a negligent attitude to one's duties;
- low efficiency of business communications;
- conflict behavior.

In order to overcome the uncertainties associated with production processes at an enterprise engaged in the sector of housing and utilities, the development of a Key Performance Indicators (KPI) system is of great importance which can be used to analyze the performance of each employee, the subdivisions and the company as a whole. Then the system of remuneration and wage incentives would be directly linked to each employee meeting the KPI which would apply to the organization as a whole and it would be transparent and understandable for all employees within the organization.

The direct correlation between emotional stability formation and the performance increase of employees in housing and utilities services by means of interiorization of psychological self-regulation methods, as well as the digestion of organizational culture norms and values should be taken into consideration while creating the personnel management system and the employees performance quality control system in the enterprises of housing and utilities services.

4 CONCLUSIONS

Emotional stability is a necessary characteristic of behavior, because this is an integral property of the psyche, expressed in the ability to overcome the state of excessive emotional arousal when performing complex activities. The development and introduction of the norms of organizational behavior adopted by the employees of the enterprise have a significant impact on the increase of emotional stability—the most important component of the communicative activity of personnel in the sphere of housing and communal services. The introduction of these norms of organizational culture on the basis of a comparative assessment of the quality of work and employee satisfaction contributes to the development of personal and professional qualities of employees.

In conclusion of the article it should be noted that managerial innovations that implement the norms of organizational behavior aimed at reducing the emotional tension of the personnel of the enterprise in the sphere of housing and communal services as a whole will contribute to improving the efficiency of the service enterprises.

REFERENCES

Berezentseva, E.A. 2014. Professional stress as a source of professional burnout. *Managing education: theory and practice*, 4, 162–170.

Bourdieu, Pierre. 1986. The Forms of Capital. *Handbook and Research for the Sociology of Education*, ed John C. Richardson, pp. 241–258. New York, Greenwood Press.

Coleman, J.S. 1988. Social Capital in the creation of human capital. *American Journal of Sociology.* n. 94, pp. 95–120.

Goleman, D. 1995. Emotional intelligence. 10th Anniversary Edition; Why It Can Matter More Than IQ. Bantam Books. ISBN 978-0-553-38371-3.

Korchagin Y.A. 2005. Russian human capital: the factor of development or degradation?: Monograph. – *Voronezh: CERA*.

Kruglov, A.F. 2013. Formation of the personnel emotional stability in an organization as a way of improving the quality of labor in the service sector. In the book of *Actual problems of modern science.* 75–78. Ufa: Bashkir State University.

Lewis R.D. Business cultures in international business. From the collision to mutual understanding: *Moscow: The Case*, 1999.

Lukashevich V.V. 2002. Efficiency of investment in human capital. *Typographer and Publisher. No. 6*, 29–35.

Mironova, M.D. 2015. Organizational behavior norms and their influence on improved emotional stability. *Leadership and management*, 2(2), 131–140. doi: 10.18334/lim.2.2.601.

Mironova M.D., Egorov D.A. 2014. Emotional stability of personnel as a factor of improving the quality of labor in the organization of the service sector. *Journal of Russian Entrepreneurship. No. 21 (267)*.289–294.

Zagidullina G.M., Romanova, A.I., Mironova, M.D. 2009. Management innovations in the system of mass service (as exemplified by housing and utilities complex). *Vestnik of the Kazan Technical University*, 5, 128–133.

Issues and Trends in Interdisciplinary Behavior and Social Science – Lumban Gaol et al. (Eds)
© 2018 Taylor & Francis Group, London, ISBN 978-1-138-55380-4

Methodology for the technical and economic analysis of a product at the projection stage

V.N. Nesterov & N.N. Kozlova
Kazan Federal University, Kazan, Russian Federation

ABSTRACT: This article presents the methodological basics of the technical and economic analysis at the stage of innovative product creation. This type of analysis allows us to optimize technical solutions by the economic efficiency criterion and to increase the quality of production at optimum costs of its production and operation that will increase the competitiveness of production.

In this study, the maintenance of technical and economic analysis at separate design stages of products is considered, as well as the methods and indicators allowing the optimization of technical solutions on the basis of the chosen economic criteria are provided. The role of functional and cost analysis as a highly effective method of the technical and economic analysis promoting considerable reduction of functional and unjustified expenses at an innovation design stage is defined.

Methods of economic optimization based on an economic assessment of multicriteria are offered. By generalizing the target criterion, either the indicator with respect to the relation of total expenses on all life cycles for unit of consumer properties and quality, or the return indicator of integrated quality as a function of the use value (usefulness) and the cumulative expenses corresponding to an object is provided. The factors influencing these economic criteria are revealed and classified, which are allowed to purposefully influence on the change of its size during both production and the operation of a product.

1 INTRODUCTION

During product design, its technical and economic characteristics are formed, which have to meet consumer needs and also be characterized by high quality and optimum expenses in both competitive production and operation.

Methods of achievement of technical and economic characteristics at design of products can be intuitive, parametrical, or functional. Experience or talent of the engineer is the cornerstone of intuitive methods. Parametrical approach is based on search of interrelations between technical characteristics of a product and economic efficiency. The functional analysis is a search of the most economically effective solutions of functions realization for which implementation of the product is intended. Technical and economic analysis (TEA) allows us to eliminate defects of intuitive methods, to make design-technology decisions and indicators for achieving an acceptable economic effect.

By means of TEA, the necessary information allowing the estimating extent of the technology solutions' influence at each design stage on technical and economic indicators of an innovation for achieving economic effect is formed. Also, this method allows us to optimize a product design proceeding from the accepted economic criteria answering to the market conditions and competitiveness of production. Such approach demands a close coordination of research and development management with formation of economic indicators, especially with prime cost of a product, and promotes identification of expenses decrease reserves and profitability increase.

The process of design is carried out in stages. Therefore, for calculations of technical and economic indicators and the choice of methods for its optimization, it is necessary to be carried out also in relation to features of each stage (Fig. 1) [1].

TEA presents creation of multilevel system of economic-mathematical models, in which, at the first level, calculations for the choice of the best design options of a new products design in general are performed.

Models of this level are applied at early development stages – the specification, the offer, and outline design – and are based on the complete technical and economic indicators characterizing a product as a research object. As criteria, generalizing indicators such as design prime cost of products, its labor input, size of initial investments, expense of the main materials, and other indicators characterizing processes of production and operation of new products are applied. Economic calculations and the analysis are based on approximate estimates, which, as a rule, have no high precision. During accumulation of information, it is specified and corrected. On the basis of these data, predesigns of economic efficiency are carried out and technical and economic indicators are proved.

At the subsequent stages of design preparation, the purpose of economic analysis is the choice of the most effective alternative design solutions of the projected product and its elements on indicators of technological effectiveness, standardization, and unification. It allows us to form more detailed information on prime cost, material capacity, labor input in a section of units, knots, and details. As a result, the accuracy of economic efficiency calculation of the chosen options technical solutions increases [9].

Choosing the functional charts of products and its elements is expedient for the application of functional and cost analysis (FCA), which represents the effective method of the technical and economic analysis promoting considerable decrease in functional and unjustified expenses by a careful research of functions of objects and costs of its implementation [4].

The technical and economic analysis of an innovation has to proceed also at a stage of technological preparation, whose purpose is an assessment of technological effectiveness of a design, its further working off. At this stage, the economic analysis is directed to the choice of the most economic options of manufacturing techniques of an innovation, organizational decisions taking into account specific conditions of its production, and preparation of the regulatory base for planning.

In Figure 1, each design stage is considered. According to these stages, the technical and economic calculations are expediently made in time for carrying out calculations upside-down

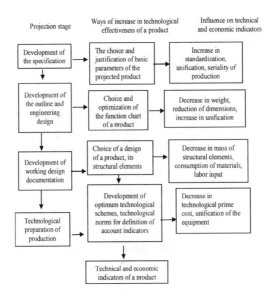

Figure 1. Formation of product indicators at each design stage [11].

as there are straight lines and feedback between separate stages of technical training, and also technical training and production is expedient.

Direct link between separate stages of technical training is shown in continuity of the results received at the previous stages for the subsequent stages. Feedback shows that by results of design of products at the subsequent stages, correction of the indicators received at the previous stages is carried out. For example, when designing separate details, changes in a design of knots and units can be made.

Between technical training and production direct link is the process of development of new products carried out on the basis of documentation developed at technical training. Feedback is shown in changes of design and technological documentation in the course of development of new products production, because of possible discrepancy of real conditions of production to design indicators.

Studying the relationships of cause and effect between technical and economic indicators of the created products in a research of the factors and conditions predetermining the level of economic efficiency holds the central position in TEA technique. The results of TEA allow us to define and prove design (standard) indicators at early stages of research and development and to estimate the actual indicators on completion of development, to define the reasons of negative deviations.

Using TEA, it is possible to mark out characteristics such as recurrence of the performed works, a variety of methods of data processing depending on a design stage, and the volume of information and decision-making on design stages. In TEA, various economic-mathematical methods, algorithms, and procedures, as well as standard reference information are applied to perform calculations of technical and economic indicators of the created objects: prime costs, material capacities, power consumption, labor input of production, expenses and productivity at operation, total expenses, and economic effect.

In this regard, it is possible to allocate and define an order of the most characteristic types of work performance:

– search and systematization of technical and economic data on analogs of the projected products;
– the choice and application of economic-mathematical models and method of calculation of the technical and economic indicators, which are most acceptable for each stage of design and conditions of obtaining necessary data;
– comparative economic assessment and analysis of technical solutions;
– preparation of recommendations to developers of innovations about the directions of increase in efficiency and quality of the projected products, optimization of its key parameters.

Consideration of maintenance of TEA and complex of methods inherent in it gives an opportunity to solve the various, but interconnected tasks:

– an assessment of expenses at various stages of creation and operation of a product necessary for summing up a research and development, planning, organization of financing, as well as analysis and an assessment of overall performance of divisions;
– the feasibility study on the created products, the choice of options of technical solution for the subsequent production and operation;
– optimization of design, technological, operational and economic indicators, and parameters of the created products;
– an assessment of economic efficiency of the investments directed to creation of an innovation.

As it was already noted, the most important problem of TEA is the economic assessment of technical solutions options. In view of the fact that the projected product does not exist in a material form, usual methods of calculation are no good. The problem of a technical and economic assessment of options is in that by means of special methods to predict the level of the expenses expected at the implementation of each option. In practice, depending on a design stage, the following main methods of definition of prime cost at a product design stage are applied: specific expenses, coefficients, dimensional coefficients, correlation modeling, the reduced standard calculation, expert estimates, and others. Dependence of expenses on change of technical characteristics of a product [3] is its cornerstone.

Competitiveness of a product considers not only technical characteristics, but also includes the product price, size of costs of operation, and other economic parameters. Therefore, to predict competitiveness of a product in the market, it is necessary to prove its future indicators including economic. Therefore, the indicator of prime cost of a new production must be set as the target criterion, which should be reached during development along with technical. The decision is made if it satisfies the established criterion size. Thus, economic criteria, along with technical, become the main, but not derivatives. Such system is applied in the concept of management on target expenses [8].

Practice shows that standards are developed proceeding from the existing level and conditions of production and last experience, and its level not always meets the requirements of production competitiveness. Target expenses force developers of new production to be guided not by last experience, but by requirements of consumers, decrease in expenses in the course of development of a product, that is to carry out cost engineering. After the main characteristics of future product are established, the main objective of the analysis consists in justification of the obtained technical solutions on achievement of the set technical and economic parameters as a solution of optimizing tasks.

In market economy, the main criteria of optimality are the following financial indicators: profit, liquidity, and cash flows. These criteria are the cornerstone of the majority of the economic efficiency calculation with the known method of innovations and investments. During innovation design, key parameters of future product are formed. In this case, the optimization of multicriteria including the use of more private and nonfinancial indicators—expenses, the income, indicators of production quality, market size, time of creation and development of a product, ecological indicators and others—is expedient. Preference is given to those indicators whose level corresponds to processes for which assessment they are intended to. Also, the accuracy and labor input of its calculation is considered. Therefore, for an assessment of the private and local decisions concerning a change of some technical parameters of products or its parts, it is better to use more concrete indicators of expenses, whereas the general indicators of efficiency are more preferable to the general decisions [7].

By generalizing the target criterion, either the indicator with respect to the relation of total expenses on all life cycles for unit of consumer properties and quality, or the return indicator of integrated quality as a function of the use value (usefulness) and the cumulative expenses corresponding to an object is provided. Furthermore, optimization of a product has to be carried out step by step with the subsequent specification of total expenses size during work performance. This indicator reflects efficiency of a product from the viewpoint of not only its producer but also the consumer [6].

In relation to separate economic entity, for adoption of the decision on the organization of new production release and expediency of investment, criteria of financial efficiency as to the investor, it is important to keep a financial solvency and to receive benefit. Therefore, calculations of innovations efficiency are based on the criteria applied to calculation of investments efficiency. Between these criteria, there are no contradictions. It supplements each other because the optimum ratio of total expenses and qualitative indicators promote market success of a product [11].

Optimizing calculations have to be performed throughout the life cycle of the product, but it is especially important at the preproduction stage as the greatest potential opportunities of decrease in costs of a product are in the sphere of production preparation, where 65–80% of cumulative expenses take place. Possibilities of expenses reduction in subsequent stages decrease in the process of product readiness [1].

2 METHODOLOGY

We will consider in more detail optimization of expenses in the course of carrying out functional and cost analyses. As it has been noted earlier, FSA is directed to elimination of functional and unjustified expenses. It is the most convenient to carry out the solution of this task by means of the structural model creation of an object submitting the scheme of the relationships among elements, which gives an idea of object material components structure. On the basis of structural model, the structural and cost model of an object with the indication of expenses on each element is under construction. The cost assessment of elements at a stage of product creation is carried using data about expenses on similar products taking into account correction coefficients. On the basis of details costs, a direct cost on assembly units is counted. On this cost, it is possible to make structural and cost schemes to carry out the analysis on a product in general.

On the basis of structural model and the functional description of a product, the combined functional and structural model is under construction, which is used for identification of unnecessary functions and elements in a product, definitions of functionality, usefulness of material carriers of a product, allocation of costs on functions, an assessment of workmanship of functions, and identification of defective functional zones in a product. It is expedient to carry out creation of functional and structural model in the form of a matrix, where in the lines, material products are put down, and on columns, functions of an object. On crossing of lines and columns, the cost sizes corresponding to an element and function (Table 1) are specified.

Cost of functions allow us to establish the volume of expenses, which is the share of separate consumer properties and the necessary level of consumer cost and useful effect when using a product. For finding decrease reserves in expenses, it is necessary to compare the limit expenses defined at early stages of development with functional expenses. Special attention is paid to those functions at which limit expenses are exceeded.

Determining the cost of function (Sf) is possible in different situations. When one or group of material objects completely provide a certain function, its production costs are defined by prime cost of these objects:

$$Sf = \sum_{i=1}^{n} Sij;$$

where Sij – costs of i object; j – function; and n – number of function object.

If the same material object performs several functions, then the related expenses are distributed between functions in proportion to a contribution (Pij) of the object to realization of these functions, which are estimated more often by the expert way. Then, costs of function are determined by the formula:

$$Sij = \sum_{i=1}^{n} Pij \times Sij.$$

Example are presented in Table 2.

After an assessment of functions, the technological level of the decisions embodied in an initial product is defined. To reveal reserves of functions cost reduction and to orient creative search of new decisions, a number of methods are available, one of which is comparison of relative costs of function and mark estimates of function importance. This method makes an assumption that the normalizing condition for allocation of costs serves the importance

Table 1. Functional and structural model of an object.

Elements	Functions				Cost of element
	A	B	C	D	
1	–	2	3	5	10
2	10	–	15	–	25
3	–	5	–	8	13
4	8	2	–	4	14
5	10	–	7	–	17
Cost of function	28	9	25	17	79 (Total cost of product)

Table 2. Calculation of cost of functions.

Name of a detail	Prime cost of a detail	Costs of function F1	F2
b ($)	60	50	10
%	100	83	17
c ($)	24	12	12
%	100	50	50
Total:	84	62	22
	100	74	26

Table 3. Basic data for the analysis.

Functions	Rank of function (importance)	Functional costs ($)
A	5	25
B	3	30
C	2	16
D	2	12
E	2	5

Table 4. Calculation of indicators [8].

Functions	Rank of function (Ki)	Value of function (Vi)	Cost of function (Ci)	Relative costs of function (Ni)
A	5	35.7	25	28.4
B	3	21.4	30	34.1
C	2	14.3	16	18.2
D	2	14.3	12	13.6
E	2	14.3	5	5.7
Total:	14	100	88	100

of functions. The importance of function characterizes a relative contribution among other functions of the level in implementation of higher function. The indicator of value of function (Vi) is determined in terms of the importance by the formula:

$$Vi = (V/\Sigma Ki) \times Ki,$$

where V – the general indicator of value of all functions in total in points (10, 100, 1000) and Ki – a rank of i function.

Relative costs of i function implementation are determined by the formula:

$$Ni = (N/\Sigma Ci) \times Ci,$$

where N – the general indicator of relative cost of all functions (10, 100, 1000) and Ci – cost of i function.

Examples are presented in Table 3.
Using the formulas given above, we will calculate indicators (Table 4):
We determine the ratio of relative cost of function to its importance (Zi) by the formula:

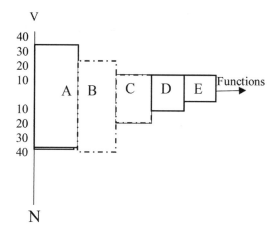

Figure 2. Functional and cost chart.

$$Z_i = N_i/V_i$$

If Z > 1, then function is considered adverse, and it is required to make the technical solution directed to function reduction in favorable compliance of cost and importance.

In our example with adverse functions, function coefficient ratio of "B" and "C" is > 1.

Za = 26.4/35.7 = 0.80;
Zb = 34.1/21.4 = 1.59;
Zc = 18.2/14.3 = 1.27;
Zd = 13.6/14.3 = 0.96;
Ze = 5.7/14.3 = 0.40.

The ratio of cost to the importance of functions is presented in Figure 2.

Other direction of product costs decrease is not minimization of the function implementation costs, but search and an exception of unnecessary functions. Unnecessary functions are found by comparison of functions that are provided with a product and demanded or by revision of technical solutions.

3 CONCLUSION

Carrying out TEA systematically in the course of development of new products allows us in due time to inform developers on the actual size of expenses; establish the main reasons, sites, and concentration of expenses; and define the main directions of decrease in expenses. Also, this type of analysis allows us to carry out calculations of economic efficiency, make rational design technology and organizational decisions, and optimize the ratio of expenses at separate design stages.

REFERENCES

[1] Nesterov V.N. The analysis of expenses in innovative development of the organization. – Kazan: KFEI publishing house, 2001. 180 pages.
[2] Maria Conceição A. Silva Portela. Value and quantity data in economic and technical efficiency measurement. – Economics Letters, Volume 124, Issue 1, July 2014, Pages 108–112.
[3] Arto Tolonen, Marzieh Shahmarichatghieh, Janne Harkonen, Harri Haapasalo. Product portfolio management – Targets and key performance indicators for product portfolio renewal over life cycle. – International Journal of Production Economics, Volume 170, Part B, December 2015, Pages 468–477.

[4] Harald Gmelin, Stefan Seuring. Achieving sustainable new product development by integrating product life-cycle management capabilities. – International Journal of Production Economics, Volume 154, August 2014, Pages 166–177.

[5] Peter C Sander, Aarnout C Brombacher. Analysis of quality information flows in the product creation process of high-volume consumer products. – International Journal of Production Economics, Volume 67, Issue 1, 10 August 2000, Pages 37–52.

[6] Katarína Teplická Comparison of Methods for Pricing of the Product and its Impact on Economic Efficiency of Enterprise. – Procedia Economics and Finance, Volume 34, 2015, Pages 149–155.

[7] Sokolov A.Y, Elsukova T.V. Using abc to enhance throughput accounting: An integrated management approach//Academy of Strategic Management Journal. – 2016. – Vol. 15, Special Issue 4. – Pages 8–15.

[8] Faizrahmanova G.R., Kozlova N.N. The system of indicators of enterprises innovative activity // Asian Social Science, Vol.11, № 11, Pages 183–188.

[9] Kulikova L.I.V.N. Nesterov, D.A. Vakhotina, I.I. Yakhin The Revision of Approaches to Innovative Analysis// Mediterranean Journal of Social Sciences, MCSER Publishing, Rome-Italy, Vol. 6 No. 1 S2, January 2015, Pages 421–425.

[10] Kirpikov, A.N., Nugaev, F.S. Relevant approaches to performing analysis of financial results of organization's activity with application of factor models//International Business Management, 10(15), 2016. Pages 2987–2991.

[11] Elsukova T.V., Lean accounting and throughput accounting: An integrated approach//Mediterranean Journal of Social Sciences. – 2015. – Vol. 6, Is. 3. – Pages 83–87.

Income inequality in Russia

R.S. Timerkhanov, N.M. Sabitova & C.M. Shavaleyeva
Kazan Federal University, Kazan, Russian Federation

ABSTRACT: In this article, we present data on income inequality in Russia. We also address the problems associated with inequality of income distribution in the country. We establish the causes of economic difference and determine the sources of economic inequality in the regions of Russia. Finally, we recommend measures to improve the socioeconomic climate in the country.

1 INTRODUCTION

Income of the population – their level and dynamics – is one of the most important economic indicators determining the population situation. In recent years, worldwide research on the inequalities in the growth of incomes of the population and the causes and structure of inequality has been gaining popularity: with special attention paid to the studies of American economists E. Stiglitz and G. Akerlof and French economist T. Picetti.

At present, in most cases, the inequality of welfare levels is considered as a socioeconomic difference, as its main reason is the difference in the incomes of the population. Income inequality is the foundation of the market economy, an indispensable condition for its effective functioning, and the strongest motivational engine inherent in human nature. The degree of income difference is an important characteristic of the quality of life, and the dynamics of this indicator can have a significant impact on the pace of socioeconomic development. The arrogant degree of difference in terms of income growth can become a serious threat to sociopolitical stability. Growth of difference in the incomes of the population depending on the level of income ultimately contributes to a slowdown in the growth of the economy. The beginning of reforms related to the transition to market relations led to the stratification of society by income, as well as an increase in the differences between strata and groups of the population. The problem of income difference has become urgent under modern conditions, not only from an economic viewpoint, but also from a social viewpoint. This phenomenon has negative consequences such as deterioration of the demographic situation, sharp socioeconomic stratification of society, increase of crime rate, and restriction of economic development. Different factors that affect income difference include territorial factors, demographic factors, moral and ethical norms of the society, taxation system, uneven adaptation of the population to the changing spectrum of economic opportunities, accumulated human capital, territorial differences in living conditions, and others.

Increasing inequality in incomes not only increases social tensions in society, but also undermines entrepreneurial activity, reducing the potential for economic growth.

2 METHODS

At present, there is a paradoxical situation: on the one hand, Russia is one of the richest countries in the world in terms of natural and human resources, and on the other hand, it lags far behind the world leaders in terms of incomes.

Despite the variety of sources of income, their main components are remuneration of labor, income from personal part-time farming, social transfers, and other sources.

The dynamics of real incomes is determined by not only the level of salaries of wage earners, the income of entrepreneurs, and transfers, but also the dynamics of taxes and the state's tax policy, trends and prices, and inflation.

In 2016, the transformation of the structure of monetary incomes slightly accelerated: the share of wages in annual terms decreased by 0.8% points (from 65.6% to 64.8%), and the share of social payments increased by 0.8% points (from 18.3% to 19.1%, which is the historical maximum for this indicator) (Table 1). Analogous, but much stronger changes in the structure of income occurred in 2009–2010, when in 2 years the share of social payments increased by 4.5% points (from 13.2% to 17.7%), and the share of wages decreased by 3.2% points (from the highest level in 15 years in 68.4% to 65.2%). The share of income from entrepreneurial activity (mainly from individual entrepreneurship) decreased from 12% in 2002–2003 to 7.8% in 2016, which indicates the preservation of a low level of entrepreneurial activity in the country. The share of income from equity (dividends, interest from deposits, payments on securities) decreased from a maximum of 10% in 2005–2006 to 5% in 2011–2012, after which it began to slowly recover, and in 2016 reached 6.3% (Table 1).

The restoration of the share of income from property suggests some improvement in the situation of the more affluent layers of the population, which leads to an increase in social inequality. At the same time, the total share of "capitalist incomes" (income from property and entrepreneurial activity) has been keeping at the level of 14% since 2011 against 22% in 2005.

There are several reasons affecting the income difference of individuals:

1. Differences in physical, mental, and entrepreneurial abilities. Some of the abilities of people are unique and therefore limited, which brings their owners an intellectual rent. The same applies to entrepreneurial abilities.
2. Differences in the profitability of professional activities. Thus, high incomes of top managers, programmers, financiers, and lawyers are conditioned by limited inelastic offer of their services due to special psychological and intellectual data that are required for mastering these professions, as well as high time, energy, and money for education and high demand for their services, on which the profitability of modern business directly depends.
3. Discrimination in labor relations. In Russia, currently, the leading place among the social and labor problems is firmly occupied by age discrimination. To get a job is very difficult for all those who are below 25 years of age and above 50 years. Most of the employed in the economy are middle-aged workers.
4. The factor of economic opportunities. There is unequal access to education, vocational training, highly paid work, and other public and private goods, which creates unequal starting economic conditions for all. Therefore, even in a highly developed market economy, the chances of getting education are distributed in such a way that some of the young people are

Table 1. Structure of incomes of the population, 2008–2016 (Russian Federation federal state statistics service).

Year	Total money-income, billion rubles	Income from business activities	Wages, including hidden salaries	Social payments	Income from property	Other income
2008	25 244.0	10.2	68.4	13.2	6.2	2.0
2009	28 697.5	9.5	67.3	14.8	6.4	2.0
2010	32 498.3	8.9	65.2	17.7	6.2	2.0
2011	35 648.7	8.9	65.6	18.3	5.2	2.0
2012	39 903.7	9.4	65.1	18.4	5.1	2.0
2013	44 650.4	8.6	65.3	18.6	5.5	2.0
2014	47 920.6	8.4	65.8	18.0	5.8	2.0
2015	53 525.8	7.9	65.6	18.3	6.2	2.0
2016	54 102.5	7.8	64.8	19.1	6.3	2.0

forced to abandon certain professions or combine their studies and work. And this directly affects the level of qualifications, prospects of employment, and, as a result, income.
5. Regional differences, caused by the natural-geographic location of the regions, and their economic and sociodemographic characteristics strongly influence the difference of the population's incomes. In March 2017, the average monthly salary in Moscow and St. Petersburg was 80,117 and 53,271 rubles, respectively, and in Russia as a whole, 38,483 rubles. This is due to the concentration in Moscow and partly in St. Petersburg of the bulk of financial, credit, and corporate structures with high and extra-high incomes. Therefore, in Moscow, the difference between the incomes of the higher and lower groups of the population is especially pronounced.

High incomes are typical for oil- and gas-producing regions such as the Tyumen Region, the Republic of Sakha (Yakutia), and areas with a highly developed external and cross-border trade – the Murmansk Region, the Khabarovsk Territory, and the Sakhalin Region.

Consider the Central Federal District, where the highest salary is inherent in the city of Moscow, that is, 80,117 rubles, and the lowest – the Ivanovo region, 22,107 rubles (Table 2).

In connection with the problem of non-equality of income distribution, the important task of the country's economy is to ensure social stability. To achieve this goal, the social policy of the state is directed in areas such as redistribution of income, provision of social protection, and provision of social benefits.

Despite different theoretical views on this problem, practice has shown that the existence of inequality has negative consequences for stable and sustainable economic growth, the implementation of the rule of law, and the formation of a culture of the population.

Various estimates are used to estimate the inequality in income distribution. The Gini coefficient or the income concentration index can be defined as a macroeconomic indicator characterizing

Table 2. Average monthly nominal accrued wages of employees for a full range of organizations by entities of the Russian Federation in March 2017, rubles.

The subject of the Russian Federation	Salary, rubles.	Level by region
Russian Federation	38,483	Middle
Central Federal District	49,226	Middle
Ivanovo region	22,107	Min
Moscow	80,117	Max
North-West Federal District	43,586	Middle
Pskov region	22,565	Min
Nenets Autonomous Okrug	71,920	Max
Southern Federal District	27,635	Middle
Republic of Kalmykia	21,751	Min
Krasnodar region	29,390	Max
North-Caucasian Federal District	23,156	Middle
The Republic of Dagestan	20,526	Min
Stavropol region	25,445	Max
Volga Federal District	28,125	Middle
Saratov region	23,960	Min
Perm region	31,884	Max
Ural federal district	43,246	Middle
Kurgan region	23,862	Min
Yamalo-Nenets Autonomous Area. Okrug	86,090	Max
Siberian Federal District	32,500	Middle
Altai region	21,807	Min
Krasnoyarsk region	37,554	Max
Far Eastern Federal District	47,686	Middle
Jewish Autonomous Oblast	33,042	Min
Chukotka Autonomous Okrug	92,136	Max

the difference of the population's monetary income in the form of the deviation of the actual distribution of income from an equal distribution among the inhabitants of the country.

In practice, the most often used is quintile coefficient, which is the ratio of the average income of 20% of the richest households to the average income of 20% of the poorest households. This coefficient is a type of coefficient of funds, and it corresponds to the case when the set of all households is divided into five groups.

3 RESULTS

Inequality can be perceived as a natural phenomenon caused by individual differences of people, if the income of 10% of the most well-off households exceeds the income of 10% of the poorest by no more than eight times. This situation is reflected in the developed countries of Western Europe. A significant excess of the above figures is called excess inequality. In modern Russia, this indicator is on average 16 times (Table 3). Many researchers believe that the official figure for Russia is understated, because it does not take due account of shadow incomes. Such a strong stratification of society is dangerous because it threatens the growth of social tension and, possibly, causes a social explosion.

The problem of inequality in Russia is more acute than in any European country. Russia in terms of inequality is closest to China, Argentina, and Venezuela (42–45%). In Africa and Latin America, for example, in South Africa, the Gini coefficient is 58%, in Brazil – 55%, which indicates a higher degree of difference. Experts believe that the reduction of excess inequality in Russia by 1% makes it possible to increase the rate of GDP growth by 5%.

In our opinion, inequality in income distribution negatively affects the rates of economic growth and development, as well as has sociopsychological consequences. Moderate income difference has a stimulating function. Namely, the segments of the population with smaller incomes will have motivation to work more and change their social status. That is, they will set themselves a clearly defined goal – to work harder and harder to increase their income and improve the quality of their lives. However, the reverse situation can arise, if the incomes of the population in the country are characterized by strong inequality. In this situation, the population will lack motivation to work, as people will consider that they are not able to influence their financial situation. Such a situation in the country negatively affects the psychological mood in society, as well as demotivates the economic activity of the population, which, on the whole, negatively affects the state of the economic situation in the country and the pace of economic development. Inequality demotivates people to invest money and time in their own development, which leads to inertia and dependency, as well as to an increase in budget expenditures while reducing economic returns from work. All these factors impede the country's economic development.

Table 3. Dynamics of income inequality in Russia.

Index	Year 2012	2013	2014	2015	2016
Monetary incomes, %	100	100	100	100	100
Including 20% of the population, %:					
The first (with the least income)	5.2	5.2	5.2	5.3	5.3
The second	9.8	9.8	9.9	10.0	10.0
The third	14.9	14.9	14.9	15.0	15.0
The fourth	22.5	22.5	22.6	22.6	22.6
The fifth (with the largest incomes)	47.6	47.6	47.4	47.1	47.1
Decile coefficient of funds, in times	16.4	16.3	16.0	15.6	15.7
The Gini coefficient (income concentration index)	0.420	0.419	0.416	0.413	0.414

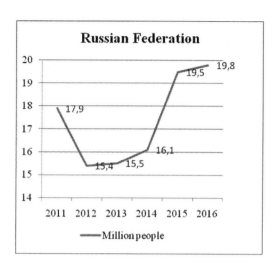

Figure 1. Russian population with incomes below the subsistence level.

Inequality in income raises the problem of poverty. In Russia, the poverty threshold, for which poverty follows, is calculated based on the subsistence minimum. In the Russian society, there were "new Russian poor" who have a high education, qualifications, and have never entered the lower strata of society. These are teachers, doctors, engineers, and other segments of the population, working primarily in the budgetary sphere. Figure 1 presents the dynamics of the Russian population with incomes below the subsistence level from 2011 to 2016.

The specificity of the current economic crisis was the decrease in real incomes of the population due to high inflation with a slowdown in the growth of nominal incomes. The problems of the poor with an average per capita income below the subsistence minimum are exacerbated. By 2013, the share of the poor has decreased to 10.8% of the population (from 29% in 2000), but from 2014, it began to increase, reaching an average of 13.3% in 2015.

According to the materials of the sample survey of household budgets conducted by the Russian Federal Service of State Statistics, 65% of the poor are economically active population (of whom only 2% are unemployed), 64% of poor families have children (33% – one child, 22% – two children, 9% – three or more), 28% of the poor live in cities with a population of less than 50,000 people, and 41% in rural settlements.

According to experts' estimates, the worsening economic situation in the country, the growth of consumer prices, and the reduction of real incomes of citizens largely influenced the growth of poverty in Russia. The fall in household incomes narrows the domestic market and aggregate demand, which causes a decline in the pace of development of the national economy.

Reduction of real incomes of the population of Russia (as well as in developed countries) leads to the fall of purchases of cars and other durable goods. However, as the crisis lasts for about 2 years, it increasingly affects the situation of the population: the level of poverty increases, the structure of demand for food products and the structure of retail trade in general change, and in particular, the share of food products grow in purchases. At the same time, the decrease in real incomes leads to an even greater compression of costs due to the inclusion of a "protective" model of consumer behavior, which complicates the prospects for restoring the growth of not only demand but also the country's economy as a whole.

4 CONCLUSION

An analysis of the difference of incomes of the Russian population testifies to the existence of a high degree of socioeconomic inequality in the level of well-being of citizens in different regions of Russia, and one of the main factors of such inequality is the disproportionality of wages.

We believe that all of the above proves the presence of a high degree of difference of the population of the Russian Federation by average per capita money income, but it should be noted that there is a tendency to reduce this indicator. To improve the current state of affairs and preserve this trend, more government intervention is needed to regulate this problem through tax and program assistance to the poor, maintaining at the subsistence level those who are not able to secure a better life, although it should not be forgotten that excessive state interference in the redistribution of income equalization processes can lead to a decline in business activity in the society and a decrease in the efficiency of economic activity, because it "extinguishes" personal enterprise.

The level of Russian income difference and the growing scale of poverty are incompatible with the requirements of a new economy, headed by a person, and the goal is its all-round development. In this connection, effective revenue policy should become one of the priorities of state regulation of the Russian economy.

In future research, it is necessary to pay attention to the impact of economic sanctions on the difference of income of the population in different regions of Russia.

REFERENCES

[1] Piketty, T. Dynamics of inequality (2014) New Left Review, (85), pp. 103–116.
[2] Grigoryev R.A., Kramin M.V., Kramin T.V.,Timiryasova A.V. Inequality of income distribution and economics growth in the regions of Russia in the post-crisis period Economy of Region Open Access Issue 3, 2015, Pages 102–113.
[3] Tatuev, A.A., Shash, N.N., Nagoev, A.B., Lyapuntsova, E.V., Rokotyanskaya, V.V. Analysis of the reasons and consequences of economic differentiation of regions, International Business Management Volume 9, Issue 5, 2015, Pages 928–934.
[4] Akerlof, G.A., Shiller, R.J. Phishing for phools: The economics of manipulation and deception. Phishing for Phools: The Economics of Manipulation and Deception 22 September 2015, Pages 1–272.
[5] Kanbur R., Stiglitz, J.E. Dynastic inequality, mobility and equality of opportunity. Journal of Economic Inequality, Volume 14, Issue 4, 1 December 2016, Pages 419–434.
[6] N.M. Sabitova, C.M. Shavaleyeva, E.N. Nikonova. Horizontal Imbalances of Russian Federation Budget System/Sabitova N.M., Shavaleyeva C.M., Nikonova E.N.//Asian Social Science – 2015. – Vol. 11, No. 11, Special Issue P.248–252.
[7] Akerlof, G.A., Kranton, R.E.Identity Economics: How Our Identities Shape Our Work, Wages, and Well-Being (Book). Identity Economics: How Our Identities Shape Our Work, Wages, and Well-Being January 21, 2010, Pages 1–185.
[8] Mareeva S., Tikhonova N. Public perceptions of poverty and social inequality in Russia Mir Rossii Open Access Volume 25, Issue 2, 2016, Pages 37–67.
[9] Russian Federation Federal State Statistics Service http://www.gks.ru/wps/wcm/connect/rosstat_main/rosstat/ru/statistics/population/poverty/#.

Parental challenges in the filial therapy process: A conceptual study

Cheng Chue Han, Diana-Lea Baranovich, Lau Poh Li & Fonny Hutagalung
Department of Educational Psychology and Counseling, Faculty of Education, University of Malaya, Malaysia

ABSTRACT: In this study, we highlight the importance of understanding parental challenges in the filial therapy process. Filial therapy is coined as golden therapy for its effectiveness across cultures, family structures, and presenting issues for over five decades. It has been demonstrated in the literature that parental challenges did emerge, but there is no clear and detailed information. Filial therapy focuses on understanding the process of parenting. This gap needs to be explored to improve the implementation and expansion of filial therapy. Here, we proposed to use Bronfenbrenner's bioecological theory to study the challenges. Besides, filial therapy's study in the context of Malaysia is nonexistent, but it is expected that the families here will benefit from this therapy modality.

Keywords: filial therapy, parents, challenges, Malaysia, process-oriented study

1 INTRODUCTION

The therapy that helps recovering people suffering from emotional distress often involves dedication and hard work. However, the hard work involved in the process can be understudied because the focus is often on the efficacy of the therapy. Hence, in this conceptual study, we aim to highlight the challenges faced by a parent throughout the journey of filial therapy.

In 1964, Bernard Guerney introduced filial therapy to help children experiencing behavioral and emotional issues. In filial therapy, parents are trained to be therapeutic agents by the therapist and have to engage in weekly 30 min nondirective play with the target child. Guerney (1964) explained the following goals of parent–child play sessions: (1) allowing the child to decide the activities within a certain limits; (2) parents get to develop empathic understanding toward the child's needs and feelings; (3) the child feels accepted by parents; and (4) the child learns to see and accept responsibility for his actions.

In the last 52 years, filial therapy has grown extensively since its outset, initial research, and development. It is considered an evidence-based treatment approach because of its efficacy to be replicable with comparable results (Guerney, 2000). Accordingly, other terms have been used to identify filial therapy, including terms such as filial family therapy (Guerney, 2000), child–parent relationship therapy (Landreth & Bratton, 2006), child relationship enhancement family therapy (VanFleet, 1994), and child relationship enhancement therapy (Bratton & Crane, 2003). These variations of filial therapy were developed to adapt to the needs of families from different places. There are two common methods of filial therapy, namely group models and individual models (Ryan, 2007). VanFleet has been credited with modifying the existing filial therapy group model to implement filial therapy with individual families and couples (Hutton, 2004).

Over the past 5 decades, filial therapy has been tested as an effective method across various cultures, family structures, and presenting issues. Filial therapy is coined as the golden therapy by Cornett & Bratton (2015) because it has yielded an abundance of valuable and long-lasting benefits for parents and children. The effectiveness of filial therapy with a variety of populations was supported by studies, such as those with Hispanic parents (Sangganjanavanich, Cook, & Rangel-Gomez, 2010), African-American parents (Solis, Meyers, & Varjas,

2004), Iranian single parents (Alivandi Vafa & Khaidzir Hj. Ismail, 2009), Jamaican parents (Edwards, Ladner, & White, 2007), foster children (Cornett & Bratton, 2014), married couples (Bavin-Hoffman, Jennings, & Landreth, 1996), and incarcerated fathers (Landreth & Lobaugh, 1998). Thus, the introduction of filial therapy in Malaysia is highly expected to benefit the families here.

However, the golden therapy comes at a "price" (challenges and difficulties in the process of filial therapy). Past researchers have discovered that filial therapy is evidently valuable and thus is referred to as "golden therapy", to describe its effectiveness. Interestingly the "price" that comes with it is understudied. In general, studies on modality of therapies are often focused on outcome to test its efficacy. It is known that filial therapy is effective, but it is not entirely understood as to "how" it works (Winek, Lambert-Schute, & Johnson, 2003) and "what" do not work. Thus, process-oriented research is equally important to highlight the necessary issues throughout the process of a therapy. Evidently, there is an abundance of literature works testifying to the efficacy of filial therapy. However, there is a lack of research with a focus on understanding the challenges faced by parents in this therapeutic journey. In reviewed studies on parental perceptions, challenges have been established without detailed explorations. For example, in a study of filial therapy with Spanish speaking mothers that uses a phenomenological approach, an emergent theme of challenges of integrating skills among the participants (Sangganjanavanich et al., 2010) was reported. It is recognized that there are challenges faced by the parents but there are no in-depth studies to provide meaningful descriptions. Another phenomenological study to collect parents' voices in filial therapy highlights that "change is hard" and deliberately applying the learned skills were exhausted (Foley, Higdon, & White, 2006). In a study conducted by Solis et al. (2004), African-American mothers indicated that they had difficulty in finding time to play with their children. Again, there is no further explanation and exploration in the area of challenges, difficulties, or barriers faced by the parents in the research.

It is anticipated that parents will face challenges when learning new skills whether in implementing change in the family or in their relationship with the target child. However, the existing literature does not provide findings that bridge the understanding of these challenges in a structured and detailed manner that would actuate improvisation in the use of filial therapy. Hence, structured and in-depth information of how these parents have responded to the challenges and consequently completed the therapy is not available in the literature.

Even so, studies on dropout in psychotherapy that essentially aimed to identify the risk factors and barrier to treatment for families are available in the literature. However, there are no studies investigating the characteristics of children and families who had successfully completed treatment (Campbell, Baker, & Bratton, 2000). For instance, a dropout study of filial therapy reported that variables such as mother's age, child's age, social support, and communication of acceptance were the predictors for dropout rate (Topham & Wampler, 2007). These factors have represented the possible challenges that parents might face in the course of filial therapy, but they have not been described in detail. The process of how these parents have faced the challenges and have eventually dropped out of the therapy has not been illustrated.

Correspondingly, because of the lack of a process-oriented research, a holistic picture in filial therapy has not been provided for researchers and practitioners. For the past 52 years, filial therapy has established for itself an evidenced-based outcome that is produced from its outcome-oriented studies. However, the focus has often been on the outcome of the therapy rather than the process. Significantly, in order to advance the practical application among a wider population and to increase the therapeutic options for children, Reed (2016) stressed the importance of a process-oriented research in play therapy. The benefits of such a process-oriented perspective are as follows. First, a process focus should enhance the understanding of the emergence of challenges from the beginning phase to the end phase of the filial therapy process. Second, the approach offers considerable insights into the processes of the evolving challenges in the process of filial therapy. Only through deeper understanding of these challenges, a practitioner may benefit from the knowledge to improvise the practice.

1.1 Research questions

1. What are the challenges experienced by the parent in the process of filial therapy?
2. How does the parent respond to these challenges?
3. How do theoretical constructs, namely process, person, context, and time (PPCT model proposed by Bronfenbrenner, 2006), help us to understand these challenges?

2 LITERATURE REVIEW

Filial therapy is not a conventional approach where parents send their children for therapy and let the child and therapist work toward the treatment goal. In contrast, parents play active role in filial therapy. Essentially, parents are the direct and main person who will receive "therapy", which, in this case, is referred to as psycho-education or training in the treatment process. Filial therapist helps the parents to help themselves and their children. Without the parents' active and committed involvement, filial therapy cannot be successful. For this reason, researchers are encouraged to focus on the viewpoint of parents because parents weight significantly in filial therapy.

The two major elements in filial therapy, which are play and parents as therapeutic agent, are meant to meet the following four goals as stated by Guerney (1964).

Goal 1: The encouragement of allowing the child to self-determine the activities fully within certain limits. Gray's (2012) definition of play is an activity that is self-chosen and self-directed; intrinsically motivated; guided by mental rules; imaginative; and involves an active, alert, but non-stressed frame of mind.

Self-directed play or self-determined activity is essential in the filial therapy process because children learn to develop sense of self, independent, self-regulate, and ability to make choices. The freedom to choose in the play session promotes creativity and self-expression (Baggerly, C.Ray, & Bratton, 2010). Unfortunately, recent generation of children are immersed in structured, scheduled, stressed, and adult-directed activities (Belknap & Hazler, 2014). In Malaysia, parents are familiar with the word of "tuition", which means extra classes to tutor students on academic subjects during the off-school hour. Most of the urban children have attended tuition classes beginning from seven years old. Tuition is one of the examples of scheduled, structured, and adult-directed activities. Furthermore, the trend of sending children to enrichment classes has become popular in the urban area of Malaysia. A child who has packed schedule has lesser time in engaging self-directed play. Hence, this has become the first goal in filial therapy to ensure that parents are with their children to engage in self-directed play for at least 30 min weekly.

Goal 2: Parents get to increase empathic understanding toward the child's needs and feelings. In the training sessions, therapists focus on increasing the parents' sensitivity to their children, acceptance of thoughts and feelings, understanding of their child's emotional needs, reflective listening, empathic responding, identification of feelings, and therapeutic limit setting (Baggerly et al., 2010). When parents grasp the idea of entering their child's world nonjudgmentally, the connection of bonding occurs and leads to the achievement of the third goal.

Goal 3: The child feels accepted by parents. Through filial play sessions, parents learn to connect to the child's feelings and needs, merit and respect the child's autonomy, and respond delicately (VanFleet & Topham, 2011). Child–parent relationship is enhanced through this unique context that parents work deliberately and consciously to create a safe space for the child to heal and grow, similar to the concept of Rogerian therapy that unconditional positive regards or acceptance enables the healing and growing progress (Topham, G. VanFleet, 2011).

Goal 4: The child learns to see and accept responsibility and consequences for his/her actions. Although self-directed play is the core activity during the play session, parents are taught to set boundary to foster appropriate and acceptable behavior (VanFleet & Topham, 2011). For example, an angry child keeps throwing the toy and nearly breaks it. It is an opportunity for parents to educate child's self-regulation through setting the limit or rule such that toys are not meant for throwing and breaking. Instead, the child can direct and release his/her anger to the pillow and others or the environment. If the child insisted the behavior of breaking the toy, he needs to bear the consequences of broken toy and the play session shall end. With the constant reinforcing of healthy boundary setting, child learns the responsibility of taking charge of his/her own actions.

Nevertheless, the overall explanation of filial therapy seems to be straightforward and simple to understand, the theoretical formation of filial therapy is comprehensive. The formation of filial therapy is built on the theoretical integration including psychodynamic, cognitive, behavioral, humanistic, interpersonal, social learning, developmental, family systems, and attachment theories (VanFleet & Topham, 2011). According to the classic attachment theory of Bowlby (1969), an infant develops emotional bonding with his/her primary caregiver during the first year of life and the level of attachment determines the capability of trust toward his/her relationship with others. The quality of the attachment is largely dependent on the sensitive responding and availability of the caregiver (Fonagy, Lorenzini, Campbell, & Luyten, 2014). The four goals of filial therapy are basically set to enable the child to experience the emotional bonding he/she deserves as a young child in order to build the trust in him/her toward his/her relationship with others. Children who do not have secure base attachment often lead to various maladaptive issues such as developmental delay, acting out behaviors, and maximizing distress cues.

These theories are incorporated fully in psychoeducation model of intervention. Parents are trained and educated on the necessary knowledge and skill about child development and child play. The bridging between the unknown and known happened through didactic training, play skill demonstration from the therapist, supervision, and group processing of experiences (Edwards, Sullivan, Meany-Walen, & Kantor, 2010). This empowerment and encouragement approach strengthens child–parent relationship in which change can occur and solve the problems faced by the target child (VanFleet & Topham, 2011). Nowadays, parents are gradually open with training and education in terms of parenting to ensure the children to achieve optimal growth.

2.1 *Conceptual framework*

To study parental challenges, Bioecological theory (Bronfenbrenner, 2006) is proposed to provide a structural understanding of parent's experiences with challenges in filial therapy. Ecological systems theory was proposed by Urie Bronfenbrenner (1977, 1979), which provides a broad perspective on human development in the aspect of accommodations made throughout the lifespan between human and environment. In the bioecological model, its development is defined as "the phenomenon of continuity and change in the biopsychological characteristics of human beings, both as individuals and as groups" (Bronfenbrenner & Morris, 2006, p. 793). The definition of development here can be applied to the parent who is undergoing filial therapy. Given that the process of learning for parents in filial therapy leads to change for individuals, as well as the target child, the implementation of developmental theory to guide the study is justified.

As depicted in Figure 1, filial therapy is described as using gears that move each other as a metaphor on how parents use child-centered play skills as a tool to "move" or enhance their relationship with their own children. In this metaphor, a healthy family relationship is depicted as a functional gear that is constantly moving. This proposed study aims to uncover the elements that affect the functionality of the gear—the challenges that are involved. It can also be seen that the size of the parent's gear is much bigger and has more gear teeth due to their capacity to make conscious change and to develop as compared to their children who are young and are going through psychological, emotional, or behavioral issues.

Process. Process is the core element in the theory. It is referred to as "particular forms of interaction between organism and environment, called *proximal processes*, that operate over time and are posited as the primary mechanisms producing human development" (Bronfenbrenner & Morris, 2006, p. 795). The main interaction lies between parent and counselor, parent and child, and parent and his/her environment. For instance, parents have reported difficulties in learning new skills in filial therapy (Foley et al., 2006), which has not been explored further and may indicate challenges during the interaction with the child or thera-

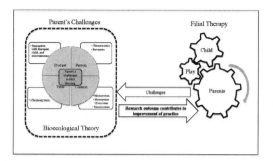

Figure 1. Application of bioecological system theory in the study of parental challenges.

pist. Correspondingly, parents have also reported the importance of establishing a respectful relationship with the therapist (Socarras, Smith-Adcock, & Shin, 2015) to enhance positive outcome of the therapy. These collaborative interactions or processes of learning with the therapist hence become an essential element in achieving a successful therapy. In addition, Winek and colleagues (2003) reported that parents who face challenges with an uncooperative child find it difficult to conduct the play session. The interaction between parent and child marks an important role to determine the success of filial therapy as such challenges that may arise need to be identified and understood further.

Person. Types of person characteristics are determined to be the most influential in deciding the course of development because they affect the direction and power of proximal processes throughout a person's lifespan (Bronfenbrenner & Morris, 2006). Individual factors such as age, race, and gender contribute to treatment attendance and adherence. For example, in a study of predicting dropout in filial therapy, older mothers have been identified to attend lesser sessions of the therapy compared to younger mothers (Topham & Wampler, 2007).

Context. There are four levels of system in the context, namely, microsystem, mesosystem, exosystem, and macrosystem. Microsystem refers to a setting where an individual has direct relation with factors such as home and workplace. For example, parents have reported that they were stressful when facing with transportation problems in order to undergo treatment, psychotherapy, and family therapy (Campbell et al., 2000; Deakin, Gastaud, & Nunes, 2012; Gresl, n.d.; Holm, 1998; Snell-Johns, Mendez, & Smith, 2004; Topham & Wampler, 2007; Werba, 2006). A direct factor such as this constitutes an immediate impact to parents' stressor in relation to time of attending, participating, and completing a therapy.

Time. Finally, chronosystem represents the characteristics of the person's changes (and continuities) over time, developmentally in environments without discounting characteristics of the environments the person is living in (Bronfenbrenner, 1994). Thus, filial therapy disputes the foremost mindset of "quick fix" over the process of making connection with the child and brings change that often takes time, which frustrates parents. One of the components in filial therapy is the supervision of parents to learn to be patient and to respect the incremental process of change. Here, parents have disclosed to have been transformed from feelings of frustration to being able to make sense of the therapy (Amy Wickstrom, 2009). Hence, the element of time plays an important role to understand parent's experiences in filial therapy.

Essentially, the relationships between and within each level determine the significance each level has in understanding the challenges parents face in filial therapy, and which needs further evaluation. As such, the structure of the bioecological theory has been identified as the framework in guiding the exploration of parental challenges in the filial therapy process.

3 CONCLUSION

Filial therapy is a golden therapy that works for various family cultures, structures, and presenting issues. Over five decades, several studies supporting the effectiveness of filial therapy have been conducted, which has resulted in the expansion of filial therapy to various types of families. Even so, this conceptual study draws conclusion that parental challenges or difficulties that emerged in the filial therapy process are understudied. It is common to have outcome-oriented study to test the efficacy of a therapy modality and thus process-oriented studies may be overlooked. The importance of the process study is to enhance the understanding of practitioner and researcher in the implementation of filial therapy. This study proposed to use the Bronfenbrenner bioecological theory as theoretical framework to study parental challenges. Process–person–context–time model is deemed to be relevant to frame the possible types of challenges a parent will face in the filial therapy process. Thus, the conceptual framework is presented by using gear as metaphor for filial therapy and the research outcome is predicted to contribute to the improvisation of application of filial therapy in Malaysia. Also, future studies stemming from this conceptual study will broaden the field of practical implementation of filial therapy.

REFERENCES

Alivandi Vafa, M., & Khaidzir Hj. Ismail. (2009). Reaching Out to Single Parent Children through Filial Therapy. *Online Submission*, *6*(2), 1–12. Retrieved from https://manowar.tamucc.edu/login?url=http://search.ebscohost.com/login.aspx?direct=true&db=eric&AN=ED504968&site=eds-live&scope=site&scope=cite.

Amy Wickstrom, M.F.T. (2009). The process of systemic change in filial therapy: A phenomenological study of parent experience. *Contemporary Family Therapy*, *31*(3), 193–208. http://doi.org/10.1007/s10591-009-9089-3.

Baggerly, J.N., C.Ray, D., & Bratton, S.C. (2010). *Child-centered play therapy research: The evidence base for effective practice*. John Wiley & Sons. Retrieved from http://web.ebscohost.com/ehost/detail?sid=033cb598-2ad0-44c3-9c23-1dc2b108057b@sessionmgr15&vid=1&hid=18&bdata=JnNpdGU9ZWhvc3QtbGl2ZQ==#db=aph&AN=9401195380.

Bavin-hoffman, R., Jennings, G., & Landreth, G. (1996). Filial Therapy: Parental Perceptions of the Process. *International Journal of Play Therapy*, (5), 45–58.

Belknap, E., & Hazler, R. (2014). Empty Playgrounds and Anxious Children. *Journal of Creativity in Mental Health*, *9*(September), 210–231. http://doi.org/10.1080/15401383.2013.864962.

Bratton, S.C., & Crane, J.M. (2003). Filial/family play therapy with single parents. In R. VanFleet & L. Guerney (Eds.), Casebook of filial therapy (pp. 139–162). Boiling Springs, PA: Play Therapy Press.

Bronfenbrenner, U. (1994). Ecological models of human development. In *International Encyclopedia of Education* (Vol. 3, 2nd Ed., pp. 1643–1647).

Bronfenbrenner, U. (1979). The ecology of human development. Cambridge, MA: Harvard University Press.

Bronfenbrenner, U. (1977). Toward an experimental ecology of human development. American Psychologist 32(7), 513–531.

Bronfenbrenner, U., & Morris, P.A. (2006). The bioecological model of human development. *Handbook of Child Psychology, Theoretical Models of Human Development*, 793–828. http://doi.org/10.1002/9780470147658.chpsy0114.

Bowlby, J. (1969). Attachment and loss: Vol. 1: Attachment. New York: Basic Books.

Campbell, V. a., Baker, D.B., & Bratton, S. (2000). Why Do Children Drop-Out from Play Therapy? *Clinical Child Psychology and Psychiatry*, *5*(1), 133–138. http://doi.org/10.1177/1359104500005001013.

Cornett, N., & Bratton, S.C. (2015). A Golden Intervention: 50 Years of Research on Filial Therapy. *International Journal of Play Therapy*, *24*(3), 119–133. http://doi.org/10.1037/a0039088.

Deakin, E., Gastaud, M., & Nunes, M.L.T. (2012). Child psychotherapy dropout: an empirical research review. *Journal of Child Psychotherapy*, *38*(2), 199–209. http://doi.org/10.1080/0075417X.2012.684489.

Edwards, N.A., Ladner, J., & White, J. (2007). Perceived effectiveness of filial therapy for a Jamaican mother: A qualitative case study. International Journal of Play Therapy, 16(1), 36–53. doi:10.1037/1555-6824.16.1.36.

Edwards, N. a., Sullivan, J.M., Meany-Walen, K., & Kantor, K.R. (2010). Child parent relationship training: Parents' perceptions of process and outcome. *International Journal of Play Therapy*, *19*(3), 159–173. http://doi.org/10.1037/a0019409.

Foley, Y.C., Higdon, L., & White, J.F. (2006). A qualitative study of filial therapy: Parents' voices. *International Journal of Play Therapy*, *15*(1), 37–64. http://doi.org/10.1037/h0088907.

Fonagy, P., Lorenzini, N., Campbell, C., & Luyten, P. (2014). Why are we interested in attachments? In P. Holmes, & S. Farnfield (Eds.), The Routledge handbook of attachment: Theory (pp. 31tledge handbook of attachm.

Gray, P. (2012). The value of a play-filled childhood in development of the hunter- gatherer individual. In D. Narvaez, J. Panksepp, A.N. Schore (Eds.), Evolution, Early Experience and Human Development: From Research to Practice and Policy (pp. 352–370). New York, NY: Oxford University Press.

Guerney Jr, B. (1964). Filial Therapy: Description and rationale. *Journal of Consulting Psychology*, *28*(4), 304–310. http://doi.org/10.1300/j085V06N03_03.

Guerney, L. (2000). Filial therapy into the 21st century. International Journal of Play Therapy, 9(2), 1–17. doi:10.1037/h0089433.

Gresl, B.L. (n.d.). Early Termination And Barriers To Treatment In Parent And Child Therapy. Retrieved from http://epublications.marquette.edu/dissertations_mu/345.

Holm, J. (1998). Examining and Addressing Potential Barriers to Treatment Adherence for Sexually Abused Children and their Non-offending Parents, (1980).

Hutton, D. (2004). Filial therapy: Shifting the balance. Clinical Child Psychology and Psychiatry, 9(2), 261–270. doi:10.1177/1359104504041922

Landreth, G.L., & Bratton, S.C. (2006). Child parent relationship therapy (CPRT): A 10-session filial therapy model. New York, NY: Routledge. Landreth,

Landreth, G.L., & Lobaugh, A.F. (1998). Filial therapy with incarcerated fathers: Effects on parental acceptance of child, parental stress, and child adjustmend. *Journal of Counseling & Development*, *76*(2), 157–165.

Reed, P. (2016). Tackling taboos: Research in play therapy. In Le Vay, D., & Cuschieri, E. (Eds.). (2016). *Challenges in the Theory and Practice of Play Therapy*. (pp. 104–124). Routledge.

Ryan, V. (2007). Filial therapy: Helping children and new carers to form secure attachment relationships. British Journal of Social Work, 37, 643–657. doi:10.1093/bjsw/bch331 Smith,

Sangganjanavanich, V.F., Cook, K., & Rangel-Gomez, M. (2010). Filial therapy with monolingual Spanish-speaking mothers: A phenomenological study. *The Family Journal*, *18*(2), 195–201. http://doi.org/10.1177/1066480710364320.

Snell-Johns, J., Mendez, J.L., & Smith, B.H. (2004). Evidence-Based Solutions for Overcoming Access Barriers, Decreasing Attrition, and Promoting Change With Underserved Families. *Journal of Family Psychology*, *18*(1), 19–35. http://doi.org/10.1037/0893-3200.18.1.19.

Socarras, K., Smith-Adcock, S., & Shin, S.M. (2015). A Qualitative Study of an Intensive Filial Intervention Using Child-Parent Relationship Therapy (CPRT). *The Family Journal*, *23*(4), 381–391. http://doi.org/10.1177/1066480715601681.

Solis, C.M., Meyers, J., & Varjas, K.M. (2004). A qualitative case study of the process and impact of filial therapy with an African American parent. International Journal of Play Therapy, 13(2), 99–118. doi:10.1037/h0088892.

Topham, G. VanFleet, R. (2011). Filial Therapy: A Structured and Straightforward Approach to Including Young Children in Family Therapy. *The Australian and New Zealand Journal of Family Therapy*, *32*(2), 144–158. http://doi.org/10.1375/anft.32.2.144.

Topham, G.L., & Wampler, K.S. (2007). Predicting Dropout in a Filial Therapy Program for Parents and Young Children. *The American Journal of Family Therapy*, *36*(1), 60–78. http://doi.org/10.1080/01926180601057671.

VanFleet, R. (1994). Filial therapy: Strengthening parent-child relationships through play. Sarasota, FL: Professional Resource Press. VanFleet.

VanFleet, R., & Topham, G. (2011). *Integrative Play Therapy*. (A.A. Drewes, S.C. Bratton, & C.E. Schaefer, Eds.)*Integrative Play Therapy*. Hoboken, NJ, USA: John Wiley & Sons, Inc. http://doi.org/10.1002/9781118094792.

Werba, B.E. (2006). Predicting Outcome in Parent-Child Interaction Therapy: Success and Attrition. *Behavior Modification*, *30*(5), 618–646. http://doi.org/10.1177/0145445504272977.

Wickstrom, A.C., & Falke, S.I. (2013). Parental Perceptions of an Advanced Filial Therapy Model. *Contemporary Family Therapy*, *35*(1), 161–175. http://doi.org/10.1007/s10591-012-9228-0.

Winek, J., Lambert-Shute, J., Johnson, L., Shaw, L., Krepps, J., & Wiley, K. (2003). Discovering the moments of movement in filial therapy: A single case qualitative study. *International Journal of Play Therapy*, *12*(1), 89–104. http://doi.org/10.1037/h0088873.

Report on the results of a monitoring study conducted in the Russian Federation on the prevention of addictive and suicidal behavior and internet risks among adolescents

A.V. Khydyrova
The Federal State Budgetary Research Institution "Centre for Protection of Rights of Children", Moscow, Russia

N.A. Mamedova
Russian Economic University named after G.V. Plehanov, Moscow, Russia

E.G. Artamonova
The Federal State Budgetary Research Institution "Centre for Protection of Rights of Children", Moscow, Russia

ABSTRACT: The article presents data on the results of monitoring studies conducted in Russia on the prevention of addictive and suicidal behavior and Internet risks among adolescents. The source of the data is the activity of the Center for the Protection of the Rights and Interests of Children in the period 2014–2017 for two interrelated projects: the identification of parents' and learners' awareness of life-threatening Internet risks and threats present on the Internet, and the monitoring of the state of work on the prevention of suicidal behavior within educational organizations. The article presents data on the methodology for conducting research and directions for using the results of the research.

1 INTRODUCTION

The modern situation of development in Russian society is accompanied by a growth of negative social phenomena, such as deviant behavior among adolescents. This trend reflects the existing trends in many countries of the world faced with defects in the development of the social arena in connection with the disproportion of the influence of the technogenic and informatization factors of society. Ostensibly, the right to adequate living conditions for children is guaranteed by the UN Convention on the Rights of the Child, the constitutions of different countries on the national level, and many other documents, but it is not always realized in practice (Artamonova, 2016). Unfortunately, children and adolescents are the most unprotected category of the population. Therefore, the educational establishment plays an important role in addiction and deviance prevention among adolescents. One of the essential conditions for work organization is the transparency of education and the efficiency of information interaction (Khydyrova, 2015, Mamedova, 2009).

2 MATERIALS AND METHODS

2.1 *Consistency of style is very important*

The research team set itself the task of developing approaches, in addition to conducting and evaluating the effectiveness of monitoring research on the prevention of addictive and suicidal behavior among adolescents. This activity was carried out on the basis of the monitoring that the Federal State Budgetary Scientific Institution, "Centre for the Protection of the Rights of

Children", had carried out on the state of addictive and suicidal behavior prevention among adolescents in educational institutions in 2014–2017. The setting and nature of the scientific and practical problem of preventing the addictive and suicidal behavior of children and adolescents led to the use of a set of special methods of research activity within the framework of the project. In the process of research, such special methods of investigation as systematization, ranking, modeling, methods of descriptive and mathematical statistics, and graphical representations of results were used. To conduct the empirical part of the research, the method of questioning was used.

This article presents summary data of the project results in two key areas of the "Protection of the Rights and Interests of Children" project of the FSBRI. The first included methodological preparation and the conducting of an all-Russian survey in accordance with the task in view. The second is the collection and statistical research of monitoring data on the state of work on the prevention of suicidal behavior within educational institutions.

The parameters of the territorial coverage, the quality of the reference base, the unconditional observance of the survey methodology, and the unity of approach in the processing of responses and the evaluation of results – all this allows us to establish the validity and adequacy of the data for completing the scientific task. The data given below on the methodology and results of the nationwide survey are the result of the activities of the Center for the Protection of the Rights and Interests of Children of the FSBSI. The goal of the monitoring study on the state of work in preventing addictive and suicidal behavior among children and adolescents was to conduct a survey of students and their parents revealing the awareness of parents and students about life-threatening Internet risks and threats present on the Internet.

As a research question we suppose to know if the awareness of the reference groups in relation to the risks of addictive behavior of adolescents has a positive effect on the effectiveness of the prevention events.

The survey was conducted in educational organizations in the 2016/2017 academic year in all 85 subjects of the Russian Federation.

3 RESULTS

The primary results of the monitoring studies conducted in 2017 are presented below. In April 2017, the Center for the Protection of the Rights and Interests of Children conducted a survey of parents[1] and school-age adolescents[2] to collect information about parents' and children's knowledge about Internet risks and threats on life that are present on the Internet.

For this purpose, two questionnaires were developed for two reference groups - schoolchildren (ages 12–17) and legal representatives of adolescents (parents, guardians), including immediate relatives (on the grounds that this reference group is often included in the nearest environment of children).

The first survey involved minors from 85 subjects of the Russian Federation. All respondents were informed that the survey was conducted for the purpose of identifying awareness and preventive measures to combat Internet risks and threats on life present on the Internet. In total, 108,631students aged 12–17 participated in this poll in the 2017 academic year. Among them the percentage of minors aged 12–13 was 32.14% (34,911 of adolescents), 14–15 – 36.53% (39,682 of adolescents) and 16–17 – 31.33% (34,038 of adolescents). The percentage in the sex-age sample was as follows: boys – 46.54% (16,248 of adolescents), girls – 53.32% (18,614 of adolescents) (12–13); boys – 46.70% (18,530 of adolescents), girls – 53.28% (21,143 of adolescents) (14–15); boys – 46.36% (15,781 of adolescents), girls – 53.62% (18,251 of adolescents) (16–17). These proportions of the age and sex of respondents are not an accidental effect, but the result of working with a sample to conduct a survey.

---

[1]Find survey on: http://xn----ttbdjacdik5e5a.xn--p1ai/surveys/view/101.
[2]Find survey on: http://xn----ttbdjacdik5e5a.xn--p1ai/surveys/view/1.

The results of the survey showed that more than half of students of all ages (12–13, 14–15, and 16–17 years old) had received information about health hazards related with a long staying on the Web: 19,733; 22,717 and 200,98 of adolescents (57%, 57%, and 59%, respectively). Information about dangerous sites was received by more than 60% of minors of all ages. Half of all the minors interviewed had received information about the dangers of communication on social networks; 40% or less of the respondents had received information about threats on life. Relatively less informed were the minors aged 16–17 (36%).

Information about the risks associated with Internet games had been obtained by more than 40%: 14,766 of 12–13 aged minors (42%), 16,903 of 14–15 aged minors (43%), and 15,223 of 16–17 aged minors (45%). The most informed were juveniles aged 16–17 with respect to the risks associated with the loss of money – 18,605 of minors (55%), while juveniles aged 14–15 – 19,836 (50%) were less well-informed, and the least well-educated were the 12–13 year-olds – 15,204 (44%).

About 9% (3,135; 3,657 and 2,849 respectively) of respondents of all ages had not received any information about any risks, and 2% (568; 664 and 638 respectively) of minors indicated that they had also been informed of other risks.

Based on the results of processing the responses of the second questionnaire, intended for legal representatives and the people closest in the environment of underage school children, the following data were obtained. All respondents were informed that the survey was conducted in order to reveal the degree of awareness of the risks and threats in the use of the Internet by their children. A total of 45,735 parents were surveyed from all 85 regions of the Russian Federation. Of these, 36,687 were mothers, 6,058 were fathers, 1,339 were grandmothers, 326 were grandfathers, and 619 were trustees.

The results of processing the respondents' answers showed that 57% of the respondents had sufficient knowledge of threats to physical health associated with a prolonged stay on the web (visual impairment, musculoskeletal issues, nervous system diseases, etc.). Sufficiency in this case was defined as an opportunity to form a personal opinion and to demonstrate one's own position on the subject of the survey. 31% of respondents could share the information, which implied the availability of special and/or professional knowledge on the subject of the survey. Only 10% of parents had a superficial view and 3% did not have any information about threats to physical health associated with a long stay on the web.

In order to complete the task of conducting and evaluating the effectiveness of monitoring studies on the prevention of addictive and suicidal behavior among children and adolescents, the Center for the Protection of the Rights and Interests of Children did a monitoring study of the state of suicidal behavior prevention among students in educational establishments (O. Efimova, L. Fonderkina, 2016).

Undoubtedly, school psychology services play a key role in the work organization of suicidal behavior prevention among adolescents. At the same time, the staffing levels of psychological services leave a lot to be desired (Table 1).

Urgent statistics about the number of suicides among adolescents in Russia show that every year about 2,500 minors take their own lives. Russia is included in the list of the leading countries for suicides. According to the available data (2014) in the 15–19 age group there are 5.9 cases of suicide per 100,000 people. In this regard, an urgent issue that should be resolved is the development of a prevention system for suicidal behavior among minors.

Table 1. Psychology services staffing level in the regions of the Russian Federation in 2015 year.

Psychology service in the school	General education	Secondary vocational education
Not organized	20622	720
1 psychologist	20395	1541
2–3 psychologists	1957	151
4–5 psychologists	500	25

Based on the results of the monitoring, proposals were developed to address the problems of human and technical equipment in the work of psychological services in educational organizations and they are presented in the report to the Ministry of Education and Science of the Russian Federation. It has also been proposed to send the results to state authorities in the fields of protecting children from information that is harmful to their health and/or development in the media (the Federal Service for the Supervision of Consumer Rights Protection and Human Welfare and the Ministry of Culture of the Russian Federation).

4 DISCUSSION

The results of the monitoring studies on the projects of the FSBSI "Center for the Protection of the Rights and Interests of Children" were presented in the form of a report on scientific and practical activities at the all-Russian and international levels and published in periodicals on the relevant subject, as well as in a scientific publication by the authors of the report on the research work. The activity of the research group received a wide public response (as confirmed by the data of Survey-Russia reports) and support from the Ministry of Education and Science of the Russian Federation, which will allow further research in 2017–2019.

The results of the research repeat the methods of sociological studies conducted earlier on similar topics. They are research on Internet addiction, various forms of autoaggression, and the study of neuropsychological patterns of various forms of deviant behavior. At the same time, the results obtained reflect the current situation in Russia and allow us to work with the results in different sections (for subjects of the Russian Federation, for a separate category of the reference group) (Pankov, Rumyantsev, Trostanetskaya, 2001; Perezhogin, 2002–2016; Egorov, Igumnov, 2005; Efimova O.I., Salakhova V.B., Mikhaylova I.V., Gnedova S.B., Chertushkina TA, Agadzhanova E.R., 2015, etc.).

The results of the research make it possible to substantiate a complex of state administrative decisions in the development of a psychological service institution on the basis of educational organizations, which will lead to a strategic approach to developing a professional standard for the activities of its employees. In this aspect, the data obtained can be considered the results of the approbation of the results of theoretical studies (Podolsky, Pogozhina, 2013, 2014; Artamonova, Efimova, 2014).

A project focused on the promotion of a healthy lifestyle in the educational environment can be successfully implemented only under the conditions of the formation of interprofessional collaboration and a constructive social network partnership of various educational institutions (Zaretskiy, Bolotnikov, Severniy, 2015).

5 CONCLUSION

The results of the monitoring studies carried out by the FSBSI "Center for the Protection of the Rights and Interests of Children" have the perspective of an approbation and development of school psychologist work models for addictive and deviant behavior prevention among minors on the order of the Ministry of Education and Science of the Russian Federation (Artamonova, Khydyrova, 2016).

Based on data research we can suppose that the research question is confirmed. As it is an ongoing research, the final data will be published in 2018.

The analytical part of the survey (the results of the survey and the collection of statistical information) is the experimentally proven baseline data for the formation of analytical and reporting materials by the authorized government agencies in the constituent entities of the Russian Federation to identify the status indicators and key factors for improving preventive work in the educational system. Systematized conclusions allow us to identify the necessary conditions for the organization of interdisciplinary and interdepartmental activities to prevent addictive and deviant behavior among children and adolescents.

ACKNOWLEDGEMENT

The data were obtained as part of the state task FSBRI "Centre for the Protection of the Rights and Interests of Children", 2017 № 29.9200.2017/5.1.

REFERENCES

Artamonova E.G. (2016) Regional experience of educational psychologists of educational institutions for the early detection and prevention of addictive behavior and deviant minors. Electronic Journal *"Obshestvo. Kultura. Nauka. Obrazovanie"*, 2.

Artamonova E.G. (2013) Prevention and correction of preschool-school exclusion – the question of physiological and psychosocial well-being of children and adolescents. In E.G. Artamonova, N.V. Zaitseva (Ed.). *Psycho-pedagogical bases of formation of the value of health, culture, healthy and safe lifestyle in the education system: Collection of scientific-methodical materials* (pp. 21–24). Moscow.

Artamonova, E.G., Efimova O.I. & Khydyrova A.V. (2016). Psychologist in the Educational System: His Role in the Prevention of Addiction and Deviance//International Journal of Environmental and Science Education (IJESE).

Belozerova M.B. (2016) Experience in preventing HIV infection in non-profit organizations. In A.I. Akhmetzyanova (Ed.). *Psychology of deviant behavior: Interdisciplinary Research and Practice: Proc. of proceedings of the First International Scientific School.* (pp. 31–34). Kazan.

Efimova O.I. (2015) Psychological help for "survivors of suicide": new perspectives prevention. In N.Y. Sinyagina, E.G. Artamonova (Ed.). *Education ID: standards and values. Collection of scientific-methodical materials of international scientific-practical conference "Theory and practice of multicultural education in the educational environment", "Actual problems of socialization of the person in modern conditions".* (153–157). Moscow.

Efimova O.I. (2015) Scientific and applied aspects of the study of suicidal minors activity. In N.Y. Sinyagina, E.G. Artamonova (Ed.). *Science for Education: Collective monograph.* (pp.163–177). Moscow.

Efimova O.I., Salakhova V.B., Mikhaylova I.V., Gnedova S.B., Chertushkina T.A., Agadzhanova E.R. (2015) Theoretical Review of Scientific Approaches to Understanding Crisis Psychology. *Mediterranean Journal of Social Sciences. 6, 2 S3,* 241–249.

Egorov A.Y., Igumnov S.A. (2005) Conduct disorders in adolescents: clinical and psychological aspects. Saint Petersburg.

Kalyagina E.A. (2009) Diagnosis and prevention of suicidal behavior in adolescents. Abakan.

Kazakova E.I., etc. (2003) Psycho-pedagogical counseling and support development of the children. Handbook for teachers, speech pathologists. Moscow.

Khydyrova A.V. (2015) Information openness of the education system as a resource for prevention of antisocial behavior: domestic and foreign experience. Electronic Journal *"Obshestvo. Kultura. Nauka. Obrazovanie"*, 3.

Kiy N.M. (2005) Pedagogical prevention of suicidal behavior in adolescents. Abstract … PhD in Pedagogy. Petropavlovsk-Kamchatsky.

Mamedova N.A. (2009) The principles for functional modeling the information society in Russia. II international scientific-practical conference "Innovative development of Russian economy": Collection of scientific works/the Moscow state University of Economics, statistics and Informatics. Pp. 337–338. Moscow.

Pankov D.D., Rumyantsev A.G., Trostanetskaya G.N. (2001) Medical and psychological problems of adolescent schoolchildren: Talk to your doctor teacher. Moscow.

Perezhogin L.O. (2016) Multimodal therapy, depending on the personal computer, the internet and mobile access to it. *Obrazovanie lichnosti, 3,* 22–28.

Perezhogin L.O. (2016) The dependence on the personal computer, the internet and mobile access to it: nosological identification. *Obrazovanie lichnosti, 1,* 45–53.

Perezhogin L.O., Emelyantseva J.V., Kryukovsky S.V. (2002) Training prevention of the dependency syndrome in children and adolescents. *Mental health and psychosocial support for children and adolescents. State and prospects. (Materials of interregional scientific-practical conference)* Kostroma.

Perezhogin L.O. (2013) Non-chemical based modern teenagers: the socio-psychiatric approach. *Obrazovanie lichnosti, 2,* 21–28.

Podolskiy A.I., Pogozhina I.N. (2014) Analysis of the psychological pressure of the phenomenon in family relations as the basis of the typology of family conflicts. *Obrazovanie lichnosti, 34,* 122–130; *1,* 90–98.

The problems of child abuse and ways to overcome them. (2008) In E.N. Volkova (Ed.) Saint Petersburg.

Zaretskiy V.V., Bulatnikov A.N., Severniy A.A. (2015) Network Innovation Marketplace: "Prevention of initiation to the use of psychoactive substances on the basis of a healthy lifestyle in the development of network interaction of educational institutions". *Profilaktika zavisimostey, 1,* 1–18.

Assessment of the quality of language education of future kindergarten teachers

I.N. Andreeva, T.A. Volkova & M.R. Shamtieva
Mari State University, Yoshkar-Ola, Russia

N.N. Shcheglova
Department of Foreign Languages and Iinguistics, Volga State University of Technology, Russia

ABSTRACT: In this study, we analyze the current state of the Russian language in the Russian society as well as the problems associated with language education of future kindergarten teachers. The objective of this study was to define the changes in the modern Russian language and the quality of language education of future kindergarten teachers. The main methods used were theoretical analysis of the literature on the problem of the research (changes in modern Russian language and language illiteracy), tests, survey, and monitoring, as well as analysis of the research results. A test on the Russian language and speech culture was carried out, which revealed some gaps in different branches of the language. We discuss about our views on the tasks that correspond to the context of teaching the Russian language and speech culture. We also analyze the types of mistakes that are usually made by students. Finally, we provide recommendations on the solution of the problems that may be applied in practice, and to monitor the academic progress of the students.

Keywords: the Russian language; lexical, grammatical, punctuation, orthographic, pronunciation mistakes; problems of the language education; intensification of the educational process

1 INTRODUCTION

The role of language in everyday life is huge, and hence the genesis and existence of a human being and the language are interconnected. The main aim of the language is to serve as the main means of communication of people. It is a form that reflects the surrounding of a person and the refection of a person himself. It is a way of getting new knowledge. The encyclopedia defines language as follows: "Language is a system that appeared naturally; it is a developing system that is expressed with the help of signs as a sign system. It enables to express some set of notions and human thoughts. It accomplishes the main goal that is communication. Language is the condition and product of human culture" (Karaulov, 1997).

This definition shows that the language is developing together with the society. Unfortunately, the current situation in the language can hardly be called development.

At present, the Russian language as well as the society itself is undergoing some significant changes. First, it is connected with the speeding of the life pace. Having analyzed some publications on the Russian language, we have pointed out the following changes that are happening with the language, and it is time to think about its future. At the same time, we tried to clarify if these problems might concern our educational establishment of higher education.

2 OBJECTIVE OF THE RESEARCH

The objective of the research is to define the changes in the modern Russian language and quality of the language education of future kindergarten teachers.

3 METHODS

The main methods are theoretical analysis of literature on the problem of the research, tests, survey, monitoring, and analysis of the research results.

4 RESULTS

1. Our language reflects the state of the society. According to Professor V.G. Kostomarov, a famous linguist, Doctor of Philological Sciences, people are not satisfied with language, as it should reflect the needs of the modern society. This means that our language has to adapt itself to meet the needs of the society (Kostomarov, 1999). Sergey Archakov, observer of the newspaper "Evening Moscow", underlines that the problems of the society are reflected in the language. In other words, when new norms appear, new words and expressions appear in order to reflect these norms. As a result, unnecessary words that the society does not need disappear. This process will always exist in the language. The language depends on many factors, such as people, development of business cooperation, quality of education, and patriotism (Archakov, 2014).

Orthographic illiteracy of people leads to grammatical and pronunciation illiteracy. Unfortunately, it is seen everywhere: in pupils' and students' essays, newspapers, and on the Internet.

Vocabulary has enlarged with the emergence of new computer, Internet, and gadget terms. Words that were borrowed from other languages have become integral part of the Russian language.

It is not a surprise that our speech is influenced by the mass media. The speech of politicians, observers, or TV presenters are full of great mistakes and fillers: if the speech of the mass media becomes cleaner, listeners and readers will speak more correctly. Giving a speech at the Russian literary meeting at People's Friendship University of Russia, Vladimir Putin, President of Russia, declared that "neglecting the rules of the native language has become the norm of Russian people, unfortunately, including the mass media and cinematography… I think that it is necessary to improve the situation…".

The language is enriching thanks to the slang used by young people. When such words appear, they are understandable only among teenagers and then they become popular among all age groups (as pointed out by a famous poet and writer Dmitry Bykov) (Medvedeva, 2014).

Some expressions that once "spoilt" our speech, as considered in the 1990s, now seem quite familiar to our hearing, though they are not used correctly ("manager", "promoter", "leasing", "marketing", etc.).

It is obvious that, on the one hand, the spoken language has become more rude, and, on the other hand, there has been excessive politeness (according to Raisa Rosina, Doctor of Philological Sciences, scientific worker of the Institute of the Russian Language named after V.V. Vinogradov, professor of the Institute of Linguistics of the Russian State University for the Humanities) (Raevskaya, 2014).

We carried out tests for full-time students and students of the correspondence course of the Institute of Pedagogy and Psychology in order to determine the mistakes that should be primarily analyzed as well as activate educational process of future kindergarten teachers. The test consisted of 13 tasks, which were elaborated according to the branches of the Russian language (Andreeva, 2012).

A total of 88 students of the first year of full-time and correspondence courses studying the majors of speech pathologists and social teachers took the test. The analysis of the works showed the following results:

The first task included a great number of mistakes with the words kvartal, logopediya, and zhalyuzi. Students tried to find the meaning of the word "kvartal", explaining the stress of the lexical meaning. A total of 145 mistakes were made by the total 88 students.

The second task included 80 mistakes, mainly in words hodatajstvovt and shchavel.

Students coped with the third task more successfully; only 68 mistakes were made, mainly with the words svekla and novorozhdennyj.

The fourth task was devoted to the norms of morphology. A total of 97 mistakes were made, mainly connected with the forms of the words brelkov-brelokov, shofyory-shofera, and dzhins-dzhinsov.

The fifth task checked the lexical norms in which students made 115 mistakes, and almost all collocations were used with difficulties. While giving explanations, the students told that they heard the expressions in the oral speech and thought that it might be the norm.

The sixth and seventh tasks were based on the knowledge of morphological norms of numerals. The sixth task included 39 mistakes, and the seventh 89. Cardinal numerals in the spoken language are used incorrectly, as was evident from the number of mistakes.

The eighth task checked the knowledge of grammatical norms, spelling of prepositions. Almost all students showed the incorrect version of the preposition "po priezdu" instead of "po priezde". The total number of mistakes was 113.

Tasks 9, 10, and 11 were connected with the correct writing of business documents, although task 9 checked the knowledge of the morphological norms of nouns, that is, the knowledge of proper names (declension of surnames). The task showed 116 wrong variants. Task 10 included numerous mistakes in grammar (117), despite the fact that the same task is used in the state exams (the correct usage of dangling participle).

Task 11 showed the ability to compile written requests correctly and more than half of all students could not find the mistakes in the test. Almost in all business letters the signature is placed to the right, date is placed to the left, and a line above the signature.

Task 12 checked punctuation knowledge, in which students made more than 138 mistakes, revealing improper school education.

The last task, 13, was based on the rules of orthography (writing of suffixes of participles and adjectives), which the students performed very badly (145 mistakes).

Analysis of the results obtained provided some distressing conclusions: 58% of students do not follow the rules of orthoepy, 26% make lexical mistakes, 29% do not know the morphological norms of students, 49% are mistaken in the use of nouns, 46% do not know the use of prepositions, and the construction of sentences with dangling participle causes mistakes in 39% of students.

The worst parameters are connected with the norms of spelling and punctuation (78% of mistakes in this area) and with making up business letters (58% do not cope with this), and spelling mistakes are made by 80% of students.

We carried out monitoring of the academic progress of the students of the full-time department (Bachelors, direction 44.03.02 Psychological and Pedagogical Education; major: Psychology and Pedagogy of Preschool Education).

Data obtained from the directions of education are presented in Table 1, from which we can conclude that the process of education directly influences the quality of language education of students.

The first-year students had an average mark of 4.3 for the winter examination period, and the second-year students had worse results in both winter and summer examination periods than other years of education: the average points were 4.1 and 3.7, respectively. These groups made the majority of mistakes in tests on the Russian language.

Table 1. Monitoring of the academic progress of students from the full-time department of groups of the major "Psychology and Pedagogy of Preschool Education".

Year	2015–2016 academic year			
	Winter examination period		Summer examination period	
	Percentage of academic progress	Average mark	Percentage of academic progress	Average mark
I	100	4.3	100	4.8
II	100	4.1	100	3,7
III	100	4.8	100	4.7
IV	100	4.4	100	4,3
Total	100	4.4	100	4.4

5 DISCUSSION

Currently, we are not concerned about the cleanliness of the language as it was 10–15 years ago but the total illiteracy. Orthographic, punctuation, and grammatical mistakes are seen in graduation projects, course works of students, students' essays, and newspaper announcements. Some examples of scientific works of students are: "Veraksa Nikilaem Evgenievichem—by a Russian psychologist, specialist in psychology of preschool education, doktorantom of psychological sciences, professor". Having studied the theoretical material on the research subject, we carried out an experiment that consisted of three diagnostics of speech development. The poor results obtained in this part created the scope of improvement in the speech development of children. By analyzing these mistakes, we come to know the importance of avoiding such illiteracy.

6 CONCLUSIONS

The recitation of students and the test allowed us to reveal the following problems of the language education of future kindergarten teachers:

1. Students do not know the rules of spelling and punctuation.
2. Students do not know the rules of grammar.
3. They cannot prepare business documents.
4. They have limited vocabulary (they do not read books, even if they do, they read low-quality literature that does not contribute to enriching vocabulary).

The analysis of examination papers and monitoring of academic progress for the period 2015–2016 (Table 1), as well as the work that was done with the students, allows us to conclude that they would not have made mistakes if they had studied well. It is also possible to conclude that the quality of language education is the indicator of the efficiency of the educational process.

Therefore, classes on the Russian language and speech culture are necessary for students in the Pedagogy and Psychology Department as they have an opportunity to bridge the gaps in their linguistic knowledge.

7 RECOMMENDATIONS

It is highly recommended to devote more hours to completely educate students, thereby improving the program and finding new forms and methods of group and individual work in order to avoid language illiteracy.

REFERENCES

Andreeva I.N. The Russian Language and Speech Culture. Андреева/I.N. Andreeva. – Yoshkar-Ola: Mari State University, 2012. – 140 p.

Archakov S. The Thread that Connects Millennia, should not be cut/S. Archakov//Evening Moscow (Vechernyaya Moskva). – 2014. – 10.02. – P. 6.

Kostomarov V.G. The Language Taste of the Epoch/V.G. Kostomarov. –3rd edited edition. – Saint-Petersburg.: Zlatoust, 1999. – 320 p.

Medvedeva T. Writer Dmitry Bykov: Three Trends of Great and Powerful/T. Medvedeva//Evening Moscow (Vechernyaya Moskva). – 2014. – 10.02. – P. 6.

Raevskaya M. Our Speech Has Become More Rude and Polite/M. Raevskaya//Evening Moscow (Vechernyaya Moskva).– 2014. – 10.02. – P. 6.

The Russian Language. Encyclopedia/Editor-in-Chief U.N. Karaulov. – 2nd edited edition– M.: Bolshaya Rossijskaya Ehnciklopediya; Drofa, 1997. – 703 p.

Developing design and construction of backspin serving skill tests to assess the learning outcomes for table tennis serving skills

Sumaryanti, Tomoliyus, Sujarwo, Bandi Utama & Herka Maya Jatmika
Faculty of Sports Sciences, Yogyakarta State University, Indonesia

ABSTRACT: The purpose of this study was to develop and design backspin serving skill tests to assess the learning outcomes of table tennis serving skills. The study employed the following research and development methods: (1) Developing the design and construction of backspin serving skill tests and testing the content validity of the design and construction of the backspin serving skills of table tennis game, (2) Searching for empirical validity and reliability of the backspin serving skill tests. The subjects of the study were students studying table tennis. The data analysis involved content validity ratios, product moment formulas and Cronbach's Alpha formula. We showed that (1) the content validity of the test was high (content validity ratio = 0.90) with the design and construction of three target marks on the table: the first area (152.5 cm × 35 cm) was scored 5, the second (152.5 cm × 35 cm) was scored 3, and the third area (152.5 cm × 67 cm) was scored 1 with a string set across the table 20 cm above the net. This string was attached according to the procedures of the test and scoring guide, (2) the empirical validity was 0.893, and the reliability was 0.938 for students studying table tennis serving. The design and construction of this backhand serving test can be used to assess learning outcomes of table tennis serving skills.

1 INTRODUCTION

Serving is a technique that is important and must be controlled by a table tennis player (Liu, 2010); (Flores, Bercades, & Florendo, 2010); (Kasai & Mori, 1992); (Hirst, 1999). Good serving skill produces precision targeting areas which the opponent finds difficult to accepted or return (Tomoliyus, 2011). When learning backspin service skills, one must keep in mind the way that the ball backspins slightly above the net, bounces on the server's side near the net with the second bounce on receiver's side close to the baseline (Tepper, 2005); (Aaron & Derek, nd). Richard (2009) stated that a target ball that is difficult to accept or return by the opponent is in the area near the net, as well those with a low ball bounce on the opponent's table. In other words, the results of serving targets that are difficult to accept or be returned by the recipient are the targets that are away from the opponent, either forwards, sideways, or to the right side of the recipient.

Based on the ball spins, there are three types of service, namely, round to the back (backspin), round to the front (topspin) and round to the side (sidespin). The technique of forehand and backhand serving includes the following: the left foot is positioned at the front, and the body slightly leans toward the table (for right-handed players), the arms position form a small corner with the arm pointing downwards. The position of bet is open when serving (Richard, 2009), turning side to the target, keeping the eyes on the ball and hitting the ball into the table at a 45-degree angle (Harrison & McCurdy, nd.). Open bet means, when hitting the ball, the front bet position is facing forward, the movement of serving with an open bet position is performed from top to bottom, resulting in backspin on the ball. Meanwhile, the serving motion with an open bet position from right to left or from left to right will produce a sidespin ball (Iizuka, Ushiyama, Yoshida, Fei, Yu, & Kamijima, 2010). Serving movement with bet open position from the rear to the front is used to create a slight backspin ball. Serving movements with a closed position bet from the rear to the front produces a topspin ball. Low arm ends the movement before the forehead. Therefore, for making a stroke, low arms form a smaller angle.

To be able to master table tennis serving, students require good learning techniques. Good learning and teaching process generate good ratings, and good judgment encourages students' learning outcomes. Good assessment of learning outcomes can be achieved by employing a good test and determining whether the test is valid and reliable.

Assessment of learning outcomes of backspin serving skills of table tennis games requires valid and reliable tests. Currently, there is a test to assess table tennis backspin serving, but the design and construction of the test still needs to be improved. Thus, the authors wanted to investigate and develop a test aimed at (1) developing skill competencies for backspin serving in table tennis (content validity), (2) testing the empirical validity and the reliability of the backspin serving skill test.

Based on the theory and objectives of the research, the research hypotheses were the following: (1) the design and construction of table tennis backspin serving test should have high content validity, and (2) the design and construction of the backspin serving skill test should have high empirical validity and reliability.

Validity and reliability are the basic requirements for measuring devices or tests to be developed or designed (Arifin, 2012); (Sukardi, 2009). Validity is the precision of a test to measure the aspects that should be measured. Tests that are valid for a specific purpose may not be valid for any other purpose. Therefore, validity is always associated with a specific purpose. The validity of the measurement is scored from low to high. The higher the level of validity is, the better the measurement is Baumgartner et al. (2007) generally divides validity in two groups, namely, rational validity and empirical validity. Rational, or content, validity is referred to as internal validity because it shows that a suitable test has to measure the contents of that which will be measured. Content validity relates to the ability of an instrument to measure the content (concept) to be measured. This means that a measuring instrument should be able to reveal the content of a concept or variable to be measured. Content validity is calculated by testing the validity of the content using a measuring instrument with rational analysis of an expert.

Meanwhile, the empirical, or external, validity (also called criterion-related validity) is the validation of an instrument by comparing it with other measurement instruments that are valid and reliable through correlations. When the correlation is significant, the instrument is considered to have the criteria validity. A criterion validity-based approach requires the availability of external criteria that can be used as a basic test for a measurement instrument score. A criterion is a variable behavior to be predicted by a measuring instrument score. To see criterion-based validity, we can use a computation of correlations between the measurement instruments scores with the criteria scores. This coefficient constitutes a validity coefficient for the measuring instrument designed, namely, r_{xy}, where x is the measuring instruments score, and y is the criteria scores (Arifin, 2012).

According to Sugiono (2007), reliability is a series of measurements or series of gauges that have consistency if the conducted measurement using that instrument are performed repeatedly. Reliability of the test is the degree of regularity or consistency of a test, namely, the extent to which a test can be trusted to produce a score that is steady and relatively unchanged although it is tested in different situations. Reliability can be obtained via test-retest. This testing is performed using the instruments twice on same respondents. In this case, the instrument and respondents should be the same, but the timing would be different. Reliability was obtained by differentiating the first test-second test, calculated with Cronbach's Alpha. If the alpha value is between 0.80 and 1.00, then the reliability is very high.

2 METHODS

This study employed research and development, with the first two stages to test the content validity of the design and construction of backspin serving skills tests; the second phase was to test the validity and reliability of the design and construction of the empirical backspin serving skills tests. The research subjects were beginner table tennis athletes. Techniques to search for content validity used seven experts in Focus Group Discussion (FGD) and the Delphi technique. Empirical validity was obtained through correlating the serving skill tests with the rating score from the seven experts due to the serving techniques used, looking for reliability by means of correlating the first test and the second test. The researchers analyzed data for content validity using the Content Validity Ratio (Wilson, Pan, & Schumsky, 2012). Empirical validity was tested using the product moment formula, and reliability was tested using Cronbach's Alpha formula.

3 RESULTS

The design and construction of backspin serving skill tests was assessed by seven table tennis experts by means of Focus Group Discussion (FGD) and Delphi techniques. Next, the results of the assessment of seven experts were calculated by a formula of content validity ratio (CVR). The design and construction of table tennis backspin serving skill tests generated a CVR value = 0.90. The design and construction of the table tennis backspin serving skill test was developed to have high content validity. This finding means that the developed test instrument was contested, and the construction tests showed a linear or accurate skill relevant to the ability of serving in table tennis. Therefore, the design and construction of table tennis backspin serving skill tests in this research was eligible to test its empirical validity. The details for the test design and construction of table tennis backspin serving skill tests were as follows: The purpose of the test was to measure the ability of serving backspin. The equipment was a table tennis ball, paddle, rope, and scoreboard. The signs table included signs for the three targets, namely, score 1, with the size of 152.5 cm × 35 cm, score 2 with the size of 152.5 cm × 35 cm, and score 3 with the size of 152.5 cm × 67 cm. Distance from rope to the net was 20 cm; the table was marked with the targets as seen in Figure 1. Test instruc-

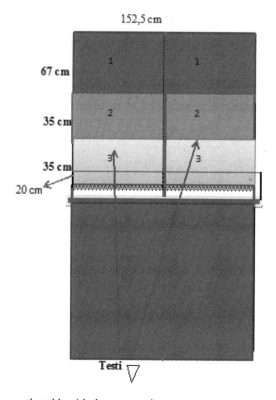

Figure 1. Target signs on the table with the score on it.

Table 1. Score results of the backspin serving skill tests (BS2T) and assessment score from the serving techniques experts (STE).

		BS2T	STE
BS2T	Pearson Correlation	1	0.893
	Sig. (2-tailed)		0.000
	N	25	25
STE	Pearson Correlation	0.893	1
	Sig. (2-tailed)	0.000	
	N	25	25

Table 2. Results of reliability test.

Cronbach's Alpha	N of Items
0.938	2

tion was as follows: the subject was asked to warm up and practice sufficiently. The subject made backspin serve towards the targets, in which the ball passes under the rope. Next, the subject serves 10 times toward the target at the right and 10 times toward the target at the left side in turns. The scoring direction was as follows: scoring carried out by two people. The first person acted as the registrar, and the second person watched the ball which was served passing under the rope and reach the target. Scores were obtained by adding the target points of the serving as many as 20 times.

The empirical validity results were gained by means of correlating the score results of the backspin serving skill tests (BS2T) and assessment score from the serving techniques experts (STE), which will be used as the criteria. The results correlation results are as shown in Table 1.

Table 1 shows the results of correlation between the backspin serving skill tests (BS2T) and of assessment score from the serving techniques experts (STE). The coefficient of correlation between serving skill test and assessment score from the experts amounted to $r_{xy} = 0893$. Based on the 0.01 significance level, $r_{xy} = 0.893 >$ r table $= 0.505$. H_1 is accepted, and H_0 is rejected. The design and construction of the table tennis backspin serving skill test developed had high empirical validity.

The reliability test results were obtained by differentiating the first test scores and second test score of the backspin serving skill tests with the same respondents, but the timing was different. The reliability test results are shown in Table 2.

Table 2 shows the results of the reliability test between test1 and test2 score is Cronbach's Alpha 0.938. The design and construction of the table tennis backspin serving skill test developed had high reliability. In other words, the test showed a degree of stable measurements made over time (stability over time).

4 DISCUSSION

The main requirement of measuring devices or test development is validity and reliability (Arifin, 2012); (Sukardi, 2013). In general, validity consists of two levels, namely, rational validity or content validity and empirical validity (Baumgartner et al., 2007). Validity of the content is obtained by employing FGD with some experts and/or using the Delphi technique. Meanwhile, the empirical validity is obtained by way of correlating between the scores of measurement instruments with the score of criteria. Meanwhile, reliability can be gained by correlating the first test and the second test with the same respondents, but with different timing (Sugiono, 2007). Based on the test development requirements above, the development conditions of construction test of backspin serving skill in table tennis must be valid and reliable. The results of this study showed that the content validity of the construction test development of backspin serving skill or content validity ratio (CVR) was 0.90. Thus, this tool of measuring service accuracy developed in this research was scientifically demonstrated to determine the target area close to the net, and produced a low ball-bounce that was difficult to accept or be returned by the opponent. This is in accordance with the opinion of Tomoliyus (2016) and Richard (2009), who stated that the goals or targets that are difficult to be received or returned by the opponent are in the area near the net, with the low bouncing ball in the opponent area. Therefore, this test can be used to test its empirical validity.

Empirical validity was sought by looking for correlations between the construction test scores of backspin servings kills with the serving technique assessment as the criterion. The result was a correlation coefficient score between the first test scores and second test score of 0.898. This finding indicated that the level of accuracy or appropriateness of the instruments used to measure the skills of serving in this research was very high. According to Guilford and Benjamin (1977), a value of the correlation coefficient between 0.80 until 1.00 has very high validity. Therefore, we can say that this construction test of the backspin serving skill has very high empirical validity. In other words, this test has the feasibility and accuracy to measure the skills of table tennis backspin serve.

Furthermore, reliability was obtained by checking the first test with the second test with the same respondent, but with different timing. The results of the variability test showed that the value of Cronbach's Alpha was 0.938. This finding showed that the level of reliability, constancy, consistency, and stability of measurements made over time was very high.

This result is in accordance with the opinion of Hair et al. (2010: 125), which stated that the value Cronbach's Alpha between 0.80 until 1.00 was very reliable. Therefore, the design and construction of the backspin serving skills test in table tennis has very high reliability. The design and construction of the backspin serving test developed has more accurate competency than the design and construction that has been previously available, because the design and construction in this test provided a limit line 20 cm above the net in order to let the ball pass between the net and the string. This finding is in line with the results of Tepper (2005) and Aaron and Derek, (nd.) who stated that learning backspin serving skills requires keeping in mind that the ball should have backspin slightly above the net, bounce on the server's side near the net, and bounce twice on the receiver's side, close to the baseline.

5 CONCLUSIONS

Based on the results and the discussion above, the following can be concluded:

The design and construction of table tennis backspin serving skill tests had high content validity. Next, the test instruments were in accordance with linear competency and are appropriate for learning the backspin serving skill.

The design and construction of the table tennis backspin serving skill test had very high empirical validity, exhibiting a coefficient value of 0.893, and very high reliability, exhibiting a value of 0.938. Thus, the test as feasible to use because it had measurement accuracy and consistency or regularity that may be applied repeatedly by the same respondents, but in a different time when performing backspin serving to measure skills.

A comparison has been made between the design and construction of this developed table tennis backspin skills test with the existing table tennis skill backspin test. The design and construction of this table tennis backspin serving test has a limit as to the path of the ball from the net height to the 20 cm rope. The test subject is forced to carry out a service wherein ball passes slightly (maximum 20 centimeters) above the net. When the ball passes over the rope, it is considered invalid. The difficulty in serving occurs when the ball bounces on the server side near the net and bounces twice on the side of the receiver, and the ball must be close to the baseline. This test is more appropriate than the existing test to measure the competence of backspin serving skills for table tennis.

ACKNOWLEDGMENTS

This work was supported by the Faculty of Sport Science, Yogyakarta State University and the Indonesian Table Tennis Association.

REFERENCES

Aaron, H. & Derek, D. nd. Table tennis. Accessed on 30 July 2017. web.uvic.ca/~thopper/Pe352/2002/Table Tennis8DerekAaron.pdf.
Arifin, Z. 2012. *Evaluasi Pembelajaran*. Bandung: PT. Remaja Rosdakarya.
Baumgartner, T.A., Jackson, A.S., Mahar, M.T., & Rowe, D.A. 2007. *Measurement for Evaluation in Physical Education and Exercise Science*, New York: Mc Graw Hill.
Flores, M.A., Bercades, D., & Florendo, F. 2010. Effectiveness of Shadow Practice in Learning the Standard Table Tennis Backhand Drive. *International Journal of Table Tennis Sciences*, no.6: 46.
Guilford & Benjamin Fruchter.1977. *Fundamental Statistics in Psychology and Education*, New York: McGraw-Hill Book Co.
Hair, Joseph F., et al. 2010. *Multivariate Data Analysis, 7th Edition*. New York: Prentice Hall International, Inc.
Harrison, J. & McCurdy, N. nd. Table tennis. Accessed on 30 July 2017. http://www.pelinks4u.org/naspeforum/discus/messages/1239/Table_Tennis_Harrison_McCurdy-2784.pdf.
Hirst, P.A. 1999. Table tennis. Accessed on 30 July 2017. www.teachpe.com/gcse/Table%20Tennis.pdf.
Iizuka S., Ushiyama, Y., Yoshida, K., Fei, Y., Yu, Z.H., & Kamijima, K. 2010. The Measuring Ball Spin at the Service in Table Tennis by Junior player. *International Journal of Table Tennis Sciences*, no.6: 122.
Kasai, J. & Mori, T. 1992. Studies of various strokes in table tennis. *International Journal of Table Tennis Sciences*, no.1: 39.
Liu, K. 2010. Reasons for the defeat of Wang Hao in table tennis men's singles final of Beijing Olympic Games from the technical and tactical point of view. *Journal of Shandong Institute of Physical Education and Sports*, 5: 73–76.
Richard Mc Afee. 2009. *Table Tennis Step to Success*. United States: Human Kinetics.
Sugiyono. 2007. *Metode Penelitian Pendidikan. Pendekatan kuantitatif, kualitatif dan R & D*. Bandung: Alfabeta.
Sukardi. 2009. *Metodologi penelitian pendidikan: kompetensi dan praktiknya* Jakarta: Bumi Aksara.
Tepper, G. 2005. ITTF Coaching Manual Level 1. As cited in USA. *Table Tennis Magazine*, vol. 76: 3.
Tomoliyus, Sumaryanti & Jadmika, H.M. 2016. Development of Validity and Reliability of Net Game Performance-Based Assessment on Elementary Students' Achievement in Physical Education. *International Journal of Assessment and Evaluation in Education*, vol. 6, no.1: 41–49.
Wilson, F., Pan, W., & Schumsky, D. 2012. Recalculation of the critical values for Lawshe's content validity ratio. *Measurement and Evaluation in Counseling and Development*, 45(3), 197–210. http://dx.doi.org/ 10.1177/ 07481756 12440286.

Marketing management of Madrasah Ibtidaiyah (MI)

Erni Munastiwi
State Islamic University, SunanKalijaga Yogyakarta, Indonesia

ABSTRACT: The purpose of this study is to review the institution of special education of madrasah institutional management. Here, we focus on marketing management of madrasah institutions. Institutional marketing is one part of the management of educational institutions. Therefore, the management of madrasah marketing must be handled properly so that the objectives of the madrasah can be achieved. We hope that well-managed madrasah institutions will be well known for marketing and then the madrasah will be known and trusted by the community. Thus, the interest of the community is high against the madrasah. In this study, we use qualitative method with five madrasah research objects, five principals, five madrasah teachers, five members of the madrasah committee, and five madrasah parents. The data collection process includes (1) identifying participants, (2) gaining access to participants and places, (3) compiling the type of information that will answer the research question, (4) designing instruments to collect and record interview results, and (5) interpreting the collected data. The data collection method involves observation, interview, documentation, and audiovisual materials. Data analysis takes place before, during, and after field study. The results of this study show that madrasah education institutions are advised to manage marketing well. Madrasah marketing strategy is done consistently. In the context of education, marketing mix strategy can be used to win the competition, in this case, known as 7P (product, price, promotion, place, people, physical evidence, and process.) Madrasah marketing management implementing the marketing mix strategy can change the image of the community to the madrasah.

Keywords: marketing management, madrasah ibtidaiyah

1 INTRODUCTION

Madrasah is a formal educational institution under the guidance of the Ministry of Religious Affairs, while the school is a formal educational institution under the guidance of the Ministry of Education and Culture. The Law of the Republic of Indonesia Number 20 the Year of 2003, Article 17, Paragraph 2 stated that primary education in the form of elementary school (SD) and *madrasah ibtidaiyah*/Islamic elementary school (MI) are equal. In its development, both institutions experience problems. The government has regulated it in the Law of the Republic of Indonesia Number 20 Year 2003 regarding National Education System, which stated that there is an opportunity for madrasah to improve its quality and introduce itself in the middle of society. Constitutional madrasah has equality with other public schools. Nevertheless, the reality is that the community positioned the madrasah as a second choice. Madrasahs are considered less qualified schools, graduates from which are less able to compete in continuing higher education. This condition proves that madrasahs are difficult to become the primary educational institution of society. We address the question of how to manage a madrasah into a community-chosen institution. Madrasahs are no longer underestimated. Madrasahs are in line with other public schools. Therefore, the intent of the researcher is to overcome the condition through improving the management of institutions in general and marketing management in particular. The low public interest in madrasah is due to the management of madrasah, which is still managed traditionally, but not yet implemented in the optimal management function. To increase community interest, madrasah management must be addressed. Madrasahs and managers must change themselves to achieve educational goals. In addition to institutional management, the leadership of the institution has not met the qualifications as an educator. Another problem is financing. Thus, managers of madrasah institutions should examine more in the management of madrasah. To attract high public interest to madrasah, it is necessary to establish policies related to marketing management program of madrasah. Philip Kotler (1997: 36) states that customer satisfaction is a function of the impression of performance and expectations. If performance is below expectations, customers will not be satisfied, and vice versa. Furthermore, if performance exceeds expectations, the customer will be very satisfied or happy. In line with that Philip Kotler, Imam Machali (2017: 226) concluded that the madrasah as an educational service industry faces challenges; on the one hand, madrasah should strive to improve the quality and competence of graduates so as to meet the expectations of stakeholders, and on the other hand, madrasah is still seen as a second-class institution. Therefore, the madrasah should meet customer satisfaction. On the basis of the results of direct observation, generally people choose public education. The reality in the field is that the results of research on marketing management madrasah are less socialized to the community, especially managers of madrasah, teachers, and stakeholders. In addition, funding or financing issues are one of the factors causing inefficiency of madrasah marketing program. Another problem, lack of knowledge about marketing management, resulted in the competence of human resources of madrasah managers. In-depth examination of the marketing management of madrasahs will deepen and expand marketing knowledge. For practitioners, managing educational institutions get institutional management guidance. Similarly, policy makers have a basis for deciding policies. Researchers can improve the results of research to obtain accurate results. Students participating in research will get new experiences so as to acquire more knowledge and experience.

2 METHODOLOGY

In this study, we use qualitative method. The data collection process involves (1) identifying participants and places; (2) gaining access to participants and places; (3) compiling the type of information that will answer the research question; (4) designing instrument to collect and record interview result; and (5) implementing the collected data. The data collection method involves observation, interviews, documentation, and audiovisual materials (John Creswell, 2015: 420). Data analysis takes place before, during, and after field study. Field study was conducted according to Miles and Huberman (1984). The data were validated using member checking and triangulation.

3 RESULTS AND DISCUSSION

Madrasah is Islamic-based educational institution, which manages various levels of educational units from kindergarten (*RaudhatulAthfal* (R.A.)). The basic education unit is called *Madrasah Ibtidaiyah* (M.I.), while the level of junior secondary education unit called *Madrasah Tsanawiyah* (M.Ts.). The upper middle level is called *Madrasah Aliyah* (M.A.). "Madrasah" educational institutions began to flourish in Indonesia in the nineteenth century. This development began with the establishment of *Madrasah Adabiyah* by Abdullah Ahmad in Padang Panjang (Malik Fajar, 1999: vii). During its emergence, the madrasah focused more on Islamic religious learning. Together with its development, various problems aroused. Madrasahs had difficulty competing with public schools and were less desirable educational institutions for the public. Madrasah became the number two institution after public school. Lack of public interest due to the management of madrasah institutions has not been optimal. However, competition in education is inevitable. Several madrasah institutions are combined because of the shortage of students. Understanding the concept of marketing management depends on understanding the concept of the market.

The marketing process is a transactional process between producers and consumers together assigning agreed rewards. As an education management institution, madrasahs should determine marketing strategies related to educational services that satisfy customers. Thus, the impact of madrasah is better known to the public. According to Sheila (1996: 25), customer satisfaction has become an important part of marketing strategy and has greater strength than an advertisement. Customer satisfaction can also be an indicator of the quality of services provided, so that the increase and decrease in income can be seen from the extent of customer satisfaction, which depends on various elements: (1) matching of customer expectations and reality; (2) satisfaction with the service received; (3) financing affordable service programs as agreed between the customer and the madrasah; and (4) the existence of a conducive madrasah atmosphere. Customer satisfaction depends on the strategy used. The marketing strategy in question is the marketing mix (marketing mix). According to Kotler et al. (2002: 9), marketing mix is a set of controllable, tactical marketing tools that the firm integrates to produce the desired result in the target market. There are seven marketing mixes: (1) product, (2) price, (3) place, (4) promotion, (5) human resources, (6) physical evidence, and (7) process. With regard to marketing management of madrasah, (1) the product is a madrasah education service that encompasses various types of madrasah institution services; (2) price is the cost of various services of madrasah institutions; (3) the place is the location of the school under question in determining the choice; (4) promotion is the promotion of madrasah; (5) human resources are those involved in madrasah services, in this case, educators and education personnel; (6) physical evidence is linked to madrasah facilities and infrastructure; and (7) the process involves all madrasah service activities. Implementation of marketing mix in madrasah institution can be modified according to condition and situation of madrasah institution. For example, the implementation of new students can be achieved before the time set by the government. The theme of the acceptance program of new students comes under the selection for entrance interests (SEI) program, which is the development of new enrollment programs. The result of the madrasah is to know the number of new students. In addition, early madrasahs can plan education services. Planning is an early stage of the marketing management of madrasah. It is a process of systematically designed marketing activities. The designed activities will be carried out to achieve the goal. The activities are (1) market identification; (2) market segmentation and positioning; (3) product differentiation; and (4) school services (Imam Machali and AraHidayat, 2016: 298). Organizing is an advanced planning in a series of marketing management. It is a marketing activity carried out by a group of people to achieve the goal. Organizing activities include the division of duties, authority, and responsibility of marketing officers of madrasah. Actuating is a series of madrasah management that realizes planning and organizing. Controlling is the process of observing and measuring marketing management activities of madrasah. Madrasahs should set marketing strategies to include the following steps: (1) identify markets to identify market conditions and map from competing madrasah; (2) market segmentation and positioning to divide the market into a group of buyers; (3) product differentiation to arrange different programs with other madrasah; (4) madrasah marketing communication involving academic and nonacademic activities; and (5) madrasah service. In the context of education, to win the competition, it can apply the marketing mix strategy, in this case known as 7P (product, price, promotion, place, people, physical evidence, and process): the product is a madrasah program that can satisfy customers; price in this context is the cost incurred by the customer; place (location) is the location of madrasah; promotion is the activity of communicating madrasah products; people (human resources) are the people involved in the madrasah program; physical evidence is the decisions; and process is the process of madrasah program. The marketing management of a madrasah that implements the marketing mix can change the image of the community to the madrasah. The role of madrasah in the society is inevitable. As image changes, madrasah constantly improves the service quality, so that it can compete with public schools, to finally become the number one educational institution of interest to the public.

4 CONCLUSION

1. Madrasah should improve madrasah management in general and marketing management in particular.
2. Madrasah should focus on madrasah marketing management.
3. Marketing strategy must be set to focus on customer satisfaction.

REFERENCES

FandyTjiptono. 2009. *Service Marketing*. Yogyakarta: Jelajah Nusa.
SofjanAssauri. 2010. *ManajemenPemasaran*. Jakarta: Raja Grafindo Persada.
Imam Machali. 2017. *AntologiPemikirandanManajemenPendidikan*. Yogyakarta: UIN Sunan Kalijaga.
Imam MachalidanAraHidayat. *The Hanadbook of Education Management*. Jakarta: Prenadamedia Group.
Philip Kotler. (1997). *ManajemenPemasaran. Marketing Management* 9e. Indonesian Language Edition, Volume 1. By Prentice-Hall. Inc. A. Simon & Schuster Company. Upper Saddle River, New Jersey 07458 (h. 36).
Fajar, Malik. *Madrasah dantantanganModernitas*, Mizan, Bandung. 1999.
Sheila, Kessler. *Measuring and Managing Customer for Educational Institution*. 2nd Edition. New Jersey: Prentice-Hall, Inc. h. 25.
Creswell, John. (2015). *Riset Pendidikan. Perencanaan, PelaksanaandanEvaluasi Riset Kualitatifdan Kuantitatif*. Fifth translation edition. Yogyakarta: Pustaka Pelajar.

Development of competitiveness as a condition of professional demand for graduates of pedagogical directions

K.Yu. Badyina, N.N. Golovina, A.V. Rybakov & V.A. Svetlova
Mari State University, Yoshkar-Ola, Russia

ABSTRACT: In this study, we explain the role of the qualitative preparation of future teachers in the development of society. The objective of this study was to define the influence of competitiveness on professional demand and further job placement for graduates of pedagogical directions. Research methods include analysis and synthesis, comparison and generalization of the results of other studies, and survey and testing of students of pedagogical directions. These methods can help examine the competitiveness of future teachers systematically and comprehensively.

Keywords: competitiveness, professional demand, graduates of pedagogical directions of preparation, modern school

1 INTRODUCTION

Professional training of teachers is one of the top priorities in Russia. It is necessary for modern schools to have skilled and professional teachers who are able to compete on the market of educational services. The competitiveness of a teacher is defined by many scientists as the ability of a certain object or subject to meet the demands of people in comparison to other similar subjects and/or objects (*Gurov, 2013, Milyaeva, 2009, Chromenkov, 2015*). For a teacher, it means the ability to solve the educational tasks efficiently, satisfying the growing needs of people and dynamically developing society, as well as to enlarge integration connections in the modern society (*Slastenin V.A., 2004*). A modern teacher should not be afraid to use the recent accomplishments of science and technology, as well as to participate not only in the educational process but also in other social processes initiating it. This is particularly important for a modern school, so that a teacher possessed some social qualities necessary for a person and developed the qualities in their followers. Consequently, schools require skilled young teachers.

Examining the question how qualitative graduates of educational directions are ready to accomplish their own skills in the practice of educational activity, what aims they have, and what motives they are ruled by, it is reasonable to talk about the degree of competitiveness development and professional need in the aims of contemporary society (*Vasilevska, Golubkova, 2014*). It should be made on the basis of the quality of graduate training, which is very important for a successful job placement in general educational organizations.

2 MATERIALS AND METHODS

Methodological bases of research are philosophical positions about universal relations and interdependence of phenomena, as well as determination of pedagogic theory and practice by social processes using the theory of systematic recognition of educational processes. Methodological complex was used during the research: analysis of literature concerning pedagogy, sociology, educational marketing, management corresponding to the problem of research; analysis of educational experience of models of competitiveness in educational sector; models of building the competitiveness in the pedagogical sector; observation; diagnostic methods; and recitation of students of pedagogical directions. The research period is 4 years based on the leading higher educational establishment of the Republic Mari El – "Mari State University". The first stage included the analysis of the research problem in the pedagogical theory and practice, as well as search for programs and research methods.

The second stage was devoted to defining and implementing professional competences as a means of competitiveness of students of pedagogical directions.

The final stage included systematization, generalization, interpretation, and calculation and reporting of the results.

3 RESULTS

The modern Russian school sets a number of requirements for graduates of pedagogical directions that are stated in the list of professional and personal requirements of a worker. Professional standard of a teacher is the instrument for implementing the strategy of education in the changing world. The standard defines the qualities of educational services that are demanded in the market. The standard envisages that a teacher should possess are qualities such as maneuverability, readiness for changes, non-typical actions at work, independence and responsibility in reaching solutions, and the ability to study and constantly develop professional skills (8). Nowadays, the labor market of the Russian Federation is formed according to the rating of professional demand. According to the site edunews.ru, the rating of a teacher is at the third place in the list of "The Most Demanded Professions in Russia" that shows the significance of this profession for the development of the society (9).

Demand is a parameter of the necessity of this profession for modern employers. On average, the statistics shows that if there are 100% of enquires for the vacancies in the sphere of education, the occupancy rate is 85%, which shows the filing of job vacancies. This trend does not mean that an employer is ready to accept any young specialist with moderate level of professional readiness. Schools need teachers capable of organizing the educational process, giving extra knowledge, and encouraging students for further self-development. A young specialist can use the most modern methods of the organization of educational process with the application of information technologies, for participation in project activity, professional skills, and constant self-development. Consequently, a teacher should be competitive in order to be demanded in the professional sphere.

In the pedagogical dictionary, competitiveness is defined as the ability of a person (specialist) to prove their competence, win in the competition in the labor market, establish the independence of educational, and attract the market of educational services (9).

The question of competitiveness as an important condition defining professional demand for a teacher was studied by many scientists who expressed different views. V.A. Adolf examined competitiveness of a graduate as an index of quality university education (*Adolf, 2007*). The essence and structure of competitiveness in the pedagogical practice are represented in the work of V.N. Gurov (*Gurov, 2013*). The main ways of increasing the competitiveness of graduates are covered in the works of Milyaeva L.G. (*Milyaeva, 2009*). The problems of professional training of future teachers and searching the best means and methods of training are examined in the works of E.V. Kondratenko (*Kondratenko, 2015*), D.L. Kolomiets, N.A. Biryukova (*Kolomiets, Biryukova, 2016*), and other modern researchers. Undoubtedly, pedagogy is responsible for quality training of modern generations, formation of spiritual and moral values of the society, and reproduction and replenishment of scientific knowledge system. Because of this competitiveness, professional demand is meaningful in the process of educational activity's arrangement.

The educational establishment, i.e., school, has to define its place in the market of educational services and enter the system of rating among the leading positions. In this connection, personnel policy recruiting highly qualified specialists on vacant places is important for lyceums and gymnasia. As a consequence, the market with professional personnel appears.

Regarding the competitiveness of a teacher, the majority of authors define it as a certain ability of an object or a subject to meet the requirements of interested people in comparison to the same subjects or objects (Gurov, 2013, Milyaeva, 2009, Chromenkov, 2015).

For a teacher, it means the ability to solve the educational tasks efficiently, satisfying the growing needs of a person, and dynamically developing society, as well as enlarging integration connections in the modern society (Slastenin V.A., 2004).

A modern teacher should not be afraid to use actively the recent accomplishments of science and technology in his/her activities, as well as to participate in not only the educational process but also other social processes initiating them according to the demand for them. For a modern school, it is important so that a teacher possessed some social qualities necessary for a person and developed the qualities in his followers.

During pedagogical training, students acquire necessary professional skills, professional training functions, readiness for pedagogical activity, professional qualities, and adequate competitiveness.

While studying the competitive abilities at the ascertaining stage of a research, we carried out a survey among students of pedagogical directions at Mari State University (including students of the first and final years of training). The survey included 273 people. The research showed that competitiveness is understood in different ways and depends on the personal qualities of respondents. The majority of students indicate the necessity to analyze a number of changing factors and conditions under which the training is carried out.

About 44% of the students surveyed think that honesty and openness are among the main qualities of competiveness of a teacher; 21% think that competiveness of a teacher is connected with the intuitiveness and ability to risk; 19% mark flexibility in accomplishing the educational process; and 16% enumerate persistence in achieving a goal. We concluded that the students do not connect competitiveness with the level of professional qualities and the formation of professional qualities. In other words, they do not draw parallels between competiveness and professional competence of a teacher, their readiness for innovative activity, and readiness to use social experience while training a child in a dynamically developing educational sphere. Consequently, to analyze competiveness, it is necessary to understand other dynamic mechanisms except for the survey.

Concerning the dynamics of competiveness of graduates, it is necessary to emphasize the models of formation of competiveness of students of the pedagogical institute as P.A. Khromenkov says. He assures that competitiveness depends not only on personal qualities but also on cultural and professional competences that are reflected in parameters such as information competence, organizational competence, managing competence, research competence, prognostic competence, technological competence, and methodological competence (P.A. Khromenkov, 2015). These qualities are formed during multisided educational activity within the period of education at the institute. On the basis of the model elaborated by P.A. Khromenkov, within 4 years, the formation experiment of professor-teaching personnel working with the students of pedagogical directions was mainly aimed at the formation of student qualities such as broadening of background; development of speech communication, sociability, rhetoric, organizational qualities, responsibility, efficiency, and objectiveness.

During the research activity of students, we encourage students to develop the philosophical way of thinking, the ability to structure experimental work, as well as desire and willingness to accomplish scientific goals.

The pedagogical practice contributed to the development of students in the following directions: skills of methodological competence; motivation for applying the best practices; and a need for working out the programs for accomplishing the educational process at school.

At the end of the experiment, the students of the fourth year carried out a research aimed to study competitiveness. A total of 280 respondents participated in the survey. The method of the research is survey. The results were calculated using automatic programs and systems. The analysis of the results revealed the following: 80–90% of graduates showed formation of parameters such as information competence, technological competence, and methodological competence. The level of formation of organizational competence equals 76%, managing and prognostic competence equals 64–72%, and research competence equals 63.8%.

Table 1. Parameters of index of competitiveness for graduates of pedagogical directions.

Parameters	Level of formation in the group
Information, communicative, technological, and methodological competence	80–90%
Organizational competence	76%
Managing and prognostic competence	64–72%
Research competence	63.8%

The results show that students face more difficulties with the formation of managing, prognostic, and research competences that are significant for the formation of competitiveness for graduates of pedagogical directions. The difficulty of formation of personal characteristics of students lies in the low level of leaders' and research qualities. The indicator of competitiveness is the assessment of job placement of young specialists.

The statistics shows that due to the effective organization of the educational process at the institute, the students form the necessary competences that are necessary to be competitive in the market of educational services.

While monitoring the employment of graduates over the last 10 years, we can conclude that 75% of students can find a job of an educational worker after graduating from the institute. Later on, about 18% of them are offered the managing positions in the sphere of education, 5% are devoted to scientific work, 20% undergo professional retraining and quit the place, and 10% of them are employed in the social sphere. These graduates possess the average level of professional readiness for pedagogical activity, and the parameters of organizational competence can hardly be more than 50 points.

The results correlate with the average level of competiveness of graduates: 74.1 points of 75% of Bachelors of Pedagogy. In general, the situation in the Republic of Mari El is the same as in other parts of Russia with a little difference depending on the geographical location of the region, as well as the social and economic conditions in it.

4 CONCLUSION

Thus, modern schools require skilled competitive graduates of pedagogical directions. About 75% of graduates are successfully employed in organizations of general education within a period of 3–6 months after graduation from the university. Noncompetitive graduates (25%) are either employed in related areas or retrained. Consequently, competitiveness defines professional demand and further job placement according to their major in the university.

REFERENCES

Adolf A.V., Stepanova I. (2007) Competitiveness—Index of Quality of VPO//Higher Education in Russia. – 2007. – № 6. – P. 77–79.

Biryukova N.A., Kolomiets D.L. Formation of Readiness of Students for Pedagogical Activity Based on the Analysis of Professional Difficulties//Modern Higher School: Innovative Aspect. – 2016. V. 8. No 2. P. 57–63. DOI: 10.7442/2071-9620-2016-8-2-57-63.

Chromenkov, P.A. Competitiveness of Graduates of Pedagogical Institute: Realities and Prospects// Teacher XXI century. – 2015. – № 1, v.1 – p. 9–14.

Gurov V.N., Daynova G.Z. Methodological Orientation and Peculiarities of Formation of the Graduate in Modern Science and Pedagogical Practice//Innovations in Education. – 2013. – № 11. – P. 15–28.

Kondratenko E.V., Kondratenko I.B., Rybakov A.V., Svetlova V.A., Shabalina O.L (2015) *Interactive Learning as Means of Formation of Future Teachers' Readiness for Self-education//Review of European Studies*. T. 7. № 8. 35–42.

Lavrentiev S.Ju., Shabalina O.L., Krylov D.A., Korotkov S.G., Svetlova V.A., Rybakov A.V., Chupryakov I.S. (2016) *Future specialists' competitiveness development: pedagogical and social-economical aspects//Journal of the Social Sciences. 11. 1855–1860.*

Masalimovaa A.R., Ivanov V.G. (2016) *Formation of Graduates' Professional Competence in Terms of Interaction Between Educational Environment and Production//International Journal Of Environmental & Science Education. Vol. 11, №. 9, 2735–2743.*

Milyaeva L.G., Borisova O.V. The Main Trends to Increase Competiveness of University Graduates in the Labour Market//Izvestiya of Irkutsk State Economic Academy. – 2009. – № 5(67). – P. 157–161.

Pedagogical Dictionary//Ed. by V.I. Zagvyazinskij, A.F. Zakirova. – M.: Academiya, 2008. – 352 p.

Professional Standard of a Teacher. Ministry of Labour and Social Development of the Russian Federation «18» October, 2013 № 544n, http://профстандартпедагога.рф.

«The Most Demanded Professions in Russia 2017» http://edunews.ru/professii/rating/vostrebovannie-Russia.html.

Slastenin, V.A. Pedagogy of Professional Education. – M.: Academiya, 2004. – 420 p.

Vasilevska D., Golubkova T. (2014) *The Analysis of Competitiveness of University Graduates in the Labor Market: The Case of Latvia//Advances in Economics and Business. № 2(1): 37–41.*

Development of students' fire safety behavior in the school education system

V.V. Bakhtina, Zh.O. Kuzminykh, A.G. Bakhtin & O.A. Yagdarova
Department of Biomedical Sciences and Life Safety, Faculty of Physical Culture, Sport and Tourism, Mari State University, Yoshkar-Ola, Russia

ABSTRACT: In this study, we present the data of annual statistical reports provided by State Fire Service agencies, which ascertain that careless handling of fire, household electrical safety violations, and child fire-play remain the leading causes of numerous fire accidents, resulting in large-scale human and economic losses. Fire protection is the responsibility of every member of a society and requires a consistent nationwide approach. In Russia, fire safety is taught to children in all educational establishments. Strong and stable fire safety skills are developed through providing systematic long-term information and guidance. The purpose of this study was to assess the level of fire safety skill development among students of grades 5–8. A survey was conducted on 54 students. Based on the survey results, it was concluded that the level of fire safety skills was higher among grade 8 students. This is because these children had a clear idea of fire safety requirements after more than a year of involvement in the educational program that aimed at the development of students' fire safety behavior. To be more effective in future work with these students, the educational fire safety program was adequately adjusted. A series of events highlighting fire prevention problems were conducted.

Keywords: fire, fire safety, fire safety behavior

1 INTRODUCTION

Over the past decade, the problem of fire disasters has become a global concern that has attracted both national and international interests. About 6 million fire accidents occur every year worldwide, i.e. one to two fire-related incidents take place approximately every 5–6 seconds. Fire causes an annual average of 50,000 deaths, with more than 6 million people being succumbed to burns and other injuries. Property losses are also enormous and exceed hundreds of billions of monetary units. In addition, fire causes air and water pollution, destroys natural resources, and thus increases environmental losses.

The works of modern authors reveal the following problems significant for our research: principles of safety culture formation (Yu.L. Vorobyev, 2006; V.N. Moshkin, 2002, etc.); sociology of fire safety (V.V. Kafidov, 2003, etc.); and student preparation for fire emergencies (V.A. Sidorkin, Yu.V. Prus, A.V. Golyshev, 2012, etc.). The problems of teaching fire safety rules to school children are also being considered in other countries (National School Code of Conduct, 2009; Fire safety trailer curriculum. Tools for Delivering Fire Safety Education Messages Using a Fire Safety Trailer. – FEMA. U.S. Fire Administration, etc.).

However, although substantial amount of research that has been conducted in this area, fire prevention and safety will remain problems of utmost importance forever.

In this study, we aim to evaluate the theory and identify empirically the most effective methods for developing students' fire safety behavior.

2 STATEMENT OF PROBLEM

In Russia, fire safety education is provided to children in all educational institutions, from kindergartens to universities. Fire safety education programs are implemented in close connection with the overall educational process, both in class and during extracurricular (after-school and out-of-school) activities. Strong and stable fire safety skills are developed through long-term systematic information and guidance provision.

In some cases, fire causes loss of human life. Therefore, fire protection is the responsibility of every member of a society and is carried out on a nationwide scale. Fire protection aims to find the most effective, economically feasible, and technically sound methods and means of fire prevention and extinction to minimize fire damage and ensure the rational use of human resources and technical firefighting equipment.

3 CONCEPTUAL FRAMEWORK

Education plays a key role in assurance of health. Here, we consider the basic indicators of the fire situation in the Mari El Republic for the period 1 January 2016 to 31 March 2016 (on the basis of the data provided by the Supervisory Activities and Preventive Work Department of the Federal Headquarters of EMERCOM of Russia in the Republic of Mari El).

In 2016, the fire situation in the territory of the Mari El Republic, compared to the same period of the previous year, was characterized by the following main indicators: 166 fire accidents were recorded (in 2015, their number was 228 (–27.2%); 14 people died in fire-related incidents (in 2015, the number of deaths was 22 (–36.4%); 14 people were injured in fire (in 2015 their number equaled 17 (–17.6%); and direct material damage from fire accounted for 21 million 381 thousand rubles (in 2015, it was 35 million 772 thousand rubles (–40.2%). Divisions of the State Fire Service rescued 113 people from fire and saved property worth more than 216 million rubles.

With regard to the days of the week, statistics of fire and fire casualties indicate that the greatest number of fire accidents occur on Sundays, Thursdays, and Fridays; the number of fire deaths is higher on Fridays. Although the difference is insignificant, it may be assumed that the number of fire-related incidents increases on a day off, or on the eve of a day off, when people go out of town and rest, their control functions being reduced. The largest number of fire accidents occur at night and in the evening (00.01, 16.00, 18.00 h), and the largest number of fire deaths occur in the morning and evening hours (08.00, 18.00 h).

The total number of fire fatalities, by months, decreased in 2016, but the winter months remain the time of highest risk, presumably, because of careless handling of fire when using stove heating or other heating equipment in residential premises. There were no fire fatalities among children during the reporting period. Men had higher death rates than women: males accounted for 57% of all fire fatalities, whereas females accounted for 43%. A proportion of 57% of people who died in fire accidents were in a state of intoxication by alcohol or narcotics (Bakhtina V.V., Kuklin E.V., 2017).

Residential fire accidents account for 70–80% of the total number of fire-related incidents that occur annually in the Mari El Republic. The main cause of residential fire accidents is due to careless handling of fire.

All too often it is the wrong actions of people that lead to fatal residential fire due to their ignorance about what to do and their inability to take the right decision in an emergency. Therefore, educating children and adolescents about fire prevention and fire safety should be one of the priority areas in the educational process.

3.1 *Research instrument*

This study was carried out in the following two publicly funded municipal schools of the Mari El Republic, Russia: Sysoyevo Secondary School named after S.R. Suvorov and Zelenogorsk Secondary School. A survey was taken among 54 students of grades 5–8.

The research methods included the analysis of literature on the problem being investigated, generalization, direct and indirect observation, testing, survey, interviews, ascertaining (checking) and teaching (formative) pedagogical experiments, and interpretation of the results of the experimental study. The empirical study covered the time period from January 2016 to April 2016.

4 FINDINGS AND SUGGESTIONS

In order to assess the level of fire safety skills among students of grades 5–8, appropriate tests were developed and conducted, each consisting of 10 questions with answer choices that suggested certain actions in a proposed situation. The student scores 1 point for the correct answer. All the tests were age-appropriate. In grade 5, for example, the question "What will you do if thick and pungent smoke is filling the room?" was answered "I will make my way to the exit" by 62.4% of students. The answer is not complete; the correct full answer is "I will cover my mouth and nose with a wet handkerchief and move forward to the exit crawling low on the floor". To the question "If fire has cut off your way to escape from an apartment on the fifth floor (no phone available), what will you do?", 72.6% of students responded "I'll make a rope from twisted sheets and go down". These actions would be risky for a child of this age. In this situation, it is recommended that students should use wet towels and bedding to plug the open space at the bottom of the door, thereby reducing the passage of smoke, then open the window and call for help. Answering the question "What will you do if oil in the frying pan catches fire while you are cooking?", 57.1% of the children chose the option "I will put out the fire with water". This is a wrong answer. In this situation, it is necessary to cover the frying pan with a wet towel. The level of the development of fire safety skills was found to be medium (61.1% of students) and low (27.8% of students) among students of grade 5. Only 16.7% of 5th graders demonstrated high level of fire safety skills. The same method was used to study the level of fire safety skills among students of grades 5–8, and the average score for each class was calculated. The results are presented in Table 1, and the comparative diagram is shown in Figure 1.

Table 1. Level of fire safety skill development among students of grades 5–8.

	Grade 5	Grade 6	Grade 7	Grade 8
Average score	5.6	5.7	5.3	7.1

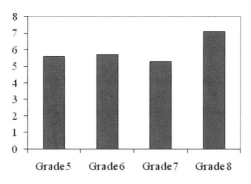

Figure 1. Level of fire safety skill development among students of grades 5–8.

Figure 1 shows that the level of fire safety skills is higher in grade 8 students, which is explained by the fact that students of this age have a clear idea of fire safety requirements after more than a year of having been involved in the educational program aimed at the development of students' fire safety behavior.

On the basis of the testing results, adequate adjustments were made to the fire safety education program. In particular, the following events and activities were carried out: practical training in fire evacuation procedures using fire escape plans; a visit to the fire-fighting department; a lecture "Fire threats"; an extracurricular activity "Fires in residential and public buildings, their causes and consequences. How people should act in case of fire in the building"; a quiz "Fire safety rules"; an extra-curricular event "Fire safety relay game: Dedication to young firefighters"; and an extracurricular event "Do not play with fire!" All of these activities were conducted in the schoolwide format.

5 DISCUSSION AND INTERPRETATION

In addition to extracurricular activities, fire safety was also taught in class when the unit "Fire safety" was studied as part of the subject "Fundamentals of Health and Safety". While planning and delivering these classes, it is necessary to take into account age-specific characteristics of students in order to choose the most appropriate method for fire prevention education.

Working on equipping students of grade 5 with fire safety behavior skills, the teacher should bear in mind that these students are in the high-risk group in relation to fire. When the topic "Causes of fire" is covered in grade 5, it is important that students are aware of the main factors of fire occurrences:

$$FIRE = fuel + oxygen + a\ source\ of\ heat\ or\ fire$$

while covering the topic "Dangerous factors associated with fires," the teacher explains that in the case of a fire, the level of oxygen in the room is sharply reduced, and, therefore, it is necessary to leave the room as soon as possible. During evacuation, one may encounter zero visibility due to heavy smoke and thus should escape the smoke-filled area crawling low to the floor where the level of smoke is lower, with the nose and mouth covered with a wet cloth. The home assignment that students receive is to make, with the help of their parents, a cotton gauze facial mask; the masks are then collected and stored in a classroom closet. When studying the topic "Actions in case of fire and evacuation from a burning building", the students learn that in the case of failure to put out the fire within the first few minutes after it has started, they must leave the premises immediately, inform neighbors of the fire, and call the fire department.

In grade 6, it is recommended that problems of fire safety education be integrated with the topic of tobacco smoking prevention, because it is at this age that children are likely to start this bad habit. The teacher draws students' attention to the fact that smoking is one of the leading causes of fires. The following various situations resulting from careless handling of fire during smoking are considered: a match or a cigarette discarded while still burning; smoking in bed; and children's smoking in attics, sheds, on balconies. Statistical data on smoking-related fires are given together with the information on the inflicted damage.

In grade 7, the unit "Forest and peat fires" is studied. Causes and consequences of forest and peat fires are discussed; statistics are provided to show that lightning strikes and spontaneous combustion of peat cause only 10% of the total number of fires, while the remaining 90% are triggered by humans. The teacher explains why forest and peat fires are called natural disasters. Considering the factors leading to forest and peat fires, findings have stated that the disasters are most often caused by unextinguished camp fires, abandoned cigarette butts and matches, and burning dry grass. The topic "What people should do during forest and peat fires" is also recommended for study.

In grade 8, students are expected to study fire extinguishing means. Encouraging students to come up with ideas, the teacher gives examples of equipment that can be used to put out a fire (crowbars, hoods, axes, shovels, bunches of green tree branches, sand, soil, washing powder, wet blankets, etc.) and briefly describes each of the latter. Different situations related to the use of fire extinguishing means in the case of a fire are discussed (e.g., fire on a TV/New Year tree/oil in the frying pan). An overview of different types of fire extinguishers (carbon dioxide, powder, air-foam, and aerosol) is given, including their general structure, instructions for use, and scope of application. Students learn about each type of fire extinguishers, as well as their principles and rules of operation. It is recommended that a simulation of using a fire extinguisher in practice, with the teachers' feedback to follow, should be organized (N.A. Gorokhova, N.I. Kolesnik, 2004).

The study aimed at students' learning of fire safety rules and acquisition of relevant skills will be successful if the students are fully aware of the need to observe the safety guidelines.

6 CONCLUSION

An important aspect of fire safety awareness work is the development of students' skills of fire safety behavior and ability to respond to emergency situations. This work includes practical training during school fire evacuation drills, excursions to fire-fighting departments, learning to use emergency fire-extinguishing equipment, and so on. This is achieved through a purposeful translation of introductory-level knowledge into skills and abilities.

Education and training should be provided at all stages of children's school life, using age-appropriate psychological and pedagogical methods for developing fire safety and fire prevention skills. Attention should be given to the danger of fire and fire extinguishing by unprepared people, as well as to explaining the consequences of such actions.

On the basis of the testimony and experience of school teachers, the organization of preventive work on fire safety in educational institutions requires extra time and extracurricular activities throughout the academic year, especially at the end of the fourth quarter preceding the summer vacation, when it is extremely important to remind students about basic safety rules, causes and consequences of fire, and things to do during a fire.

Education on fire safety problems should meet the following requirements:

– continuous increase in the complexity of the learning process, starting with the simplest game activities followed by a smooth transition to holistic views on the fire safety system and
– systematic revision of the most significant problems of fire safety while making the consideration process more complex and in-depth moving on from imposing to convincing explanation.

With the purpose of developing fire safety skills among students of grades 5–8, a series of events were held, including thematic classes on fire safety, in which the following teaching methods were used: studying the conditions of burning; explaining the causes of fires; persuading of the need to observe fire safety rules; analyzing cases of wrong actions leading to fire risks; giving the rationale for the algorithm of behavior in case of fire using visual aids, including photo and video materials; practical training through school fire and evacuation drill; case study and problem solving in the form of a quiz game; participation in sports competitions dedicated to the work of firefighters; and visiting a fire department. Children are likely to keep for years their memories of seeing the firefighters' professional body armor and gear, firefighting equipment, and the fire engine, as well as talking to the firefighters. This means that such events raise students' fire safety awareness and encourage adequate behavioral changes.

Thus, we may conclude that purposeful systematic work aimed at the development of fire safety skills is efficient under the condition of ensuring the integration of theoretical knowledge acquired by students in class along with their practical skills that they develop during extracurricular activities.

REFERENCES

Bakhtina V.V., Kuklin E.V. Characteristics of the fire situation in the Republic of Mari El//Bulletin of scientific conferences. 2017. N 2–2 (18). P. 19–21. http://ucom.ru/doc/cn.2017.02.02.pdf.

Fire safety trailer curriculum. Tools for Delivering Fire Safety Education Messages Using a Fire Safety Trailer. – FEMA. U.S. Fire Administration. –371 p.//https://www.usfa.fema.gov/downloads/pdf/publications/fire_safety_trailer_curriculum.pdf.

Methodical recommendations for teaching children how to fire-safe behavior/Comp. N.A. Gorokhova, N.I. Kolesnik/ – Ekaterinburg: GOU DPO IRRO, 2004.– 45 p.

Moshkin, V.N. Decelopment of the safety culture of schoolchildren. – Barnaul: BSPU Publishing House, 2002. – 318 p.

National School Code of Conduct. Ministry of Education St. Clair Port of Spain Republic of Trinidad and Tobago. – May 2009. – 57 p. http://moe.edu.tt/general_pdfs/National_Schools_Code_of_Conduct.pdf.

Sidorkin, V.A. About problems of the formation of a culture of fire-safe behavior of children/V.A . Sidorkin, Yu.V. Prus, A.V. Golyshev//Technologies of technospheric security .– 2012. – № 3 (43). – P .1–8.

Sociology of fire safety/Ed. by Kafidova V.V.– Moscow.: VNIIPO. – 2003. – 362p.

Vorobyev, Yu. L. Basic principles of the formation of culture of life safety of the population/Yu, L. Vorobyev, V.A. Puchkov, R.A. Durnev; Under the Society. Ed. Yu.L. Vorobyov. Russian Emergency Situations Ministry. – Moscow: Delovoi Express, 2006. – 316 p.

Development and enhancement of communication for junior school children with vision impairments

E.Y. Borisova, O.V. Danilova, T.A. Karandaeva & I.B. Kozina
Department of Special Pedagogy and Psychology, Mari State University, Yoshkar-Ola, Russia

ABSTRACT: In this study, we reveal the specific characteristics of communicative development for junior school children with vision impairments. The experimental results for the process of communicative formation based on self-cognition and personal quality development are presented. The correlation and interconditionality of communication defects and personality transformations in children with vision impairments are analyzed. We consider the ways of enhancing communication and the directions of special correctional and developmental work with children, with the aim of facilitating active communication among junior students of educational orphanage, as well as training emotional expression and comprehending other peoples' emotional state.

Keywords: communication, communication skills, adaptation, diagnostic tools, independence, communication, communication skills, social responsibility, junior school children, visually impaired children, correction, self-esteem, social status

1 INTRODUCTION

Development and implementation of various methods and means of communication for different types of human activity result from similar principles of ontogenesis for children with normal vision as well as with vision impairments.

Communication is a complex and multidimensional process of establishment and development of human contacts. It designates human activity and determines fully functional development of human as a personality and as a subject of activity. Problems in communication faced by children with vision impairments were dealt with by L.S. Vygotskiy, V.Z. Deniskina, A.G. Litvak, L.I. Solntseva, P.R. Cox, M.K. Dykes, and Ian Bell, to name a few.

2 STATEMENT OF PROBLEM

Vision impairment adversely affects communicative and personal development. The necessity for enhancing communication for a child with vision impairment is proven by the fact that communication is an effective means of correction and compensation for secondary deviations of mental development resulting from vision impairment. This fact has stimulated us to study the development of communication among children with vision impairment.

Our research was aimed at elaborating the methods and techniques facilitating the development of communication, as well as the expansion and conduct of communicative activities.

The research had been conducted from 2015 to 2017 within the framework of State Budget-funded Institution of General Education "Savinskaya Orphan School" involving grade 4 students aged 10–11 with vision impairments such as amblyopia, strabismus, myopia, and hypermetropia.

The diagnostic period revealed the following findings: developmental delay in children, including underdevelopment of their communicative abilities; a number of specific characteristics of mental process development; and an influence exerted by the time of defects' emergence on communicative development.

3 CONCEPTUAL FRAMEWORK

According to V.Z. Deniskina and A.A. Leontiev, emergence of earlier defects is a significant mental disorder in children. Such disorders impede communicative development, especially nonverbal communication, for example, facial gestures and body language (Deniskina V.Z., 2007).

Communication represents human cognition of other human minds. In the process of communication, people self-identify and reveal their own mental properties and character traits. Communication favors formation and development of human skills, serving as a means to acquire knowledge and experience of cultivated behavior, and represents the basic means of nurture and socialization. Only the process of communication enables shaping communicative experience. These provisions can serve as a methodological base for analysis of the process of fashioning communication among children with vision impairments (Karandaeva T.A., 2013).

4 FINDINGS

The correctional and developmental work was conducted due to the specific characteristics of personal and communicative evolution of children with vision impairments. The goal of this study was to shape the skills of self-sufficient activity, social responsibility, empathy, as well as cognition of self and others. It was also aimed at forming and correcting basic mental processes, as well as pursuing the enhancement of communicative means, self-esteem, and social status.

The study of personal and communicative characteristics of children with vision impairments was conducted using the following techniques:

- "What am I?"
- "Evaluate Yourself"
- "The Choice in Action" or "The Gift"
- S. Rosenzweig's technique of pictorial frustration.

The following are the basic criteria for selecting a technique:

- correspondence with age-related and individual abilities;
- the diagnostic value of identifying specific characteristics belonging to the components of children's communicative activity;
- the abilities of the technique to adapt to the conditions of group work with children with vision impairment.

The adaptation of the techniques included the following focuses: adaptation of impetuses, involving the use of clearer hard images, elimination of insignificant details, and increase of the images to facilitate the process of perception of planar images); individual work with exact description was implemented in the function of the procedure of conducting the adaptation.

The "Choice in Action" technique enabled researching the structure of personal interrelations. During the research, the experimental group had demonstrated low level of advantageous interrelations among children. The number of mutual choices is 1, which constitutes 12.5% of students. One student (12.5%) is isolated. The number of students with low sociometric status constitutes 50%, whereas, in contrast, those having high sociometric status constitute 37.5%. As to those having middle sociometric status, these students constitute 12.5% of the group.

The "Evaluate Yourself" technique contributed to evaluation of students' self-esteem and accentuation of the leading qualities, which constitute the base of communication. Analysis of the examination revealed that 37.5% of students have inflated self-esteem, 37.5% have deflated self-esteem, and 25% have adequate self-esteem.

The "What am I?" technique helped to determine students' self-esteem and confirmed the research results of the "Evaluate Yourself" examination. In the course of the study, the following three groups of the students were revealed:

I – The group with middle level of development and self-esteem (25%);
II – The group with low level of development and self-esteem (37.5%);
III – The group with high level of development and inflated self-esteem (37.5%).

The research revealed the following most significant qualities chosen by the students: kind, polite, and honest.

Rosenzweig's "Pictorial Frustration" technique enabled providing qualitative and quantitative characterization of relationship maintained in various communicative situations by junior students of the specialized orphan school.

The examination demonstrated that extrinsic denunciation is a prevailing direction of reaction in communication for this group of children. The index of self-sufficiency is equal to 0, which indicates dependence as well as lack of self-sufficiency and active social position. The characteristic of reactions that occur during conflicts with parents, other adults, and same-age peers has drawn the interest of researchers. This interest is substantiated by the fact that child's communication is mostly influenced by parents, whereas the group includes children from one-parent families, disadvantaged families, and even those whose parents are deprived of the custody rights. The extrinsic denunciation (extrapunitive) reactions toward same-age peers constitute 55%; those aimed at adults, 52.5%; and those aimed at parents, 37.5%.

5 DISCUSSION AND INTERPRETATION

During this study, the greatest attention was paid to creating an advantageous climate in the group, as well as individual approach to the children. To achieve this goal, the recommendations of S.Y. Benilova, a clinical psychiatrist, were applied (Benilova S.Y., 2006).

The next necessary factor of positive communication is initiation of conditions favoring communication. It suggests arranging various life activities to let every student manifest his/her strengths, individual interests, and hobbies.

The following basic kinds of activity shaping personality and communicative experience were chosen:

- Exercises correcting basic mental processes. In the course of educational work, including walks, unsupervised study, excursions, and classes, the children were given various tasks, which encouraged them to develop attention, observation skills, different memory types, creative thinking, and imagination.
- Free and theme-based drawing. In the course of preparation and exercises, the children were made to draw some kind and wicked characters of fairy tales, a magic mug or a cup of kindness, and the child's mood or his/her friends' mood. These tasks allowed eliminating distortions of emotional response, as well as developing empathy and benevolence.
- Simulation and analysis of given situations and creating problem situations. The exercises offered various situations to be resolved by children individually and they also investigated children's conduct in everyday life. They enabled accumulating various communicative experiences and contributed to activity, communicative initiative, and adequate expression of thoughts.
- Participation in schoolwide events enabled developing qualities such as self-assurance and self-esteem.
- Reading fiction with subsequent discussion and composing stories, enabling teaching children to analyze their deeds, express thoughts, and expand vocabulary.
- Conversations, games, and correctional and developmental exercises. The topics of conversations and exercises were elaborated on the basis of analysis of existing problems among children with vision impairment.

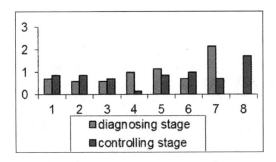

Figure 1. Sociometric status change of students (diagnosing, controlling stages).

– Specialists' support: speech therapist, psychologist, physical therapist, rhythmic gymnastics instructor, and facial mobility specialist. The 35–40 min exercises were conducted three to four times per week and enabled children to retain their experience and knowledge.

During this study, positive results were revealed. Students' self-esteem increased by 50%. The controlling stage of the experiment demonstrated that 12.5% of students have deflated self-esteem; 12.5% have inflated self-esteem, and 75% have adequate self-esteem. As has been demonstrated by the analysis of this study, low sociometric status correlates to self-esteem level: students with inadequate self-esteem have the lowest sociometric status.

Figure 1 shows the dynamics of students' sociometric status.

The verification analysis of reactions on frustration demonstrated that the percentage of extrinsic denunciation reactions decreased from 48.31% to 15.82%. This indicates the fact that students are aimed at optimization of collaborative activity. The social activity of students increased, and the index of self-sufficiency equaled 1, whereas it became 0 at the diagnosing stage. Communication with adults acquired positive characteristics.

6 CONCLUSION

In this study, we revealed that communication of children with vision impairments and development of their personalities are interdependent and require specially organized nurturing and education. Productive communication of children with vision impairments becomes possible when there is advantageous climate in the group and the personality-oriented approach is implemented in education and nurturing. This study accelerated the processes of development and evolution of communicative space in class, as well as the formation of interpersonal relationships. The indicators of position adopted by children in communication approached the optimal one.

REFERENCES

Benilova S.Y. Special children—special communication./S.Y. Benilova//Upbringing and teaching of impaired children. 2006. № 2, p. 77.

Deniskina V.Z. About the classification of visually impaired children and children with secondary deviance./V.Z. Deniskina/All-Russian pedagogical lectures, Moscow, November, 2007. – 17 p.

Karandaeva T.A. Bases of teaching and upbringing of visually impaired children: textbook/ Mari State University; T.A. Karandaeva—Yoshkar-Ola 2013. – 172 p.

Neuropsychological approach to the correction of developmental deviations of preschool children with health limitations

E.Y. Borisova & I.B. Kozina
Department of Special Pedagogy and Psychology, Mari State University, Yoshkar-Ola, Russia

ABSTRACT: This article is devoted to the study of applying neuropsychological diagnostics and correction for providing psychological and pedagogical assistance to preschool children with health limitations. The effectiveness of correctional and developmental work has experimental grounds in favor of the neuropsychological status of the child.

Keywords: neuropsychological diagnostics, psychological and pedagogical assistance, multilingual and monolingual environment, children with disabilities

1 INTRODUCTION

Currently, there is an increasing number of children with mental deviations and disorder variations without acute pathologies, whose learning problems sometimes can hardly be solved with traditional psychological–pedagogical methods involving learning and correctional work (*Lilian de Fátima Dornelas, Neuza Maria de Castro Duarte & Lívia de Castro Magalhaes. 2015*). Because of the increasing number of children with developmental delay; complexity, stability, and variability of the disorders; and the uncertainty of differential diagnostics, the attention is being increasingly focused on the problem of the neuropsychological approach to the children's development assessment (Akhutina T.V. & Pylaeva N.M., 2008; Glozman J.M. 2012, Semenovich A.V. 2002).

In addition to identifying the reasons and mechanisms of children's backwardness and difficulties at school, studying the underdevelopment or delay related with higher mental functions and their structural incompleteness, and revealing the insufficient functionality of certain areas of brain, the neuropsychological approach shows a wide range of opportunities (Akhutina T.V., Vizel T.G., Kamardina I.O., Korsakova N.K., Mikadze Y.V., Polonskaya N.N., Semenovich A.V., Tsvetkova L.S. and others). The key tasks of a neuropsychologist are to reveal the special aspects of deficiency of psychic functions at various stages of ontogenesis, study the dynamics of their development, identify the reasons and mechanisms of the disorders occurred, and develop corresponding preventive and correctional methods (E.G. Simernitskaya, 1991,1995; Y.V. Mikadze, N.K. Korsakova, 1994; T.V. Akhutina and others, 1996; N.K. Korsakova and others, 1997; L.S. Tsvetkova, 1998, 2001; A.A. Tsyganok, 2003; T.V. Akhutina, N.M. Pylaeva, 2003; A.Y. Potanina, A.E. Soboleva, 2004; L. Kiessling, 1990; D. Tupper & K. Cicerone, 1991).

The neuropsychological method focuses on the special aspects of forming higher mental functions which determine the human's conscious activity and behavior during personal development. This method allows us to analyze the reasons and the quality of brain disorders for various types of psychic activity at different stages. It determines a wide range of opportunities for its application in diagnostic, prognostic, correctional-pedagogical, and preventive and advisory practices.

2 STATEMENT OF PROBLEM

At present, the neuropsychological approach is considered to be one of the most appropriate approaches for the diagnostics of mental development of children. Several diagnostic, preventive, and correctional techniques are being continuously developed within the framework of this method, as the complex neuropsychological diagnostics of higher mental functions allows us to reveal weak cognitive mechanisms that can cause the most common learning difficulties and to plan an effective correctional work. A variety of developed diagnostic suites were considered as children-friendly versions from the battery of tests conducted by A.R. Luriya. Among these suites, some focused on both general neuropsychological examination and some versions of instant diagnosis. The first group of techniques, which allow us to diagnose the statement of various mental functions, may include those developed by T.V. Akhutina and others (1996, 2008), J.M. Gloseman (2006), A.V. Semenovich (2002), L.S. Tsvetkova (1997), and other researchers.

The effectiveness of the application of the neuropsychological method is increasingly acknowledged by psychological and pedagogical community working on deviant development problems. In the recent decades, school neuropsychology has been contributing more into the modern pedagogical practice, involving both diagnostic and correctional methods. However, most of the developed programs are for elementary school age only; besides, there are no programs considering the special aspects of development and interaction of brain systems in terms of the child's multilingual acquisition, which is a challenging issue.

Within the framework of the study "The neuropsychological analysis of the structure of psychoverbal pathology of modern children in multilingual regions (in terms of the Mari El Republic)" conducted in the Mari El Republic with support of Russian Foundation for Humanities (project No. 14-16-12004) in 2014–2015, there is an analysis of the results of neuropsychological examination of preschoolers (3–7 years old) with conditionally normative and deviant development who are raised in multilingual and monolingual environments (Borisova E. Y. & Kozina I.B., 2014, 2015). The comparative analysis of the results has allowed us to determine the group of mental functions that have higher values for children with conditionally normative development and lower values for those with various forms of dysontogenesis, as well as the group of functions expressed equally for both samples.

3 CONCEPTUAL FRAMEWORK

As a rule, the special aspects of development and interaction of brain systems in terms of the child's multilingual acquisition are rarely considered for developing programs and techniques. Because of the significant degree of multiculturalism throughout Russia, some percentage of preschoolers are raised in multilingual environment, which determines the importance of considering neuropsychological and psychological mechanisms and the development factors of higher mental functions in terms of linguistic diversity.

All the aforementioned determines the importance of development based on the neuropsychological approach of the technology of psychological–pedagogical and psychological follow-up for preschoolers being raised in multilingual social environment.

The goal of psychological-pedagogical follow-up of children with health limitations is to provide them with social and psychological adaptation in the educational environment as best as possible, to assist with the personal development in various activities with a glance to the type and degree of disorder and personal psychological constitution. The subject matter of the follow-up provided to a child with health limitations is methodologically developed in the works by I.I. Mamaychuk, M.M. Semago, and N.Y. Semago. Most of the authors are of the same opinion that follow-up is a dynamic process that includes the following components: systematic monitoring of the child's clinical–psychological and psychological–pedagogical status in terms of the dynamic of psychological development; creating social–psychological conditions for effective development of children in social environment; systematic psychological assistance provided to the children as consulting, psycho-correction, psychological support; systematic psychological assistance of parents of children with developmental disorders; and the organization of the child's life in social environment considering mental and physical abilities.

3.1 *Research instrument, respondents of the study*

This study focused on analyzing the regulatory and speech indicators' dynamic, which is unique to preschoolers (the fifth year of life) from different nosological categories, who are raised in monolingual and multilingual environments. The dynamic was monitored with psychological–pedagogical follow-up including neuropsychological diagnostics and correction.

The main goal of this study is the development and approbation of psychological follow-up technique on the basis of the neuropsychological approach on preschoolers with health limitations who are raised in multilingual social and cultural environment. The hypothesis of the study represents the idea stating that the correctional and developmental learning program based on the findings of differentiated and system neuropsychological diagnostics can be effective enough..

The tests were conducted on 70 pupils with general speech underdevelopment of the 3rd level and mental developmental delay from the middle groups of preschool facilities. This includes 40 preschoolers from the experimental group and 30 from the control group. The groups were formed in accordance with the findings of psychological–pedagogical and medical commission as well as on the basis of expert evaluation obtained from teachers and logopedists.

Among the participants of the program implementation are teachers-logopedists, educational psychologists, and nursery teachers.

The diagnostic program of our study included particular units of child-friendly battery of tests by A.R. Luriya (T.V. Akhutina and others, J.M. Glosman and others). The evaluation of the values of programming and control unit was conducted in accordance with results of the following tests: dynamic praxis, reciprocal coordination, correction test, composition a story on the series of pictures (from 4 years old), "the odd one out" task, dialogue, naming of objects and acts, composition of phrases, and composition a story on a picture.

The forming stage included the comprehensive correctional and developmental work as a mixture of motor and cognitive correction on individual and group sessions. The correctional and developmental programs were implemented mostly with gaming techniques, including the ones related with the mediation of cognitive processes. The gaming tasks, their number and complexity, were selected personally for each testee, as determined with the neuropsychological status of the child. The expected result of the correctional work was the improvement of the preschoolers' skills to assimilate and execute movement programs, set the main units of meaning of a story in a regular string, and focus on the correction test execution that would reflect in conjunction positive dynamics in the formation of voluntary regulation, programming, and control. The speech status assumes fully-formed, consistent, and correct speech.

4 FINDINGS

In order to evaluate the dynamics of indicators, some intermediate diagnostics for particular tests (dynamic praxis, reciprocal coordination, deferred motor memory, correction test, and composition a story on pictures) was conducted after implementing the first stage of the program.

Improvements in most of the studied regulatory sphere and coherent speech parameters were detected in both experimental and control groups. The number of testees requiring assistance for the task accomplishment in the dynamic praxis test decreased from 78% to 54%. The number of testees accomplishing this task step by step or in batch mode decreased from 81% to 68%, and the number of pupils making no mistakes increased from 6% to 16%. Although the successful accomplishment of the reciprocal coordination test at the primary diagnostics stage was available only for 16%, in the repeated diagnostics, it resulted as 36%. These positive trends expressed to different extents are noted for all analyzed tests that probably reflects the age dynamics of regulatory function formation.

In order to evaluate the significance of differences in the dynamics of studied indicators in the control and experimental groups, a comparison of changes by the Mann–Whitney U criterion was conducted. The statistical analysis allowed us to highlight significant changes in improvements between the groups by parameters such as program execution ($p = 0.036$) and serial organization errors in the deferred motor memory test ($p = 0.008$), productivity in the correction test ($p = 0.043$), and composition a story ($p = 0.003$).

5 DISCUSSION AND INTERPRETATION

The detailed analysis of the structure of changes allowed us to assume that the improvement in the dynamic praxis results (at least for one point) has appeared to be specific to more than half of the children in the experimental group and only one-third of preschoolers in the control group (Table 1).

However, the changes of program acquisition indicators in this test appeared to be almost identical in both groups, that is, about 40% preschoolers did not improve their results. The difference of positive changes in the composition a coherent story is more appreciable: 40% of testees from the experimental group decreased the number of missed units of meaning and unreasonable repeats of linking words, whereas 90% of children in the control group did not improve their results. As a whole, the experimental group more significantly improved the results of the program acquisition, in both movement and speech tasks, although the process of task acquisition was accompanied by some difficulties, and in most cases, there was a necessity to read the instruction again.

Let us now review the changes in speech test parameters. In the "naming objects" test, all children from the experimental group identified the items with no mistakes. In the control group, 20% of testees showed the search of nominations, and the increase of latency period was noted. It is characteristic that both groups do not contain children with multiple paraphasias and perseverations. The analysis of low-frequency word pronunciation test showed that 80% of children from the experimental group did not have any difficulties in pronouncing words of this type. The remaining children showed the search of nominations, the increase of latency period, as well as single paraphasias. There were testees who required the first letter prompt. About 50% of the control group testees showed some difficulties in naming objects, and the rest accomplished successfully. The greatest interest is attracted by the analysis of "Composition a story on the series of pictures" test (Bastrakova O. P. & Kozina I. B. 2017) The children's stories have become more coherent, do not contain the signs of inadequateness anymore, and the plot was retold more correctly. Children from the experimental group who required prompts at the first stage accomplished the task themselves in the next tests. A few of testees of the control group kept on repeating the linking words. Thus, the dynamic

Table 1. Shifts of indicators of dynamic praxis and reciprocal coordination tests at the control diagnostic stage (the percentage of children shows the corresponding shift).

Indicator Group	Program accomplishment (dynamic praxis) 0	1	Serial organization errors (dynamic praxis) 0	1	Program accomplishment (reciprocal coordination) 0	1	2
Experimental	46.3%	53.7%	46.3%	53.7%	34.1%	63.4%	2.4%
Control	65.2%	34.8%	65.2%	34.8%	52.2%	47.8%	0%

indicators of speech development were more significant in the experimental group than in the control group.

6 CONCLUSION

In general, the analysis of the study findings confirms the difference in the quality of correctional and developmental interventions performed in accordance with the child's abilities determined by means of the neuropsychological status. The identified trends require further study, and show the need for providing psychological–pedagogical assistance to preschool children with health limitations by taking into account their neuropsychological status.

ACKNOWLEDGMENT

This study was funded by the RFBR and the Government of the Mari El Republic based on the research project No. 16-16-12004.

REFERENCES

Akhutina, T.V. & Pylaeva, N.M. 2008. Overcoming the difficulties of teaching: a neuropsychological approach. St. Petersburg: Peter.

Bastrakova O. P. & Kozina. I. B. 2017. Features of correction of phonemic processes in children of preschool age with a phonemic dyslalia. Proceedings of the II International Scientific and Practical Conference, Modern tendencies and innovation in the field of humanitarian and social sciences (p. 94–98).Yoshkar-Ola: Mari State University.

Borisova, E. Y. & Kozina, I.B. 2015. Neuropsychological approach to the analysis of peculiarities of mental development of children of preschool age living in the multicultural region Yoshkar-Ola: Mari State University.

Borisova, E. Y. 2015. A neuropsychological approach to the assessment of deviations in the development of regulatory functions in junior preschool age. Psychology of learning, 11, 77–87.

Borisova, E. Y. 2015. Age dynamics of formation regulatory functions in the preschool age: Neuropsychological aspect. Messenger of Mari State University.

Borisova, E. Y., Kozina, I.B. & Chernova, E.P. 2015. Age dynamics of indicators of function block programming and control in preschool children with mental and speech pathology. News of Southern Federal University. Pedagogical sciences. – Rostov-on-Don: «Southern Federal University», 1, 123–127.

Borisova, E. Y., Kozina, I.B. Chernova, E.P. 2014. Features coherent expression of children of multicultural region of the senior preschool age with mental and speech pathology. Messenger of Mari State University, 15, 25–29.

Glozman, J. M. 2012. Neuropsychological screening: a qualitative and quantitative assessment of the data, Moscow: Meaning.

Lílian de Fátima Dornelas, Neuza Maria de Castro Duarte &.Lívia de Castro Magalhaes. 2015. Neuropsychomotor developmental delay: conceptual map, term definitions, uses and limitations. Revista Paulista de Pediatria (English Edition) 33, 88–103.

Kozina, I.B. Borisova, E. Y., & Chernova, E.P. 2014.Some features of speech of children 5–7 years multicultural region. SOCIETY, 3-4, 33–36.

Neuropsychological diagnosis, examination of writing and reading to younger students, Eds. T.V. Akhutina, O.B. Inshakova. 2012. Moscow: V. Sekachev.

Semenovich, A.V. 2002 Neuropsychological diagnosis and correction in children: a Training manual. – Moscow: ACADEMA.

Active and interactive methods for teaching psychological counseling

I.E. Dremina, L.V. Lezhnina, I.A. Kurapova & S.V. Korableva
Department of Developmental Psychology and Education, Mari State University, Yoshkar-Ola, Russia

ABSTRACT: In this study, we focus on active and interactive methods for teaching psychological counseling. We consider future psychologists' competences, skills, and techniques applied to psychological counseling. We also describe the main stages of psychological counseling, aimed at developing skills for confidential conversation and acquiring certain competences necessary for counseling a client. The course consists of four stages and involves various active and interactive teaching methods, which make it possible to master basic skills, as well as techniques and procedures of counseling.

Keywords: students, teaching methods, active methods, interactive methods, psychological counseling, nondirective approach, counseling skills and techniques

1 INTRODUCTION

Currently, because of the transition to the two-level system of higher education and the sharp increase in information flow, the traditional educational pattern is inefficient, and the role of active and interactive methods of training bachelors and masters is becoming crucial. University professors face a difficult problem of developing students' professionally important qualities and competences in a relatively small number of course hours. Theoretical, methodological, and practical bases of using innovative technologies and teaching methods, cardinally different from traditional and familiar methods, have widely been developed recently in the Russian education system (Grudzinskaya, E.Y., Mariko, V.V. 2007; Griffiths, S. 2009).

The goal of this study is to establish the ways of teaching psychological counseling more effectively through active and interactive teaching methods.

We suppose that active and interactive methods for teaching psychological counseling help students to develop basic competences and skills of a counselor, as well as to gain experience in counseling during training.

The following two types of teaching methods have been accepted so far: passive methods and active methods. Passive teaching methods are teacher-centered, where the teacher controls the whole process and students rarely participate. From the viewpoint of modern pedagogical technologies, passive methods (e.g., a lecture) are least efficient; however, these methods have some advantages: it is easy to adapt to them and more information can be presented in a short time. These advantages are significant, and they make a traditional lecture a steady and useful teaching method.

Active teaching methods are student-centered, and they involve cooperation between students and teachers. Interactive teaching methods are the most modern form of active methods, and they focus on students' broader interaction not only with teacher but also with one another. We consider this characteristic feature to be the most important, as it allows to involve not only intellectual processes, but emotional and communicative processes, as well as to double down on deep personal self-development mechanism while learning. In our opinion, it is especially important for training future educational psychologists and it plays a discrete role in the formation of their professional competences (Dorogushina, O.A. 2010; Rutten, A. & Hulme, J. 2013).

Our experience in working with students shows that many of them have a true interest in the work of a psychologist, but at the same time, they are afraid of it especially of its communicative aspects. In this case, active and interactive teaching methods enable students to prepare for true-to-life situations better in their future career, and, to some extent, experience them, succeeding or failing.

Psychologists are responsible for providing psychological counseling. It is the most difficult aspect of their work. Psychological counseling refers to an interaction between a psychologist and a client, it involves application of psychological knowledge and special communication skills for solving clients' problems. Psychological counseling requires particular competences, including the ability to build a relationship of trust and confidence with the client, the ability to feedback, the ability to follow the client, the ability to ask questions properly, the skills to structure the client's information, the use of special techniques (silence, normalization, self-disclosure, confrontation), and others. Active and interactive teaching methods contribute to the full acquisition of these competences.

2 CONCEPTUAL FRAMEWORK

Currently, there are many approaches to psychological counseling, each contributing to its development. The general objective of providing psychological support determines a special purpose in each approach. The psychoanalytic approach (S. Freud) suggests bringing repressed experiences and conflicts into the conscious mind. The behavioral approach (A. Bandura) suggests affecting non-adjustive behavior and teaching clients to behave correctly. The Adlerian approach (A. Adler) suggests transforming life goals and shaping more socially beneficial ones. The rational–emotive approach (Ellis A.) suggests the application of scientific methods for solving behavioral and emotional problems. The existential approach (V. Frankl) suggests responsibilities of the client and helps in searching the meaning of life. Finally, the client-centered approach (C. Rogers) suggests a favorable environment for counseling, providing client's self-enquiry and insights, as well as recognition of factors that hinder personal growth (Kochunas, R. 1999; Rogers, C. 1975).

We adhere to the person-centered, client-centered, nondirective approach, which was proposed by Carl Rogers in the framework of a humanistic therapy and then improved and developed by other researchers (Rogers, C. 1975; Brammer, L.M., Shostrom, E.L. 1982; George, R.L., Cristiani, T.S. 1990; Bugental, J.F.T. 1987).

3 RESULTS

With due regard to this approach, we have developed and implemented a psychological counseling course, which aims at developing skills to conduct a confidential conversation for acquiring various psychological competences necessary for counseling (Lyutova-Roberts, E.K. 2007). The course involves four stages, which are discussed as follows.

In the first stage, students are introduced to the basic principles of psychological counseling. The conditions for successful counseling and the techniques for trust-based relations between the counselor and the client are described. Much time is devoted to the development of students' active listening skills. We ask students if they can attentively listen to a person who shares his/her concerns. Most students usually answer in the affirmative. Then, the students are encouraged to prove it: in pairs they share their life experience for 10 minutes, appearing as the psychologist and the client in turns. Moreover, there will be a student observing the participants' body language (postures, facial expressions, gestures). In 20 minutes, they gather in a circle and discuss their experience playing the roles of the client and the psychologist, whether they were better at speaking or listening, or if they were distracted by their thoughts while listening.

Such exercises may produce the following results: many students find it difficult to listen to the client and to obtain further insight into the client's concern. They usually want to comment on the client's words, give advice or ask a question (questions are not allowed so far). In a word, it was characteristic of students to have some difficulty in perceiving the psychological content of the client's story. However, by the end of the course, the atmosphere in the group changes: each pair seems to create its own space and conducts a quiet conversation, the students are attentive and concentrated.

The second stage is devoted to the development of basic skills and techniques of a psychologist such as the ability to use nonverbal means of communication, give and receive feedback, structure information, follow the client, ask questions, and so on. Students acquire some of them quite easily and quickly (e.g., information structuring, using and monitoring nonverbal means of communication, echo technique, summarizing), while other skills require much time and effort (following a client, paraphrasing, asking efficient questions). Students learn the following main counseling techniques: silence technique, self-disclosure technique, normalization technique, confrontation technique, counsel technique, and others.

In the first and second stages, active and interactive teaching methods, such as discussion of topical issues in groups and in a circle, 10-min pairwork with one student as a psychologist and the other as a client, and exercises aimed at developing new skills, done in pairs or in groups of three or small groups, are used. At the beginning of each lesson, there are warm-up activities that create a positive mood and introduce students to the subject.

The third stage deals with the counseling structure and its staged development and completion. There is not a consistent approach to counseling stages and their number [Rogers, C. 1975; George, R.L., Cristiani, T.S. 1990; Bugental, J.F.T. 1987). We take into account R. Kochunas's approach, who distinguished six main stages (Kochunas, R. 1999). Knowing the stages helps students realize that counseling is a structured process, and students have some difficulty in understanding the process. To help students, the teacher uses his/her counseling notes, which include up to six sessions. While listening to them, students take on the roles of learners and supervisors: they distinguish counseling stages, as well as determine and analyze the techniques.

The fourth stage involves developing psychologist's main competences and more difficult counseling skills. The following active and interactive methods are used:

- Students' pairwork: one of the students is a counselor and the other is a client, who speaks about a problem situation in his/her life, in turns. It takes them 20–30 min (compared with the first two stages, when sessions last 5–10 min). By counseling clients, students realize how deep and profound a 20-min session can be.
- Work in groups of three: the first student is a client, the second one is a counselor, and the third one is a supervisor. There are three short sessions where students change their roles. The supervisor watches the psychologist's work with the client, gives him/her feedback, and comments on his/her work. It takes students 60 min to act as clients, psychologists, and supervisors (20 min for each role).
- Work with junior students. Other students are invited to take part in such sessions. They take on the role of the client and can tell the psychologist about a real problem, or about something that surprised, upset, or made them angry. Such sessions last from half an hour to one hour. Students develop their skills of working with clients in situations close to reality.
- Short counseling sessions in pairs and group supervision are held if any of the students require psychological counseling. Any student can play the role of the psychologist and hold a counseling session. The group observes it and gives feedback (only with the client's permission).
- Work with real clients: the group invites a real client, as a rule, a junior student. One of the students counsels a client. The group observes the process and notes the techniques used during counseling. If the student has any difficulty, the teacher can interfere with the counseling session and help him/her to achieve the task.
- The teacher's master-class: the teacher counsels a real client (clients are usually students of the same group or another one).

These teaching methods together with short lectures, containing a little theory, implemented in practice, contribute to the integration of abstract and theoretical and practice-oriented knowledge, and students' personal contribution to its development transforms information into knowledge, which is important and significant. Students become comprehensive and aware of this knowledge. The session with real students sharing their real problems and the group supervising the process and the teacher or a more advanced student counseling is very important for developing students' competences.

4 CONCLUSIONS

Although active and interactive teaching methods are not offered by every subject, they enable students' interest to be increased in the subject, their independence, and their critical thinking. Consequently, they help students develop their creative and communicative skills and oral speech, and reveal their inner potential.

Therefore, a positive result in teaching psychological counseling through active and interactive teaching methods is achieved by means of the following principles and technologies:

- phased acquisition of basic techniques of counseling and their gradual improvement;
- development of practical counseling skills (the ability to listen, ask questions, paraphrase, summarize, give feedback, inform, etc.);
- development of practical counseling skills, when students work in pairs and counsel each other;
- active discussion of the practical exercise results in counseling, students' feedback;
- opportunity to exchange the roles of clients, counselors, and supervisors;
- opportunity to share personal problems and discuss them with the counselor;
- prolonged counseling sessions from 5 min to 30–40 min;
- analysis of the observed sessions and self-analysis of one's own work as a counselor or a client.

The implemented course helps to master basic counseling skills, develop certain techniques and procedures, broaden the range of students' methods, relieve their failure anxiety, and develop their professional skills, competences, and personal growth.

REFERENCES

Brammer, L.M., Shostrom, E.L. 1982. Therapeutic Psychology: Fundamentals of Counseling and Psychotherapy. 4th Ed. Englewood Cliffs, New Jersey: Prentice-Hall.
Bugental, J.F.T. 1987. The Art of Psycho-therapist. New York: W. W. Norton.
Dorogushina, O.A. 2010. Active methods in teaching core subjects [Russian language]. Proceedings of the 2nd International Conference on Psychology and Practice 2010. Nizhny Novgorod: Nizhny Novgorod State University of Architecture and Civil Engineering.
George, R.L., Cristiani, T.S. 1990. Counseling: Theory and Practice, 3rd Ed. Upper Saddle River, New Jersey: Prentice- Hall.
Griffiths, S. 2009. Teaching and learning in small groups, 3rd Ed. New York: Routledge.
Grudzinskaya, E.Y., Mariko, V.V. 2007. Active teaching methods in higher education institutions. Nizhny Novgorod.
Kochunas, R. 1999. Basic psychological counselling principles. Moscow: Academic Project.
Lyutova-Roberts, E.K. 2007. Training young counselors: making a confidential conversation. Saint-Petersburg: Speech.
Rogers, C. 1975. Empathy: an unappreciated way of being. The Counseling Psychologist, 5, p. 2–10.
Rutten, A., Hulme, J. Learning and teaching in Counselling and Psychotherapy https://www.heacademy.ac.uk/system/files/counselling-report-final.pdf.

Superstitions as part of the frame "pregnancy and birth" in the English, Russian, and Mari languages: Postliminary level

E.E. Fliginskikh, S.L. Yakovleva & K.Y. Badyina
*Chair of Foreign Language Communication, Faculty of Foreign Languages,
Mari State University, Yoshkar-Ola, Russia*

G.N. Semenova
*Chair of Chuvash Philology and Culture, Faculty of Russian and Chuvash Philology and Journalism,
Ulianov Chuvash State University, Russia*

ABSTRACT: This article represents the second part of the study devoted to the frame "pregnancy and birth" in the superstitions of the English, Russian, and Mari languages. Following A. van Gennep, we distinguished three top-level groups: preliminary, "liminary", and postliminary. Preliminary and "liminary" levels are described in the first part of the study. The second part describes the postliminary level. This level is the most expanded as it includes two actors (a mother and a child), but most attention is paid to the child. It includes the following slots: (1) predicting the future life of a child based on different factors (date of birth, weather, child's look, etc.); (2) choosing the name for a child; (3) protecting a mother and a child from evil spirits; and (4) baptismal ceremony. We describe all the aforementioned slots by presenting examples and revealing universal and unique peculiar features of every language and culture under study.

Keywords: superstitions, birth, rites of passage, the English language, the Russian language, the Mari language

1 INTRODUCTION

Birth is the first rite of passage experienced by every person. It is a separation from a neutral asexual world or the world preceding the human society and inclusion into the world divided according to the gender, into a family, or a tribe.

As for a woman, giving birth to a child allows her to become a full-fledged member of a society and a new family, confirm her status as a wife, and give her rights in the house.

Hence, this event has always played a very important role in the society and has always been given a special attention. At the same time, it is the most vulnerable period for both of them as it is believed that they could be attacked by spirits and demons, which has been described in several studies (Lori, 2009; Biddle, 1996; Barber, 2004; Kendall, 2013).

Because of this fact, all folks created several traditions, customs, as well as omens and superstitions connected with it (cultural dimension).

The goal of this study is to investigate, discuss, and compare superstitions related to birth of children in Britain, Russian, and Mari cultures, which are reflected in the corresponding languages and language worldviews. This study will help find universal features and features peculiar to specific language and culture.

2 MATERIALS AND METHODS

Superstitions are paroemias in the form of sentences with prognostic functions. They appeared several years ago and accumulated the experience of previous generations. However, they include irrational belief based on unreasonable arguments on something that is invisible (Fliginskikh, 2014). Superstitions are based on magic, that is, on the possibility to control people's lives by executing special actions and rituals. Superstitions and rituals of the postliminary level connected with them are intended to include a woman and her child into the society.

This linguistic study is based on 54 English, 40 Russian, and 34 Mari superstitions. The limited number of examples is explained by the limited field of research. To select the material for this purpose, we carried out continuous sampling from monolingual and multilingual dictionaries and books describing customs and traditions of the folks (quantitative research methods), which are listed as follows:

1. English superstitions: Black Cats and April Fools by H. Oliver; Strange and Fascinating Superstitions by Cl. De Lys; The Encyclopedia of Superstitions by R. Webster; Multilanguage Dictionary of Superstitions and Omens by D. Puccio; The Story of the World's Most Notorious Superstition by N. Lachenmeyer; Superstitions: 1,013 of the World's Wackiest Myths, Fables & Old Wives Tales by D.J. Murrel; and Oxford Dictionary of Superstitions by I. Opie, M. Tatem.
2. Russian superstitions: Everyday Life of a Noble Class of Pushkin Epoch. Omens and Superstitions by E.V. Lavrentyeva; Superstitions or Prejudices by Yu.V. Shcheglova; Multilanguage Dictionary of Superstitions and Omens by D. Puccio; Black Cat with an Empty Bucket. Folk Omens and Superstitions by E.G. Lebedeva; and Functions of Family Members in the Rites Connected with the Birth of a Child among Russian Citizens of Kazan Volga Region in the Second Half of the XIX Century – The Beginning of the XX Century by N.V. Leshtaeva.
3. Mari superstitions: Ethnography of the Mari people by G.A. Sepeev; Mari Mythology: Ethnographic Reference Book by L.S. Toydybekova; Studies in Cheremis: the Supernatural by Th.A. Sebeok and Fr.J. Ingemann; and Musical Instruments in the Family Life of the Peoples of the Middle Volga Region by E.P. Busygin and V.I. Yakovlev. Because the information about Mari superstitions is very limited and not properly presented in the literature, we conducted field studies in the villages of the Mari El Republic questioning local residents (interviews as qualitative research methods).

The analysis presents data from two viewpoints: statistical analysis with data presented in figures, and qualitative analysis of superstitions of three languages in the form of sentences that are grouped according to their meaning.

3 RESULTS AND DISCUSSION

During this study, we singled out four slots in the postliminary level: (1) predicting the life of a child based on different factors (date of birth, weather, child's look, etc.); (2) choosing the name for a child; (3) protecting a mother and a child from evil spirits; and (4) baptismal ceremony.

Slot 1. Predicting the future life of a child
(ENG 24, RUS 22, MARI 10)

Subslot 1. Predictions based on physical appearance
Superstitions with predictions based on physical appearance contain the names of different parts of the body of a child as the main component (*caul* and *umbilical cord* (as tissues connecting a mother and a child), *speckles, veins, teeth, face, hair*, as well as *resembling* and *cry* that are also connected with the body).

The common superstition for all languages contains component *caul*, having the same meaning: *It was considered lucky to be born with the caul*. It ensured protection against evil forces for the baby's entire life. Such babies were also thought to have special psychic and prophetic powers. Cauls were often used as amulets against drowning and shipwrecks.

It is necessary to mention that A seventh child, particularly a seventh son, was thought to have special powers and the ability to see ghosts as well (ENG).

Superstitions with component umbilical cord were found in the Russian and Mari languages. In Mari, it predicts the duration of a child's life: *If the umbilical cord dries out for 5–6 days, the child will live a long life; the less number of days reduces life.* Both Russian and Mari parents wanted to destine child's future by doing special rituals: *Boy's cord was cut on the ax; girl's cord was cut on the spinning wheel to make them good workers* (RUS). *Children's cords were kept tied all together to make them live in friendship* (MARI). These superstitions reveal values specific for the culture: good workers in Russian, family relation in Mari.

Russian superstitions include components *speckles, veins,* English and Russian – *teeth. A speckle is a sign that a baby will live in poverty. Veins* and *teeth* in both languages have negative connotation. Veins are connected with death because of water: *To be born with blue veins on the side of the nose means death will come by drowning.* Teeth are just the bad omen or they could mean death: *If the first tooth appears in the lower jaw, the child will die in infancy* (ENG). *To be born with teeth is a bad omen* (RUS).

Components *face, hair,* and *resembling* are involved in Russian superstitions. A baby's future depends on the process of birth: *If it is born with its face down, it will die soon; if it is born with the face up, it will live a long life.* Hair always means wealth and happiness: *babies born with long hair will be rich and happy.* Resembling with the parents is a sign of happiness as well: *If the son resembles the mother, and the daughter resembles the father, they will be happy.*

Component *cry* exist in Mari superstition: *Newborn's frequent cry means fast death, rare cry means long life.*

Subslot 2. Predictions based on the date, day, and time of birth
This subslot includes superstitions with the components naming months, dates, lunar phases, days of the week, holidays, parts of the day, and sharp time. There are several English superstitions in this group, which show the great meaning of the time of birth in British culture.

English people based their predictions on the date of birth (good dates are the first day of the month, February 29; bad days are Friday 13, December 28), holiday (on the one hand, Christmas is good – it means talented child; on the other hand, it is bad – it means just 33 years of life before death like Jesus Christ; Halloween as the day of spirits predicted the powers of communication with the dead), day of the week (according to the English poem, the worst day is Saturday as the child shall work hard for living), part of the day (it was luckier to be born during the day, but not night), and time ("chime" hours were thought to be unlucky).

Russian superstitions include the names of holidays (New Year and Christmas are good for birth), seasons and months (winter means happiness, but July and September mean absence of harmony in the family and bad health), and moon cycle (new moon means long life).

In the Mari language, there are two superstitions containing the parts of the day as components: *Birth in the morning gives hard life, birth in the evening gives easy life.*

Subslot 3. Predictions based on nature description
This subslot contains Russian and one Mari superstition, which show the meaning of nature in the life of these peoples.

Compass points predict different spheres of life: *To be born with the head to the North means good health, to the West – long life, to the East – bad luck, and to the South – bad health.*

Nature predicts future life as well: *Bad weather and strong wind predict hard life, while polar light was a good omen.*

Both the Russian and Mari superstitions contain the fauna images as components: *If the nightingale is singing near the window at the birth time, child will study music* (RUS), *whereas dog's howling or raven's note were bad omens* (MARI). These images are bad omens in different cultures (Fliginskikh et al., 2016c).

Subslot 4. Predestination of future
There were special rituals that could predestine the future of a baby.

In the British culture, *it is very important for an infant to go up in the world before it goes down* that is why it was lifted upstairs.

In the Russian culture, a good sign was *dough on a newborn's body* (meant happiness and wealth), *a piece of baked apple in the mouth* (it will not drink alcohol in future), *a present on the day of birth or the next day* (it will be the beloved of all).

In the Mari culture, *placenta burial meant healthy life of a child and mother's fertility*.

It is interesting to note similar superstitions in the Russian and Mari languages with the same components: *sheepskin, father's shirt*: *A newborn was put on the pulled out sheepskin* (to make it happy, healthy, and wealthy as fur was a symbol of wealth). *A newborn was wrapped into fathers' shirt* (Mari superstition specifies to make it a favorite child).

Slot 2. Choosing the name
(ENG 6, RUS 5, MARI 11)

Choosing the name is a special ritual. As the analysis shows, there are several Mari superstitions in this slot. It tells about the special meaning of the name in the life of Mari people.

In all three cultures, there was a fear of bad spirits, fairies, and evil eye, and hence the baby's name was kept in secret or it was sometimes given two names: before and after the baptismal ceremony. Mari pagans had a special ritual of "selling" if the baby falls sick very often. *The mother went to the street and "sold" her baby to a woman, and then she "bought" the baby back through the window and gave it another name very often.* It was done to deceive evil spirits.

In the British and Russian cultures, *it was a bad omen to name a child after the dead relatives*, especially if their death was not natural. In the Russian culture, *it was a bad sign to name a girl after a mother or mother's mother, but a good sign to name a son after a father*.

In the Mari culture, it is quite the opposite; they tried to keep sounds from parents' names in the names of children. And there was a special ritual of naming when they pronounced the names of forefathers: when the baby cried, it was its name.

In the British culture, it was a bad omen to give the name of a pet, while Mari had names derived from the names of animals, plants, or connected with snow and rain (if it was snowy and rainy during birth time).

English people named their children after some celebrities to give them their luck, or after saints to get their protection. Mari people named their children after Mari heroes.

There are interesting rituals in the Russian and Mari cultures connected with giving the names.

In Russian, they went to the street after the baby was born and asked the first person they met to say his/her name. And they name the child after that name.

In the Mari culture, it is interesting to note three rituals. The first was connected with baking loaves. Three loaves were given the names before baking. The best loaf gave the name to a baby. Other two rituals were performed by a pagan priest – kart. He stroke fire from the firestone or picked up a baby in his arms when that was crying and pronounced different names. When the fire was stroked or the baby stopped crying, it was its name.

To conclude, in different cultures, there are different rituals to choose a name for a baby.

Slot 3. Protection from evil spirits
(ENG 12, RUS 3, MARI 10)

Forty days after the birth were the most dangerous for both the mother and the baby. There is a hypothesis that these days were given for the soul to settle in a baby's body. These days were the days when the mother could not go to the church as well. It was believed that all evil spirits gather around the mother and the baby. That is why, there are several rituals and superstitions for protecting them.

First, it is necessary to mention that *babies were often welcomed into the world by being spat at three times as protection against evil spirits* (ENG). Component *three* is often present in the superstitions as it is one of the sacred figures of many peoples. It contains three worlds (upper, intermediate, and lower), three time parallels (past, present, and future), and in Christianity it is Trinity and unity of human body, soul, and spirit (Fliginskikh, 2016a)

It is interesting to note the opposed superstitions in the English and Russian languages: *Weighing or measuring babies was thought to tempt the Fates to do something horrible* (ENG) while *they painted an icon of the size of a baby and depicted four Christmases to give a baby a long life* (RUS). Here, we can see that the English concealed intentionally the size and the weight of the baby as they thought the more knowledge that fairies, elves, and witches had of human affairs, the more likely they would interfere while the Russians sized the baby up on purpose.

In the Mari culture, it is necessary to note several rituals intended to protect a mother and a baby after birth: *The mother got a piece of brown bread and salt. The mother and the baby were washed in banya (washhouse) during three days* (here, we meet component three again). *The midwife put hot ash, silver coins, corns, and nails into the water* (silver coins is a symbol of health and wealth, as well as a good protective metal from evil spirits, corns is a symbol of prosperity, nails were made of iron – one more protective metal from bad spirits). *The baby was sweated with mountain ash switches* to protect it from different deceases and evil spirits as mountain ash was thought to be a magic tree.

In English culture, the washing process is mostly connected with luck, health, and beauty, but not with protection: *They did not wash baby's right hand for the first three days not to wash away its luck. Especially after a baby's first bath, the water would have to be carefully poured over a tree in leaf to help the child grow healthy. If the water was poured over a tree at the time of flowering, it was considered particularly good for the child who would grow to be exceptionally beautiful.*

To protect a baby parents took different measures. In the English culture, they tied red ribbon, thread, or wool on baby's cradle (red color had an apotropaic/protective function in different cultures (Fliginskikh, 2016b)), put silver or iron amulets, knives, or necklace made of wolf teeth inside the cradle. Brown bread in the cradle meant health.

In the Russian culture, it was advised to put a veil on the cradle to avoid evil eyes.

In the Mari culture, they put scissors, knife, gunpowder, brown bread into the cradle, sewn red biggins for babies.

As for the clothes, both English and Russians thought that *baby clothes must not be left out to dry after dusk* as bad spirits could settle on them. In the British culture, *when babies were dressed for the first time, it was important that they were dressed only in used clothes and it was common to wrap an infant in clothes belonging to its parents, especially the mother* as new clothes would attract evil spirits and witches and contact with an extension of the mother's body would protect the child. At the same time, it was a tradition *to dress little boys in girls' clothes and to let them grow beautiful long curls, as well as to dress them in blue – the color of the Virgin Mary* so as to confuse devil and avert the evil eye. Girls were inferior to boys and they were thus less likely to be attacked by demons and witches.

The Mari sewn bright ribbons and silver coins on the clothes (silver coins were very popular among Mari people and were present on clothes of all women as well).

We can mention metal things in one more Mari superstition: *First food serving devoted to the birth was accompanied by the noise of dampers, scissors, and other metal articles.* It was done to drive bad spirits out of the house.

In the British culture, *until the baby was baptized, it was important for all visitors to the household to make a toast and to eat something to protect the health of the child.* Especially, if it was a girl, not to carry her beauty away.

In the Mari culture, it was bad to leave a mother or a baby alone. If they had to, they gave her an iron piece.

Therefore, we can find the same popular components in the superstitions of all three languages as protective sings: *silver, iron, and red color*.

Slot 4. Baptismal ceremony
(ENG 10, RUS 10, MARI 3)

The lack of Mari superstitions connected with baptismal ceremony is explained by the fact that the Mari are pagan people, and they were baptized a lot later. Nowadays, there are still pagans in the Mari El Republic on the territory of Russia.

First, it is necessary to mention that in the British culture it was thought to be good to *baptize a baby on the same day of the week it was born*, and during double ceremony, *the boy was baptized first or he will not have a beard*.

Other superstitions are connected with the godparents. In the British culture, *the godparents should come from three different parishes if you want your children to attain a long age*. For the Russian culture, it was important that *godparents should not have love or any relations*. And after the ceremony, *they shall eat buckwheat porridge in order the baby not to be spotted* and *try everything on the* table *to make it rich*. In both cultures, it was thought *to be dangerous to have a pregnant woman as a godmother*: it could kill an unborn or both unborn and born babies.

As for the clothes, both in English and Russian cultures, it should be *white*, but in English, *they should not be new* not to attract evil eye, and in Russian, *it was considered to be good to baptize all children in the same clothes to make them friendly*.

The baptismal ceremony straight after a funeral was considered unlucky, but very lucky after wedding (ENG).

English superstitions note lucky signs during baptismal ceremony: *It was considered lucky for a child to cry during the baptism ceremony, especially at the time of receiving the holy water on their heads. It was advised to let the holy water dry naturally on the baby's head and not to dry it so as to allow it fully to permeate the little one's body. Wetting the eyes of the baby with holy water would protect it from seeing the apparitions of the dead later in life.* Russian superstition notes an unlucky sign: *If wax with baby's hair is drowned in the baptismal font, the baby will die soon.*

As for the Mari superstitions, they are mostly connected with the time after the ceremony. First, *it was necessary to go back home by different road*. Then at home, *the baby was lifted up to grow high, brought to all four corners of the house and around the* table, *wrapped into the sheep skin to make it a strong host*.

It is interesting to mention the role of money after this ceremony. In the Russian culture, it was good to *bring a golden coin for a baby* and *a silver spoon put into the dough* means wealth. In the Mari culture, *the guests were given the right to look at a newborn only after they presented a silver coin*. Usually, it was a man who wished wealth, happy future, clear, and shiny life as silver.

Therefore, we can see that parents should take into account many factors for the baptismal ceremony.

In this study, we distinguished four big groups (slots) of superstitions connected with the afterbirth period. The first group predicts the future of a baby based on different factors. The second group deals with naming a baby. The third group gives pieces of advice to protect a mother and a baby from evil spirits. The fourth group is connected with baptismal ceremony.

4 CONCLUSION

It should be noted that the sphere of superstitions presents a multidisciplinary study of different sciences, especially linguistics and folklore studies because superstitions reflect culture through language and show the worldview of people distinguishing universal and unique characteristics.

We showed that in three languages under study, there are definite similarities in superstitions connected with the birth of a baby. We could underline the universal nature of superstitions mostly connected with protection of a mother and a baby. As it was mentioned, fear of evil eye, bad spirits, and fairies were common for all the three cultures as well as for many other cultures in the world. However, at the same time, there are unique peculiar superstitions that deal with different specific rituals. They were found in different slots, but the most unique ones are those connected with naming the babies.

REFERENCES

Barber, G.D., 2004. Giving Birth in Rural Malawi: Perceptions, Power and Decision Making in a Matrilineal Community. Thesis ... of Doctor of Philosophy in Social Anthropology. Goldsmiths College, University of London.

Biddle, J.M., 1996. The Blessingway: A Woman's Birth Ritual. Thesis ... of Master of Arts in Interdisciplinary Studies in Speech Communication. Speech Communication. and Anthropology.

Busygin, E.P., Yakovlev, V.I., 1990. Musical Instruments in the Family Life of the Peoples of the Middle Volga Region (end of XIX – beginning of XX). Family Ritualism of the People of the Middle Volga Region (p. 3–28). Kazan: Kazan University.

Cultural Dimension of Pregnancy, Birth and Post-natal care. https://www.health.qld.gov.au/__data/assets/pdf_file/0035/158669/14 mcsr-pregnancy.pdf.

Fliginskikh, E.. 2014. Towards a Definition of a Term 'Superstition'. Vestnik Chuvashskogo Universiteta (p. 153–157). Cheboksary: Ulyanov Chuvash State University.

Fliginskikh, E.. 2016a. The Comparative Characteristics of Colour Naming in the Ritual superstitions of the Russian, English, and Mari Languages. Innovative Technologies in Teaching Foreign Languages: from Theory to Practice (p. 110–114). Yoshkar-Ola: Mari State University.

Fliginskikh, E.. 2016b. Numerical Symbolism in the Superstitions of the Rites of Passage of Peoples from Different Language Families: Russian, English, and Mari Peoples. Vestnik Bashkirskogo Universiteta (p. 211–216). Ufa: Bashkir State University.

Fliginskikh, E.. Yakovleva, S., Kudryavtseva, R., Badyina, K., Akeldina, S., 2016c. Linguistic Analysis of Fauna Images in the Ritual Superstitions of the English, Russian and Mari Languages. The Social Sciences (p. 1634–1640). Medwell Journals.

Kendall, E., 2013. Fa'amatala Lau Tala: Samoan Pregnancy and Childbirth Narratives. Independent Study Project (ISP) Collection. Paper 1563.

Lachenmeyer, N., 2006. The Story of the World's Most Notorious Superstition. Moscow, 212 p.

Lavrentyeva, E.V., 2006. Everyday life of a Noble Class of Pushkin Epoch. Omens and Superstitions. Moscow, 516 p.

Lebedeva, E.G., 2007. Black Cat with an Empty Bucket. Folk Omens and Superstitions. Moscow, 543 p.

Leshtaeva, N.V., 1990. Functions of Family Members in the Rites Connected with the Birth of a Child among Russian Citizens of Kazan Volga Region in the Second Half of the XIX Century – the Beginning of the XX Century. Family Ritualism of the People of the Middle Volga Region (p. 83–103). Kazan: Kazan University.

Lori, J.R., 2009. Cultural Childbirth Practices, Beliefs and Traditions in Liberia. A Dissertation ... of Doctor, the University of Arizona.

Lys, Cl. De, 2005. 8,414 Strange and Fascinating Superstitions. Northfield, 494 p.

Murrel, D.J., 2008. Superstitions: 1,013 of the World's Wackiest Myths, Fables & Old Wives Tales. Plesantville, NY, 256 p.

Oliver, H., 2006. Black Cats and April Fools. London, 306 p.

Opie, I., Tatem, M., 2005. Oxford Dictionary of Superstitions. Oxford University Press, 494 p.

Puccio, D., 2013. Multilanguage Dictionary of Superstitions and Omens, 384 p.

Sebeok, Th.A., Ingemann, Fr.J., 1956. Studies in Cheremis: the Supernatural. New York, 357 p.

Sepeev, G.A., 2001. Ethnography of the Mari people. Yoshkar-Ol, a 184 p.

Shcheglova, Yu.V., 2007. Superstitions or Prejudices. Rostov-on-Don, 234 p.

Toydybekova, L.S., 2007. Mari Mythology: Ethnographic Reference Book. Yoshkar-Ola, 312 p.

Webster, R., 2009. The Encyclopedia of Superstitions. St. Petersburg, 352 p.

Functional problems of the tourism development institutions

G.F. Ruchkina, M.V. Demchenko, N. Rouiller & E.L. Vengerovskiy
Financial University under the Government of the Russian Federation, Moscow, Russia

ABSTRACT: The Article explores the sources of formation of assets owned by development institutions that are aimed at the advancement of the tourism industry. It examines the problems of development institutions corporatization and public funds allocation. It puts forward the measures coming from the changes in the Russian laws to ensure that financial resources of development institutions are generated and used in a transparent way. This paper contains recommendations for further reform of the public funding system by imposing legislative requirements for reporting on the effectiveness and appropriate use of budget spendings.

1 INTRODUCTION

The concept of the Russian Federation's long-term socio-economic development appeals to enhancing of the investment potential and development of innovation-driven economy. The mentioned benchmarks are impossible to be implemented by leveraging purely administrative measures in terms of current changes: sanction wars, geopolitical tensions and other adverse events. The problem lies in the point that investment attractiveness is to be improved together with the private sector operating in the market and making decisions based on whether it is possible to obtain economic benefits. Within this framework, special importance is attached to institutional transformations, generation and evolution of the mechanisms to address market economy issues at the intersection of interests and strengthening of partnership relations between the state and businesses that still maintain really incompatible positions.

Given the situation, a key direction of institutional transformations is generation and further enhancement of development institutions that act as the organizational and economic entities and encourage the allocation of financial funds, fund-raising for high-priority projects, generation of new technologies and improvement of the Russian Federation's competitiveness under such tough circumstances.

The Russian legislation does not reflect the status of the development institution as a specific business entity with a set of certain functions, duties and powers. This greatly complicates an objective assessment of their operational efficiency, comparison of their performance indicators and providing of recommendations for further development. Harmonization of the requirements for the development institutions operating in the Russian Federation should ensure their establishment as a form of business entity which, under the Civil Code of the Russian Federation, applies to non-profit organizations that do not pursue profit-making as its main goal. The formation of such an entity is to be accompanied by the corresponding legal documentation that gives an entity the status of the development institution. Development institutions operating pursuant to uniform legal requirements and the charter designed in accordance with these requirements and specifying all the conditions of its activities arrangement must become an essential condition for their formation.

We would like to highlight open joint-stock company special economic zones (fez Oscar) among the development institutions operating directly in the field of tourism. This entity was established as a participant in the creation of the tourist cluster in the north Caucasian federal district, Krasnodar krai and the republic of Adygea.

In western countries, it is possible to transform a development institution into a joint-stock company only if all set goals and objectives are met. The authors believe this legislative decision to be extremely reasonable as after the set objectives are met and a joint-stock company is formed, reorganized entities are deprived of state funding and continue to operate in the new context.

Up to this time, the Russian Federation has witnessed none of its previously established development institutions stop its operation as it reached its goals. However, a range of development institutions successfully completed corporatization, while the concerned special economic zones was originally established as a joint-stock company.

2 MATERIALS AND METHODS

The methodology of the research is based on the application of the dialectical method of cognition, which allows us to study objective economic laws and patterns in their interrelations and interdependencies. Theoretical constructions in the article are considered using general scientific and particular methods of selective research. The factual base is based on the legislative and regulatory legal acts of the Russian Federation, statistical and analytical reports on the research topic.

We believe that commercial entities whose activities are aimed at profit-making should not gain the status of a development institution. Moreover, corporatization of state entities, without transferring them to private business, will result in monopolization of the market, sluggish competition and adversely affect the investment climate.

The authors believe that the political tendency to transform development institutions into joint stock companies while preserving their status puts the concept of a development institution at risk of commercialization. In this context, the state capital will weaken competition and push out private business, rather than stimulate innovation and attract private investment in the tourism industry, inter alia.

In the Russian Federation, transferring of the functions of the development institution to the private sector after the former has fulfilled its key mission is complicated by the fact that there is still no unified approach to evaluate the productivity and efficiency of development institutions. Scientific papers provide a really piecewise coverage of this topic and suggest no proven techniques.

In light of the above, we propose to approve the method of efficiency estimation for the activities of development institutions and allow current institutions to switch to joint-stock companies in terms of their business form only if they achieve their goals.

3 RESULT AND DISCUSSION

Corporatization should be accompanied by deprivation of the status of a development institution and respective budget funding. We believe that development institutions should generate no commercial results. They should not support personal interests of their management or implement projects that are more attractive in terms of profitability. We propose to reorganize development institutions operating as joint-stock companies into public companies in order to prevent the Russian legal framework from offering development institutions an opportunity to function with the aim of profit-making.

Apart from the problems associated with corporatization of development institutions and approval of business entity forms, the Russian legal field experiences the problem of their funding arrangement. The Russian legislation specifies that the sources of formation of assets owned by development institutions in the area of tourism may be money, personal belongings, shares, government and municipal bonds, exclusive or other intellectual property rights, etc. The federal law "on joint stock companies" contains no individual provisions on the sources of assets formation. At the same time, this law contains no restrictions. Therefore, funds can come to a joint-stock company from any non-prohibited sources, including the budget system, donators, financial market transactions, owners, income-generating activities, etc.

However, the practice has shown that the greatest proportion of development institutions funding is formed by the budgetary system of the Russian Federation. This study found that planning of budgetary funds allocation in order to finance open joint stock company special economic zones lacks transparency and a unified approach to the planning procedures. Also, there is no explanation for the needed amount of budget investments which contribute to the authorized capital of development institutions.

The data presented in the report of the Russian Federation's accounts chamber speaks for the unreasonable increase of the fez Oscar authorized capital in 2016 in the amount of 5.2 billion rubles, of which 476.8 million rubles were transferred to Lipetsk fez, 500 million rubles to lotus fez (astrakhan region), 3.3 billion rubles to Kaluga fez and 1 billion rubles to northern Caucasus resorts jsc.

According to the conclusion of the same agency, the federal budget is to allocate 13.1 billion rubles from 2017 to 2019 in order to establish fez Oscar and increase the authorized capital of northern Caucasus resorts jsc. However, 6.7 billion rubles of the federal funds on the corporate accounts were not used as of January 1, 2017.

While studying available materials provided by development institutions and legal documents, the authors concluded there was no economic explanation of the needed asset contributions to sustain them in certain areas.

The amount of allocated budget funds is determined based on plans and applications submitted by development institutions. Moreover, after budget allocation, the funds received shall be placed on the financial market and aimed at ensuring profitability of development institutions by using state financial resources.

To resolve the identified violation, we propose to further reform the public funding system of development institutions by imposing legislative requirements for reporting on the effectiveness and appropriate use of budget spendings.

Also, we offer to ensure legal regulation of the requirements that enable the transparency of constituent documents, annual accounting and financial reports, information on the activities of branches and representative offices for all organizations with state participation, apart from those representing a state secret.

In this regard, article 32 of the federal law "on non-profit organizations" should be supplemented by the following text: "non-profit organizations that form their assets by using budget funds are to ensure transparency and availability of the documents specified in paragraph 3.3. Article 32."

These changes will coordinate strategic documents of the Russian Federation and will comply with the provisions of subprogram no. 3 "ensuring openness and transparency of public finance management", state program "public finance management and financial markets regulation" and the basic principles of budgetary system construction.

4 CONCLUSION

The analysis of the development institutions aimed at the development of the tourism industry in the Russian Federation showed the absence of a single document that allows to determine the features of their functioning and the organizational and legal form of the institution. In addition, in the Russian legal field there are no regulations on the provision of publicly available information on the activities of development institutions receiving budgetary funds.

Proposals formulated in the article by a team of authors aimed at eliminating the above-mentioned problems, with their successful implementation into practice, will improve the effectiveness of the development institutions and the effectiveness of using the budgetary resources of the budgetary system of the Russian Federation.

REFERENCES

Budget code of the Russian Federation dated July 31, 1998 no. 145-z.
Civil code of the Russian Federation dated November 30, 1994 no. 51-z.
Kudelich M.I., 2016. Improvement of legal regulation of development institutions activities in the Russian Federation. Financial magazine (finansovy zhurnal) 2: 83.
Melnikov R.M., 2016. Productivity and efficiency of the Russian Financial development institutions: assessment approaches and ways to improve.
Evaluation of performance efficiency of Russian and foreign development institutions. Edited by I.N. Rykova. Scientific report. Moscow: Scientific and Research Financial Institution. Industrial Economics Center.
Decree of the Russian government dated April 15, 2014 no. 320 "On Approval of the Russian State Program 'Public Finance Management and Financial Markets Regulation'".
Federal law dated December 26, 1995 no. 208-z "On Joint-Stock Companies". http://сп.рф/press_center/news/29969 – Report of the Russian Accounts Chamber collegium on the results of expert and analytical event "Efficiency Analysis of Special Economic Zones in 2016".

Formation of students' readiness for safe behavior in dangerous situations

M.N. Gavrilova, O.V. Polozova, S.A. Mukhina & I.S. Zimina
Department of Biomedical Sciences and Life Safety, Faculty of Physical Culture, Sport and Tourism, Mari State University, Yoshkar-Ola, Russia

ABSTRACT: The formation of safe behavior is generally the process of forming readiness for various types of activities (cognitive, physical, communicative, etc.), performing various social functions (citizen, producer, consumer, etc.), and appropriating various types and fragments of culture (ideological, moral, aesthetic, etc.). This study was conducted to investigate the formation of students' readiness for safe behavior in dangerous situations. Modern schools must pay special attention to the development of safe-type behavior personality using various means and methods. Each individual must be aware of the nature of occurrence and development of dangerous situations; know their strengths and capabilities to overcome the danger; and be able to make a correct diagnosis of the situation. The general purpose of the safe-type personality development is to reduce itself to acquisition of skills and abilities, allowing people to properly shape their behavior and thus reduce the level of outgoing threats, as well as prevent the danger surrounding them in the modern world. This experimental study included a total of 129 students studying in 5th–9th grades of the comprehensive schools of the Mari El Republic (5th grade – 21students; 6th grade – 17 students; 7th grade – 23 students; 8th grade – 44 students; and 9th grade – 24 students).

Keywords: safe behavior, types of personality response in various situations, knowledge of health and safety fundamentals, secondary education

1 INTRODUCTION

Currently, crisis phenomena are related to different spheres of people's life, which contributes to the development of insecure feeling in both children and adults. Safety assurance has become an integral part of each individual's life. In this context, a concept of safe-type behavior personality and its development is one of the key methods for teaching health and safety fundamentals. Safe behavior in everyday life must be natural and seamless. This is because most accidents do not originate from natural disasters: children and adolescents die in road and fire accidents as well as become victims of crimes. Therefore, it is very important and even vital for schoolchildren to acquire knowledge, abilities, and skills of safe behavior (Ahmadullin, 2016).

2 STATEMENT OF PROBLEM

A significant role in the formation of safe-type behavior personality is played by the school course "Fundamentals of health and safety", which is aimed at achieving the following goals:

1. Acquisition of knowledge about an individual's safe behavior in dangerous and emergency situations of natural, technogenic, and social origin; about health and a healthy lifestyle; about the state system of population protection from dangerous and emergency situations; and about citizens' duties to defend the state.
2. Development of a sense of responsibility for personal safety.

3. Development of personality's qualities necessary for living a healthy lifestyle.
4. Acquisition of abilities to foresee potential dangers and act in the right way upon their occurrence.

The syllabus of this school subject provides for the formation of students' general academic abilities and skills, universal ways of activity, and readiness for safe behavior in their life (Ayzman, 2012). Therefore, the study of pedagogical conditions of development of safe behavior skills within the scope of health and safety course becomes a vital task.

3 CONCEPTUAL FRAMEWORK

To develop safe-type behavior personality, students must completely understand the rules of safe behavior at a reproductive level. In modern schools, the course of health and safety fundamentals is not always an object of close attention, especially in final grades, as it is not a major and is not on the list of exams to be taken when entering universities. The low level of students' competence in the field of health and safety suggests the necessity of more effective organization of the process of forming adolescents' readiness for safe behavior, and use of new methods to select interested students. Despite the fact that we do not need this knowledge when entering a university, it is vital in dangerous life situations and can help save human life and people around us.

Thus, students leave school with a wealth of theoretical knowledge of various school subjects, but they do not have skills of safe behavior in a social, technospheric, and natural environment. Pedagogical theory is searching for methods and means of students' preparation for safe behavior. The research issue is prioritized in the education system of the Russian Federation. Work syllabi, practice books, course schedules and lesson plans, and methods of teaching health and safety fundamentals (Kazin et al., 2013) are developed.

It is impossible to assure absolute safety from various external and internal threats: there is always some risk. Risks can be managed, which is assured by goal-oriented activity and safe behavior.

Safe behavior is behavior that assures safety of personality existence and does not do harm to people around them.

Thus, safe behavior is a goal-oriented system of actions performed in sequence, which bring about a relatively safe contact of an individual with surrounding conditions, ensure satisfaction of vital interests, and achieve significant goals.

Nowadays, the state, society, and education system pay special attention to childhood safety and safe behavior of a child. An integral part of the education system development strategy is the objectives of children's health preservation and formation of children's healthy and safe lifestyle, responsible behavior with respect to their life and health.

Education plays a key role in the assurance of health and the safety of a personality, in particular, of a child. The new Federal State Educational Standards (FSES) of basic education introduced the term "safe lifestyle" and provided a detailed description of a safe lifestyle grace during the learning process—from execution of rules to the formation of attitudes and values.

The Federal law "On education in the Russian Federation" states the need for developing students' culture of a healthy and safe lifestyle.

The development of students' systematic knowledge, skills, and abilities will allow them becoming fully aware of health and safety on the basis of a differentiated approach to health and safety education adapted to the specificity of educational activity and the formation of the content of educational fields.

The following fields of education must be studied in more detail:

– foundations of a healthy lifestyle (health-promoting factors; health-destroying factors);
– safety in a social environment (safety upon the occurrence of terrorist attacks; safety upon the occurrence of regional and local armed conflicts and mass riots);
– safety upon the occurrence of emergency situations of military nature;
– fire safety and rules of behavior in case of fire;

- main areas of activities of state organizations protecting population and territories from emergency situations in the time of peace and war;
- state services protecting citizens' health and assuring their safety;
- legal foundations of organization of population safety assurance and protection;
- issues relating to state and military construction of the Russian Federation (military, political, and economic foundations of military doctrine of the Russian Federation, the Armed Forces of the Russian Federation in the structure of state institutions);
- war-historical training (military reforms in the history of the Russian state, the Days of War Glory in the history of Russia);
- war-legal training (legal fundamentals of the state defense and military service, military duty and preparation of citizens for military service, legal status of a military servant, record of military service, military discipline); and
- state and military symbols of the Armed Forces of the Russian Federation (symbols of the Armed Forces of the Russian Federation, rituals of the Armed Forces of the Russian Federation) and the state.

There are several publications devoted to the formation of students' readiness for safe behavior that consider the issue from different viewpoints (N.P. Abaskalova, 2008; L.A. Mikhailov, 2008; V.N. Latchuk, 2014, and others).

3.1 Research instrument

The experimental study includes a total of 129 students studying in 5th–9th grades of the comprehensive schools of the Mari El Republic (5th grade – 21student; 6th grade – 17 students; 7th grade – 23 students; 8th grade – 44 students; and 9th grade – 24 students). All the students took the test. For each grade, we chose questions corresponding to the course syllabus. Each test consisted of 15 questions, which suggested one or more answer options.

According to the test results, the following assessment criteria were determined:

"5" – 14–15 of correct answers (high level)
"4" – 11–13 of correct answers (above-average level)
"3" – 8–10 of correct answers (average level)
"2" – 0–7 of correct answers (low level).

The rating system in the Russian Federation does not provide for the allocation of points 0 and 1.

4 FINDINGS

The test results showed the following distribution by the level of students' knowledge of safe behavior rules:

5th grade: "2" – 43%, "3" – 57%
6th grade: "2" – 29%, "3" – 71%
7th grade: "2" – 39%, "3" – 61%
8th grade: "2" – 55%, "3" – 45%
9th grade: "2" – 42%, "3" – 58%.

According to Figure 1, students have average and low levels of knowledge of health and safety fundamentals. Most students with an average level of knowledge study in the 6th grade. The lowest indices were detected in schoolchildren of 8th grade. This can be related to schoolchildren's transition into a critical adolescent period of development, when they change life orientations and studies take last place in their system of priorities. At the same time, adolescents become more independent and they more often have to face and cope with situations on which their safety level depends. Therefore, the teacher organizing fundamentals of health and safety and the grade master must pay special attention to the issues of developing safe-type behavior personality.

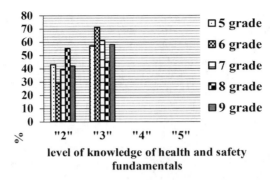

Figure 1. Test results among 5th- to 9th-grade students by the level of knowledge of health and safety fundamentals.

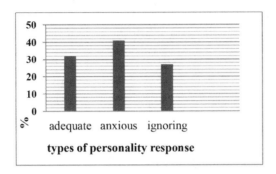

Figure 2. Results of the questionnaire used for testing students' sensitivity to threats.

Also, we offered a questionnaire test of sensitivity to threats (V.G. Maralov, I.A. Tabunov) containing a number of questions relating to the life of adolescents and their attitude to various situations. For each question-statement, four answer options were provided, from which they were to select only the option that corresponds to their personal opinion to the fullest extent.

The results of diagnostics demonstrated dominating types of personality response in various situations: adequate, anxious, and ignoring. However, we can face certain difficulties here: the test subject can respond to different situations in different ways: in some cases, they can make an adequate choice, and in others overstate or understate the threat. Hence, distribution scales were defined separately by each parameter; the average score was computed; and high, average, and low levels were distinguished. This provided an opportunity to subsequently determine the dominating response type in dangerous situations for each test subject.

Among adolescents with an average level of knowledge of health and safety fundamentals, most of them adequately (32%) respond to various dangers.

Students of anxious response type (41%) tend to overstate the importance of dangers. This can be related to the fact that such schoolchildren are not aware of behavior rules in various dangerous situations, and hence they try to avoid them. In some situations, such behavior can be considered justifiable, but when facing dangers, such children will hardly be able to assure their safety and can be a threat for people around them.

The students of ignoring response type (27%) are schoolchildren understating the importance of threats, flaunting their courage (or sometimes stupidity). However, such attitude to dangers can be explained perfectly well. First, they admit their laziness or reluctance to perform their academic duties, in particular, their home tasks; furthermore, they explain their laziness by the fact that they completely lost interest in their studies. Second, they suppose that teachers have a grudge against them, that is, disrespect them and refuse to treat them like individuals. Third, parents accuse them of inadaptability, and in addition, make them perform certain household duties.

It can be assumed that, facing the danger, such children who are not aware of personal safety rules will be more vulnerable. First, they do not know how to behave themselves upon the occurrence of any danger. Second, they will be too anxious, which can prevent them from making a right decision, or too self-confident, overstating their abilities. Such children can become a source of danger not only for themselves, but also for people around them. Therefore, we can assume that, when the level of knowledge and skills of safe behavior increase, their response type will change as well in various situations.

5 DISCUSSION AND INTERPRETATION

Thus, the formation of safe behavior ways is a continuous process lasting throughout the entire human life. The most important aspect of this process is the period of school studies because it is at this age when foundations of health and personality are laid. The school course "Fundamentals of health and safety" provides massive opportunities for developing safe-type behavior personality.

To lay foundations of schoolchildren's safe behavior in the institution, it is necessary to organize the education system integrating all activities. To develop students' readiness for safe behavior within the scope of the course "Fundamentals of health and safety", it is reasonable to use active teaching methods stimulating practical and mental activity without which it is impossible to move forward in the process of knowledge acquisition.

To ensure effective process of teaching health and safety fundamentals, it is necessary to use all types of organization of academic studies and academic activities of students, as well as optimal means and methods of evaluation of education process quality and the level of competence of people leaving school.

During the classes, most diverse teaching methods should be used. A teaching method is a well-ordered activity of a teacher and students aimed at achieving the specified learning goal. In teaching practice, the following teaching methods are distinguished:

- verbal methods: narration, lecture, conversation;
- bookwork;
- illustrative methods: various types of illustrations, schematization, symbolization, demonstration, video method, etc.;
- practical methods: experiment, practical work, laboratory work, observation, demonstration experiment, exercise, measurement, sampling;
- interactive training methods: nongame methods—discussion, case study, problem lecture, etc.; game methods—didactic, computer, role, simulation, business games; and
- control methods: oral, written, programmed.

Practical methods are applied when teaching fundamentals of health and safety should include practical works, exercises, and training sessions. The practical work is a students' activity related to the object of study, familiarization with the technique, and action in which application of acquired knowledge and ability to use theory in a real-case scenario predominate. In most cases, the practical method is used after studying theoretical foundations of education material and includes several steps as follows:

1. Actualization of knowledge aimed at the transfer of knowledge and abilities from the latent state into the manifest (acting) state. Using conversations, the teacher helps students theoretically comprehend the work purpose, relate its performance to the acquired knowledge, skills, and life experience.
2. Instructions are of great importance for achieving independence of performance and awareness of actions performed by students. The teacher orally explains the work performance process as a whole or step by step. Methodological cards, tables, texts from a textbook, or written text on the blackboard can be used as instructions.
3. Work performance. Students do their work independently, individually, or in a group. The teacher makes sure that it is being performed correctly, helping those students who find it difficult to cope with the task.

4. Summing up the work results. The students' work is accepted and assessed. The work results can be summed up in the form of a conversation or group reports on the work performed.

During the educational process, a wealth of knowledge, abilities, and skills are acquired in practical classes. During practical classes, the following should be ensured:

- correct demonstration of ways of actions in dangerous and emergency situations, as well as during first aid treatment, and watching their drill;
- uninterrupted lessons;
- detection and timely correction of errors during the drill of practical issues by listeners;
- encouraging listeners to achieve best results;
- observation of safety instructions.

6 CONCLUSION

In this study, we found that low and average levels of knowledge of safe behavior rules were detected in students studying in 5th–9th grades of the comprehensive schools of the Mari El Republic. The percentage of children with anxious (41%) and ignoring (27%) types of behavior is high. Such children will be more vulnerable when facing dangers, having no knowledge of personal safety rules at the same time.

Thus, modern schools must pay special attention to developing safe-type behavior personality using various means and methods. Each individual must be aware of the nature of occurrence and development of dangerous situations; know their strengths and capabilities to overcome the danger; and be able to make a correct diagnosis of the situation. The general purpose of developing safe-type personality is to reduce itself to acquisition of skills and abilities, allowing people to properly shape their behavior and thus reduce the level of outgoing threats, as well as prevent the danger surrounding them in the modern world.

It follows from the above findings that the development of safe-type personality must be based on universal human values, kindness and delicacy of feeling, tolerance, and responsibility. Such personality must first be developed by the educational institution. The education system is a key aspect of forming adolescents' readiness for safe behavior. The method used for developing safe-type personality is based on a comprehensive approach to learning through subjects which are directly related to safety issues.

REFERENCES

Abaskalova N.P. 2008. Theory and methods of health and safety fundamentals teaching: study guide a. Novosibirsk.
Ahmadullin U.Z. Safe behavior: theoretical prerequisites for investigation of the safe behavior problem [Electronic source]. – Available at: www.work.vegu.ru (accessed date: 7.12.2016).
Ayzman R.I. & Korolyov V.A. 2012. The importance of «Fundamentals of health and safety» training course in modern training of schoolchildren. Siberian Pedagogical Journal (p. 210–215). Novosibirsk.
Kazin E.M. et al. 2014. Peculiarities of organization of health and safety fundamentals teaching under the conditions of implementation of Federal State Educational Standards based on a differentiated approach. Siberian Pedagogical Journal (p. 269–272). Novosibirsk.
Latchuk V.N. Mironov S.K., Mishin B.I. 2014. Fundamentals of health and safety. Planning and organization of schoolwork. 5–11th grades: study guide. Moscow: Drofa.
Maralov V.G. & Tabunov I.A. 2013. Psychological structure of sensitivity to dangers. *Cherepovets State University Bulletin* (p. 122–126). Cherepovets.
Mikhailov L.A., Ahmadullin U.Z., Vassilev E.S. 2008. Psychological and pedagogical approaches to the development of safe type personality qualities. *Bulletin of Humanitarian Academy of Economics and Law* (p. 103–107). Moscow. 1880.

A poetics of Mari verbal charms

N.N. Glukhova, E.V. Guseva, N.M. Krasnova & V.N. Maksimov
Mari State University, Yoshkar-Ola, Russia

ABSTRACT: This article shows poetical stylistic characteristics of Mari verbal charms 'shüvedyme mut' which constitute an integral part of the spiritual culture of the nation. Texts of charms represent a rare phenomenon due to the long history of their existence and their peculiar language features. At the same time the tradition of practicing magical texts can be observed still today. An informational content of ca 500 verbal charms from different sources allowed dividing them into six groups. They differ in compositional structure, temporal characteristics, expressive and stylistic devices. The latter is an object of investigation in this paper. The carried out research showed that among expressive means and stylistic devices the investigated charms are rich in syntactic stylistic means and folklore tropes. This choice of means can be explained by the 'laws' of oral presentation of magical texts and their pragmatic aims.

1 INTRODUCTION

The main goal of this research is to describe some poetical features of esoteric texts in Mari culture. The article gives a classification of stylistic means which create expressiveness of Mari verbal charms (incantations). Mari charms are a separate folklore genre, which is an integral part of Mari oral tradition. It should be noted that the tradition of using magic texts in everyday life, in different rituals (weddings, childbirth, funerals) is a 'non-dying' cultural phenomenon among the Mari.

An overview of the currently available literature on the chosen subject in Mari culture shows the absence of a unified and generally accepted interpretation of a charm, or incantation. Here we offer a working definition of this folklore genre. A Mari incantation – 'shüvedyme', 'shüvedyme mut' – is an oral, rhythmically organized verbal formula of considerable length, containing a wish, a will or a command. It is employed in a ritual situation and is believed to produce a desired effect under certain conditions because of the magical power both of the word and the person who uses it with definite pragmatic goals (cf: Glukhova 1995; Glukhova, 1997; Glukhova & Glukhov 2009; Glukhova et al. 2016).

The majority of the Mari population lives in the Republic of Mari El. The Republic is situated in the eastern part of the East European Plain of Russia in the basin of the middle flow of the Volga River. The results of the population census in 2010 showed that out of 571 382 Maris in Russia 290 900 live in the Republic. Despite strong influences of the neighboring ethnoses during centuries of contacts the Mari nation has succeeded in preserving its cultural traditions remaining faithful to its magic and religion. Mari magic charms present a considerable interest in understanding of contemporary Maris' spiritual culture. They are also important for understanding the Mari language.

The research goal of the present paper is to define a stylistic potential of Mari verbal charms, showing stylistic devices which make charms poetical texts and a unique phenomenon in spiritual Mari culture.

A poetics of Mari verbal charms has never been in the focus of researchers' attention. Summary of stylistic features of this folklore genre enriches a stylistic description of the Mari language, its system of functional styles. This fact shows originality of the research as texts of Mari charms have interested scholars only by their compositional and informational characteristics (Sebeok & Ingemann 1956; Sebeok 1974). The central focus of the study is placed on the tropes which are also called lexical expressive means. The study significance can be explained by the fact that it closes the language lacuna and produces the revealed system of stylistic devices. It represented a research problem for scholars of Mari stylistics.

The pragmatic goals of verbal charms are manifold. They aim at healing people and domestic animals; protecting people and domestic animals from witchcraft, countering bad effects, bringing good, benefit or profit to people, inflicting evil on people, animals, plants and objects. As we see the goals are quite straightforward. Therefore, as the hypothesis goes, there cannot be too elaborate stylistic devices typical of the author's or the belle-lettres style such as metaphor, metonymy, irony, periphrasis, oxymoron etc. Stylistic approach together with semantic analysis will prove that there should be 'simpler' stylistic means which could provide the fulfillment of the wish, will, or command contained in the verbal charm.

The interest in esoteric components of peoples' spiritual life has never ceased since the 1980s. There are many original works analyzing peculiar aspects of charms in different cultures (Kõiva 1990; Koiva 1996; Napolskikh 1997; Roper 2003a, b; Kropej 2003; Charms, Charmers, Charming 2009; Misharina 2011). In all of them charms were treated as magic texts.

The notion of "text" has been the centre of attention of many schools of thought since the 1980s. Nevertheless the approaches to the analysis of text and attempts to define it differ greatly. A short message, or a piece of information, encoded into a linguistic form, consisting of several phrases, a whole book or a novel, can be equally called texts by representatives of different schools of text linguistics, pragma-linguistics or linguo-stylistics. Despite the controversial treatment of this notion by different scholars, a good consensus exists among them as to text communicative functions and pragmatic goals.

Text possesses certain features which permit it to be treated as a whole unit, an entity. These features differentiate it from a string of separate sentences. Despite multiplicity of approaches to text its dominant characteristic traits are: informational content, integrity, inter-dependence of the components, coherence, uniting idea and one general aim (Galperin 2011; Galperin 2012). Text, therefore, is regarded as a semantic unit which forms a unified and coherent whole united by a pragmatic aim.

Mari esoteric texts have clear-cut pragmatic goals. This view is shared in this paper though in the focus of attention is a poetic aspect of texts, their expressivity.

According to pragmatic aims Mari charms are classified into six types: 1. *Incantations for healing:* a) people, b) animals. 2. *Incantations protecting from witchcraft aimed at:* a) people, b) animals, c) plants and objects. 3. *Counter incantations and incantations freeing from already inflicted witchcraft on:* a) people, b) animals, c) plants and objects. 4. *Incantations changing interpersonal relationship among people, spoiling or changing them for the better.* 5. *Incantations bringing good, benefit or profit to:* a) people, b) animals, c) plants. 6. *Incantations inflicting evil on:* a) people, b) animals, c) plants (Glukhova 1996).

Different types of texts are marked by different composition parts, by their structure. In text linguistics the constituent parts of text are called supra-phrasal units. They represent the combinations of sentences, producing a structural and a semantic whole with one topic sentence which determines the subject matter of each of them. Mari incantations differ from each other structurally within the wide range of patterns—from complete (having all components of the ritual text) to partial texts (having the most necessary magic text elements) (Glukhova 1997; Glukhova et al. 2016). Both complete and incomplete patterns of texts are in everyday use in Mari El.

2 METHODS AND MATERIAL

The algorithm of the research, combining methods and stratagems from stylistics, text linguistics with elements of statistical analysis, consists of several stages.

In the initial stage of the research circa 180 folkloristic texts were collected during the folklore and ethnolinguistic expeditions of the author to different regions of the Republic of Mari El between 1992–1997 and 2008–2016.

During field expeditions the main methods used were the observation and interviewing of a considerable part of the village inhabitants in every region of the Republic, as well as participation in some magic rituals aimed at improving health and relieving stress.

The interviewing was carried out both with the help of questionnaires that concerned different spheres of Mari traditional culture and without, during free talks on different topics in an informal atmosphere. The interviewing of the informants, mainly local 'tradition bearers', was carried out according to a common pattern in order to make the comparison of collected texts.

The next stage included the enrichment of the gathered collection with the already published texts or texts from the archives.

All in all ca 500 incantations were analyzed syntactically, semantically and statistically.

For this paper the results of stylistic analysis were taken into consideration as the main goal is to determine expressive means and stylistic devices—folklore tropes—in Mari charms.

Stylistic devices or *folklore tropes* were determined and analyzed with the help of componential and contextual types of analysis.

3 RESULTS AND DISCUSSION

3.1 *Tropes*

The quantitative approach applied to investigated material showed that dominant lexical expressive means were *tautological epithets, folklore hyperboles, as well as sustained* and *simple similes*.

Epithets constitute a group of leading tropes in the analyzed verbal charms (82%). Examples of hyperbole (11%), and simile (7%) are less numerous in comparison with epithets.

3.1.1 *Epithet*

The epithet as a lexico-syntactical trope is based on the correlation of logical and emotive meaning in the adjective expressing quality or attribute.

Epithets are employed to characterize specific properties of the objects evaluating them. In the investigated charms there are two groups of epithets: *tautological* (traditional) and *explanatory*. They are expressed by: 1) qualitative adjectives; 2) relative adjectives; 3) nouns playing the role of relative adjectives or words of undifferentiated semantics. All these groups are equally represented in the investigated material.

Qualitative adjectives in the function of epithets are used mainly in the positive degree: *black (red, yellow, green)* water; *black (red, yellow, green)* pebble:

'*Shem (joshkar, narynche, užarge) βüdyshtö shem (joshkar, narynche, užarge) shargüm shinchapunžo dene kunam nalyn kertesh, tunam iže shincha βochsho!*' (Shaberdin 1973)

'When he (the sorcerer) is able to take a black (red, yellow, green) pebble with his eyelashes in the black (red, yellow, green) water, only then let him bewitch me with the evil eye!'

Qualitative adjectives are also used in the next extract: the *white* man; the *white* sea; the *white* sand:

'*Osh jeng, os tengyžyshke kajen, osh oshmam konden, sholagaj kandaram punen kertesh gyn, tunam iže … shincha βožen kertshe!*' (Glukhova 2008)

'If the white man is able to go to the white sea, to bring some white sand, to twine the rope of the left twining, only then let him be able to bewitch me with the evil eye!'

Structurally, epithets in the analyzed charms are classified into *simple single epithets* and *compound epithets*. Besides, in some texts, chains of epithets may occur.

Simple epithets are expressed by qualitative adjectives, as was shown in the previous examples.

Relative adjectives are also often used as epithets: *choman ßül'ö* 'a mare with a foal'; *pachan shoryk* 'a sheep with a lamb':

'*Choman ßül'ö kuze shke chomažym jörata, tuge myjym jöratyže! Pachan shoryk kuze shke pachažym jörata, tuge myjym jöratyže!*' (Gorskaja 1969)

'As a mare with a foal likes its foal, let him love me too! As a sheep with a lamb likes its lamb, let him love me too!'

Epithets expressed by nouns (*chodyra, shörtn'ö*) are numerous in the examined material:

'*Chodyra shordo shörtn'ö shinchyr dene shörtn'ö mengesh kylten shogalten, merang üj dene omdykten, shörtn'ö kümažysh lüshten, shörtn'ö kümyž-soßla dene 77 türlö kalykym ik minutyshto pukshen-jükten kunam kertes, tunam iže ushkalyn onchylym nalyn kertshe!*' (Shaberdin 1973)

'When he is able to tie a moose cow with a golden chain to a golden pole, to soften its udder with hare's butter, to milk it into a golden dish, in a minute to feed 77 different peoples on golden plates with golden spoons, only then let him be able to take my cow's first milk!'

Compound epithets are built on a more complicated pattern, as is seen from the following examples: *shij-tükan üskyz* 'a bull with silver horns', *shörtn'ö-punan ushkal* 'a cow with golden hair'.

In the analyzed corpus of charms epithets may be organized in certain chains: *tale tütan mardež* 'strong hurricane wind', *püsö pulat kerde* 'saber made of Damascus Steel', etc.

3.1.2 *Hyperbole*

Investigated charms are also rich in hyperboles. It is a stylistic device commonly known as a deliberate overstatement or exaggeration. Therefore any described feature inherent in the object, different phenomena or even whole situations can be exaggerated.

Texts may contain *simple hyperboles* as seen in the following passage:

'*Mündyr üžaram kunam posharen kertesh, tunam iže posharen kertse! Er-kechym kunam posharen kertesh, tunam iže posharen kertse!*' (Porkka 1895).

'When he (the sorcerer) is able to bewitch a far-away dawn, only then let him be able to bewitch me! When he is able to bewitch the morning sun, only then let him be able to bewitch me!'

Hyperbole in this text appears due to the understanding of apparent discrepancy between the normal flow of events and imaginary situation depicted in the incantation: *to bewitch a far-away dawn; to bewitch the morning sun*.

Hyperbolical situation in the texts can include two or more exaggerated conditions which are practically impossible to accomplish:

'*Nylle ik pushengym kunam shinchapunzho jeda keryn, shinchazhe dene onchal kertesh, tunam izhe onchalyn kertshe! Nylle ik vaksh-küm kunam shinchapunzho jeda keryn, shinchazhe dene onchal kertesh, tunam izhe onchalyn kertshe!*' (Gorskaja 1969).

'When he is able to open his eyes after having hung 41 trees on each eyelash, only then let him be able to see! When he is able to open his eyes after having hung 41 millstones on each eyelash, only then let him be able to see!'

In the examined material the most typical overstated circumstances are connected with the *time* of specific actions and the *amount of conditions* to be carried out. It is also seen in the extract of the charm *'Vash ushnash'* 'Love spell':

'*Shymlu shym türlö mlande ümbalne nylle ik choman βül'ö shke chomažym ik minutyshto shke pomyshkyžo kuze pogen nalesh, tugak tudynat shüm- mokshyžo ik minut žapyshte myjyn mogyrysh saβyrnyže!*' (Gorskaja 1969)

'As in a minute in 77 different countries 41 mares gather their own foals, let his heart/liver stick to me in a minute!'

In the next example – "*Poshartyshym shörymö*" "The removal of witchcraft" – the stress is laid on the amount of conditions:

'*Kö myiym posharen gyn, shörtn'ö kurykysh kayen, shörtn'ö imn'ym yshten, shörtn'ö orzhan lupshym yshten, shörtn'ö imne ümbake küzen shinchyn, shörtn'ö lupsh dene lupshal kolten kunam ik shagatyshto kertesh, tunam izhe posharen kertshe! Ik shagatyshto ok kert gyn, shke dekyzhe pörtylzhö!*' (Petrov 1993)

'Only then you can bewitch me when, having climbed up a golden hill, having made a golden horse, having woven a golden whip, having mounted a golden horse, having whipped a golden horse with a golden whip, you can wave a golden whip for an hour! If you can't do it for an hour let the evil come back to you!'

Enumerating several conditions, hyperboles, therefore, does not only serve as a simple separate stylistic device in charms. Hyperboles appear to be a *text constituent factor*.

In addition to simple and sustained hyperboles *numerical hyperboles* are of a certain interest in the examined material, for example in the incantation "*Razymym shörymö*" "Against rheumatism":

'*77 türlö βüdysh kajen, 77 türlö küjym kudalten, 77 türlö rožym shüten küeshyže, tushan 77 türlö mengym keryn, 77 menge βujyshto 77 türlö imym kerlyn ... ik shagat, ik minutyshto tache kechyn imyshke shogalyn, shüsken-muren kunam kertes, tunam iže (tide ajdemym) kochkyn-jüyn kertshe!*' (Petrov 1993)

'When it is (the spirit of rheumatism) able to get into 77 waters, throw 77 different stones, make 77 different holes, put 77 poles into these holes, put 77 needles into the bases of these 77 poles .... and then, staying on these needles for an hour and a minute, to whistle and sing, only then let the spirit of rheumatism be able to drink and eat (this man)!'

Such numerals as "one", "seven", "nine", "eleven" and "77" are amply used in the texts. They are considered to be magical among Finno-Ugrians (Petrukhin 2005).

Complex numerals are also used in the analyzed charms. A good example is a passage from the incantation "Shulymo" ('Evil dissolving'):

'*...77 türlö mlande βalne er pokshym kuze shulen kaja, tugak shulen kajyze!*

77 türlö mlande βalne er tütyra kuze shulen kaja, tugak shulen kajyze!

77 türlö mlande βalne er lups kuze shulen kaja, tugak shulen kajyze!

100, 99, 98, 97, 96, 95, 94, 93, 92, 91. Nimat uke!

Loktysh, poshartysh, ovylymo, shinchavochmo, jylmepuryltmo, marii cher, rush cher, koshtyra, nimat uke!' (Петров 1993).

'...How the morning hoarfrost disappears on 77 different lands, let his sorcery disappear!

How the morning mist disappears on 77 different lands, let his sorcery disappear!

How the morning dew disappears on 77 different lands, let his sorcery disappear!

100, 99, 98, 97, 96, 95, 94, 93, 92, 91.

Nothing is here. There is no sorcery, no damage, no evil eye, no evil tongue, no Russian disease, no Mari disease, no skin disease, nothing!'

The last passage is repeated nine times. The numerals are also reiterated nine times each time being reduced by ten units.

3.1.3 *Simile*

The investigated charms contain similes, too. It is such a device whereby two concepts are imaginatively and descriptively compared. This comparison leads to the intensification of one of the features which becomes prominent.

Structurally similes are figurative language means with an explicitly expressive referent and agent—the first and second components of the device. Simile has formal elements showing comparison in its structural pattern: they are connectives, usually post-positions or conjunctions. In the Mari language they are *gaj, semyn; kuze ... tuge /tugak*: *Müj gaj, üj gaj, sakyr gaj, shinchal gaj, lum gaj shulen kaizhe!* 'Let the disease melt away like honey, like butter, like sugar, like salt, like snow!' (Petrov 1993)

In the majority of the investigated charms similes are based on comparison but not on the contrast of different phenomena of reality.

Analysis of the incantations showed that on grounds of expediency it is necessary to classify figurative comparisons into *simple similes* and *sustained*, or *situational similes*. This taxonomy reveals a qualitative difference in the complex character of these objects of investigation, on the one hand. On the other, it corresponds to grammatical differences in their syntactic structures.

A large group of the analyzed charms is structurally and semantically based on *the situational simile*, which has a specific linguistic form. It has grammar indicators—two conjunctions: *kuze* (as) in the subordinate clause and *tuge/tugak* (so) in the main clause: *Shochmo keche kuze nöltesh, tuge tide eŋyn kapshe-kylže, kidshe-jolžo nöltshö!* 'As the sun on Monday rises, so let this man's body, arms-legs also rise!'

The *situational simile* retains some features of a simple simile. It becomes apparent when a verb (predicate) in the subordinate clause is not semantically independent but is closely connected with a verb (predicate) in the main clause. Manifestation of such semantic ties is the repetition of one and the same verb – *to rise* – in both parts of the complex sentence:

'...*Kushkyzhmo keche kyze nöltesh, tuge tide eŋyn kapshe-kylže, kidshe-jolžo nöltshö!*

Vÿrgeche keche kuze nöltesh, tuge tide eŋyn kapshe-kylže, kidshe-jolžo nöltshö!

Izarnya keche kuze nöltesh, tuge tide eŋyn kapshe-kylže, kidshe-jolžo nöltshö!' (Petrov 1993).

'... As the sun on Tuesday rises so let this man's body, arms-legs also rise!

As the sun on Wednesday rises so let this man's body, arms-legs also rise!

As the sun on Thursday rises so let this man's body, arms-legs also rise!'

Sometimes the verbs in both parts can be contextual synonyms as is seen in the following extract:

'*Shopke lyshtash kuze kushtylgyn kaja, tugak myjyn imnem kushtylgyn modyn kynel koshtsho! T'fu!*' (Evsevjev 1994).

'As the aspen leaf lightly rises in the air falling from the tree, so let my horse lightly rising, playfully go! Pah!'

It is interesting to note that simple and sustained similes, hyperboles create the concepts of impossibility (*improbability*) and *inevitability* of events and actions (Glukhova & Glukhov 2008). Their usage reveals the intuitive nation's knowledge of several concepts from contemporary probability theory.

4 FINDINGS

The present paper has treated the salient textual features of Mari charms, revealing stylistic markers on the lexical level of this folklore genre. The combination of known approaches and techniques has made it possible to obtain valuable information on the expressive character of folklore style of these ancient esoteric texts.

Folklore text, being a complex phenomenon, demands different lines of approach, because it can be viewed as an oral performance, on the one hand, and as a written text, on the other. This dual character has required a certain algorithm of analysis and the implementation of the theory of composition.

Stylistically significant features in Mari charms, constituting its poetic nature occur at each of the levels, traditionally distinguished in linguistics.

Lexical expressive means and stylistic devices characterize specific properties of objects, actions and phenomena, showing an evaluating attitude of the community towards them.

Evaluation is expressed in different types of epithets. The research showed such groups of epithets as: traditional and explanatory; simple single and compound epithets. They were expressed by two types of adjectives, by nouns playing the role of relative adjectives and by 'words of undifferentiated semantics'. All the revealed groups were represented in the investigated material.

Both simple and sustained hyperboles showed different situations and things in their untrue dimensions, described the unreal character of events. Numerical hyperboles stressed the number of events or actions the magical practitioners should undertake to achieve the positive or evil goals.

The analyzed verbal charms showed the existence of simple similes and sustained, or situational similes. This structural classification revealed a qualitative difference in the complex character of figurative comparisons, throwing an unexpected light on the compared objects.

The revealed and described stylistic devices have one point in common: they bear an imprint of the collective imagination. Consequently they can be termed folklore epithets, folklore similes, and folklore hyperboles. These stylistic means add to our knowledge of Mari functional folklore style.

The above enumerated folklore stylistic devices of lexical level, constituting the poetic aspect of the Mari charms, demonstrate the high linguistic culture of their creators, and the highly elaborated and developed Mari people's verbal art.

The obtained results may be successfully used in teaching practice as well as in compiling certain parts in the books on the Mari folklore stylistics.

REFERENCES

Charms, Charmers and Charming in Europe. International Research on Verbal Magic. 2008. Edited by Jonathan Roper. New York: Palgrave Macmillan.
Galperin, I. R. 2012. English Stylistics. M.: Librokom.
Glukhova, N. et al. 2016. An Esoteric Part of Mari Culture. The Social Sciences, 11 (8): 1808–1812.
Glukhova, N. 1995. Rhythm in Mari Charms and Its Geometric Interpretation. Finnisch-Ugrische Mitteilungen. 16, 17: 115–122.
Glukhova, N. 1997. Structure and Style in Mari Charms. Szombathely: Berzsenyi Dániel Tanárképző Főiskola.
Glukhova, N., Glukhov, V. 2008. Expressions of Impossibility and Inevitability in Mari Charms. Charms, Charmers and Charming in Europe: Materials of the International Conference of the International Society for Folk Narrative Research. New York: Palgrave Macmillan: 108–118.
Kõiva, M. 1990. Estonskiye zagovory. Klassifikatsiya i zhanrovyie osobennosti: aftoref. ... diss. kand. filol. nauk. Tallinn (Kõiva, Mare. Estonian charms. Classification and genre characteristics. Abstract of Candidate Dissertation on Philology).
Kõiva, M. 1996. The transmission of knowledge among Estonian Witch Doctors. Folklore. Electronic Journal of Folklore 2: 12–35. www.folklore.ee/folklore/vol2/ web_PDF 1996 vol. 2.

Kropej, M. 2003. Charms in the Context of Magic Practice. The Case of Slovenia. Folklore. Electronic Journal of Folklore 24: 62–78. https://www.folklore.ee/folklore/vol24/.

Misharina, G. 2011. Funeral and Magical Rituals among the Komi. Folklore. Electronic Journal of Folklore 47:155–172. http://www.folklore.ee/folklore/vol47/misharina.pdf.

Napolskikh, V. 1997. Seven Votyak Charms. Folklore. Electronic Journal of Folklore 5:141–144. http://www.folklore.ee/folklore/vol5/napolskikh.pdf.

Petrukhin, V. 2005. Mify finno-ugrov. M.: Astrel': ACT Tranzitkniga, (Petrukhin, V. 2005. Finno-Ugrian Myths. M.: Astrel': ACT Tranzitkniga).

Roper, J. 2003a. English Orature, English Literature: the Case of Charms. Folklore. Electronic Journal of Folklore 24:7–50. https://www.folklore.ee/folklore/vol24/.

Roper, J. 2003b. Towards a Poetics, Rhetorics, Proxemics of Verbal Charms. Folklore. Electronic Journal of Folklore. 24:50–62. https://www.folklore.ee/folklore/vol24/.

Sebeok, A. T. 1974. Structure and Texture. Essays in Cheremis Verbal Art. The Hague-Paris: Mouton.

Sebeok, T. A., Ingemann F. J. 1956. Studies in Cheremis. The Supernatural, New York: Viking Fund Publications in Anthropology 22.

Timotin, E. 2013. Folk narrative in the modern: unity and diversity. Charms symposium, the 16th Congress of the International Society for Folk Narrative Research. Vilnius, Lithuania, June 25–30, 2013. Incantatio. An International Journal on Charms, Charmers and Charming 3.

Zagovornyie teksty v strukturnom i sravnitel'nom osveshenii, 2011. Materialy konferentsii Komissii po verbal'noi magii Mezhdunarodnogo obshestva po izucheniuy fol'klornykh narrativov. 27–29 oktyabrya 2011, Moskva/redkollegiya T.A. Mihailova, J. Roper, A.L. Toporkov, D.S. Nikolayev. M.: PROBEL-2000. (Oral Charms in Structural and Comparative Light, 2011. Conference materials of the Committee on Charms, Charmers and Charming of the International Society for Folk Narrative Research's. 27–29 October. Moscow. Editors: Mikhailova T.A., Roper Jonathan, Toporkov A.L., Nikolaev D.S. – M.:PROBEL-2000).

LIST OF SOURCES

Evsevjev, T. 1994. Kalyk oipogo. Yoshkar-Ola: Marii kniga savyktysh (Evsevjev T. 1994. Folk's wisdom. Yoshkar-Ola: Mari Publishing House).

Glukhova, N. 1992–1997. N. Glukhova's notes from field expeditions into New-Toryal and Morki regions. Informants: Kolgasheva Anastasiya Ivanovna (1922), Shabalina Alevtina (1972), Vasilyeva Anastasiya (1935).

Glukhova, N. 2008. N. Glukhova's notes from field expeditions into Volzhsk region: Informants: Sokolova Marina (1989), Dubnikova Galina (1948).

Gorskaya, V. 1969. Nauchnyi rukopisnyio fond. Zametki OP 1. N 942. Mariiskii Nauchno-issledovatel'skii institute yazyka, literatury i istorii imeni V.M. Vasilyeva (Gorskaya V. 1969. Scientific manuscript foundation. Notes OP 1. № 942. Mari Scientific Research Institute of Language, Literature, and History named after V.M. Vasilyev).

Petrov, V. 1993. Mari ü: Türlö loktymo, cher, muzho vashtaresh shüvedymash. Yoshkar-Ola: Marii kniga savyktysh. (Petrov, V. 1993. Mari magic: different charms against witchcraft, diseases and prevention of evil. Yoshkar-Ola: Mari Publishing House).

Porkka, V. 1895. Volmari Porkka's Tscheremissische Texte mit Übersetzung. Herausgegeben von Arvid Genetz. JSFOu XIII1. Helsingfors.

Shaberdin, I. 1973. Nauchnyi rukopisnyio fond. Zametki OP 2. № 830. Mariiskii Nauchno-issledovatel'skii institut yazyka, literatury i istorii imeni V.M. Vasilyeva (Shaberdin, I. 1973. Scientific manuscript foundation. Notes OP 2. № 830. Mari Scientific Research Institute of Language, Literature, and History named after V. M. Vasilyev).

The trends in the dynamics of Mari ethnic values

N.N. Glukhova, G.N. Boyarinova, G.N. Kadykova & G.E. Shkalina
Mari State University, Yoshkar-Ola, Russia

ABSTRACT: The main objective of the paper is to show the results of some changes in Mari ethnic identity represented in the traditional system of values. They were reconstructed from Mari proverbs and sayings (10 690) and compared with the data of sociological surveys which were carried out for five years among the Mari young people, students of the Mari State University whose mother tongue is Mari. The aim of the research dictates the combination of investigation techniques used in linguistics, systems theory and mathematical statistics – that is semantic and factor types of analysis within a framework of a systemic approach. The most important stage includes the process of discerning values called 'factors' in the paroemias with the help of componential and contextual types of analysis. Statistics helps to organize the revealed values according to their frequency. Special attention is paid to ethics. It is just in this sphere one can see the trends in the dynamics of the axiological paradigm.

Keywords: paroemias, ethnic values, Mari culture, factors, systemic approach, semantic analysis, statistical method, ethics, dynamics, axiological paradigms

1 INTRODUCTION

The Mari people are one of the several Finno-Ugric nations. The greater part of the ethnos lives in the Republic of Mari El, situated in the European part of Russia. According to the results of the population census in 2010, there are 547 605 Maris. Out of all the Maris in Russia 290 900 live in the Republic. The Mari represents one of the Russian Federation rural nations. In the cities there are ca 26.1% of the Mari population, and in Yoshkar-Ola this figure is even smaller – 23.4%. The sociological investigations carried out in the 1970s, 1980s, 1990s and at the beginning of this century by Mari scholars showed basic features of the Mari ethnic consciousness which the interviewed population stressed in their answers. Crucial parameters of a national identity were: the language (75%–77.2%); traditional culture (61.6%); common historical past (21.6%); religion (15.7%); character features (15.4%) (Shabykov 2000).

The national language and traditional culture constitute the most important parts of the national identity structure. The common language helps identify oneself closely with a person or group. But the term "ethnic consciousness" includes a greater variety of notions incorporating ideas of typical national features: ethnic stereotypes, historical past, behavioral norms, etc. All of them can be explained by a type of economy, life on a certain territory, and even geographic climatic conditions. If the concept of identification on the personal level takes a center stage in the formation of self and its result is the individual's identity, then ethnic identity is considered a basis of ethnic consciousness. It represents a definite set of behavioral or typical distinctive traits by which an individual is recognized as a member of a group. Conclusion on identity is made if investigated characteristics in their totality and relationship among themselves are relatively unique. The core of ethnic identity is represented in traditional values of an ethnic group.

The research problem discussed in the paper is changes in the axiological paradigm of the Mari youth.

The research goals are a) to analyze Mari proverbs and sayings with the aim of discerning traditional values expressed in them; b) compile a list of selected values highlighting the dominant ones; c) to compare the obtained list of values as evidenced in Mari paroemias with the values shared by the students; d) to mark trends in the dynamics of the ethnic value system.

In hypothesis testing we are interested in the trends of values system changes, especially in the dynamics of the evaluation of ethical and moral norms.

2 METHODS AND MATERIAL

In modern cultural anthropology ethnic identity is generally viewed as collective ethnic consciousness, which is predominantly manifested through the following psychological functions: thinking and feeling; sensation and intuition; volition and will (Jung 1981).

Ethnic cultural identity is defined by its most conservative components accumulated by the ethnic group on the territory of its living during centuries (Shkalina 2012). These components can be obtained from folklore texts of various genres, mainly from proverbs and sayings as texts embodying nation's wisdom. In the paper the system of values constituting the core of national identity (hereinafter referred as a system) is reconstructed from 10 690 proverbs and sayings (Ibatov 1960; Grachyova 2001; Kitikov 2004). We analyzed Mari proverbs and sayings as the main source of people's wisdom.

The investigation includes methodological basis of the systems theory which, in its turn, involves a decision making theory and a theory of values. A dichotomy method, factor and statistical analysis are also applied to the material together with componential (semantic) and contextual analysis.

The term "value" is defined in dictionaries both as certain abstract notions, and material objects. The values expressed in Mari paroemias express concepts and qualities referred to a spiritual sphere of life not to its material realm.

An algorithm of the research was already described in several of the authors' papers (Glukhov & Glukhova 2008; Glukhov & Glukhova 2013; Glukhova et al. 2015). So here we will only recall some of the most important steps in defining the values. The Mari paroemias as proverbs and sayings in other cultures are used in their direct and figurative meanings. To reveal ethnic values we should use componential and contextual types of analysis, to define the components of value meanings. But as there are literally thousands upon thousands of people's aphorisms semantic analysis should be supplemented by a statistical data method. This combination helps to reduce the number of values. By this step we distribute paroemias into semantic groups and further calculate their mention. We assume that the more important the value is the more often it is used in the proverbs and sayings (Glulkhov & Glukhova 2012). So, we compiled the list of values; calculated their estimation of the incidence in the paroemias and the probability of their usage; ranked values-factors in descending order of probability; divided all the discerned values into four groups (1) main, 2) auxiliary, 3) additional, 4) insignificant) using the method of consecutive dichotomy by the criterion of a simple majority; built a histogram of probabilities; made some conclusions and summaries.

3 RESULTS AND DISCUSSIONS

3.1 *Mari ethnic values as evidenced in paremiaes*

Systemic approach to 10 690 Mari proverbs and sayings allowed to reveal eight basic values and arrange them in the following order of priorities: *ethics, family knowledge, speech, wealth, food, labor* and health.

The results of statistical analysis of the introduced factors for the easiness and simplicity of perception are depicted in the form of a differential factor diagram (Fig. 1) which is based on the data in Table 1.

Table 1. Averaged probability distribution of values mentioned in Mari paroemias.

№	Values	Probablity of usage (positive evaluation)	Probablity of usage (negative evaluation)	Sum total of probabilities
1	Ethics	0.077	0.147	0.224
2	Family	0.097	0.054	0.151
3	Knowledge	0.107	0.038	0.145
4	Speech	0.051	0.057	0.108
5	Wealth	0.068	0.038	0.106
6	Food	0.082	0.02	0.102
7	Labour	0.05	0.038	0.088
8	Health	0.038	0.038	0.076
9	Sum total	0.570	0.430	1.000

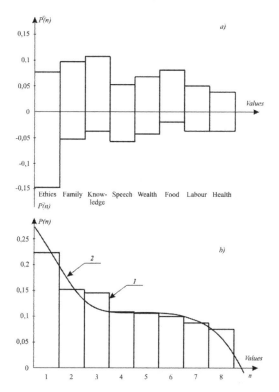

Figure 1. Diagrams of values probabilities mention in Mari paroemias: a) differential b) summarizing.

On the diagram (Fig. 1) the x-axis shows the values and the y-axis illustrates their average probability distribution. Analyzing the table and the diagram, we draw a conclusion that the principal values for the Mari are: *ethics, family, knowledge*. They constitute 52% of the usage of all values. Of secondary importance, least vital are: *speech* and *wealth*.

Family, knowledge, wealth and *food* are mainly positively assessed. *Ethics* and *speech* are evaluated more negatively than positively. Thus, the diagram analysis shows that in Mari paroemias we can find a high evaluation of the family, ethics and work and absence of striving for wealth and money, a stoic attitude towards health.

Here we shall discuss the most vivid examples of all the factors.

The contents of the majority of the proverbs and sayings are quite clear and do not need to be commented on. But when a proverb does not have clear meaning then there is an explanation of it.

By *ethics* people's wisdom expressed in the proverbs and sayings imply strict hierarchy of moral qualities (with a strong accent on a respect to elder men) (22.4%). In the proverbs describing ethics the positive assessment is given to *bravery* ('People respect the brave') and *kindness* ('Beggars should be given alms without calculation'). *Greed, envy, malice* are evaluated negatively: 'A mean person returns the debt only when the hare's tail gets longer', 'The wealthier the rich man becomes, the more difficult it is to satisfy him', 'Who doesn't follow the path of evil, will not reach it', 'He is as greedy as he is rich' (Kitikov 2004).

The next factor – *family* – usually describes big extended families (15. 1%). The contents of the analyzed paroemias reflect the time when the Mari community belonged to a peasant civilization based on a patriarchal family structure. The family consisted of peasants united by blood kinship and by closely knit economic ties. There were big undivided families and small nuclear ones. A nuclear family consisted of two parents and their children. Big extended units might include several married couples from one to three generations living together according to a patrilineal pattern. As such a subsystem combined agrarian production with elements of hunting and gathering and living in a family and working hard for it were main requirements for the survival ('Fingers are many, but you will not cut any', 'It is easy to live on one's parents store', 'A man is a head of the household' (Ibatov 1960)). The peasant culture taught family members to conform to rigid rules of unwritten law and ethics. Respect for parents was one of the most important ('Even if you make an omelet on your own palm still you will be in debt to your mother', 'You should not be bigger (more important) than your parents' (Kitikov 2004)). Thus, another most significant prerequisite for living in a family was obeying the moral norms and rules of existing together. The proverbs set apart *ethics* as one of the vital values of a peasant life.

The next factor is *knowledge of nature* and *climatic conditions* which was useful in the organization of an everyday life of an extended family (14.5%). Judging by the contents of the paroemias it is interesting to note that knowledge and skills are associated with "natural" intelligence and work and not always with bookish education. ('Without work you will not be clever', 'Skills are more precious than work', 'A bird is beautiful by its feathers, a man – by his mind', 'He who learns much knows much, but not he who has lived much (long)', 'The head without brains is like a lamp without light', 'Good work lives for two centuries' (Kitikov 2004)).

Next in a descending order of priorities is the factor *communication, speech* (10.8%). The negative assessment is given to *boasting* ('An empty barrel makes much noise'), *lies* and *hypocrisy* ('The tongue is smooth, the soul is loathsome (repulsive, cruel))'. The proverb 'A good rope is long, a good word is short' praises people speaking to the point. People speak about the wounds which can be inflicted by words: 'The evil tongue is like an arrow' (Ibatov 1960).

The factor *wealth* occupies the next place (10.6%). The examples are: 'Big money might crash a person', 'If you want to be with money, get up early', 'It is difficult to be friends with the rich', 'The rich man's greed is bottomless' (Kitikov 2004).

Next comes the factor *food* (10.2%): 'If there is bread, there is cattle', 'You will eat an old bast shoe if it is fried in butter', 'You will not spoil porridge with butter', 'The beer is good but the cup (ladle) is small', 'If there is no meat even lungs are a treat' (Grachyova 2001).

Mari proverbs and sayings give a positive evaluation to the benefit of agrarian *labor*, which constitute another value (8.8%): 'One gets rich from work, dies from sorrow', 'If you work hard your festive table will be rich', 'Who works hard eats butter and honey'(Kitikov 2004; Ibatov 1960).

And the last factor in the list of values belongs to the factor *health* (7.6%). The examples show a lack of serious health care: 'One will not put on a kerchief if there is no headache', 'An aching spot draws a hand', 'He became so thin that only body and skin are left', 'He is fit only for a coffin', 'A sick person in a fever is cold even on the stove', 'Illness comes in heaps, goes away in grains', 'Sauna cures all diseases' (Grachyova 2001; Kitikov 2004).

The fundamental system of traditional values esteemed by the Mari community and reflected in paroemias has been stable for a long time.

It might be well to point out that the reconstructed system of values contradicts to the existing direction of contemporary social processes of globalization, transition to a market economy and postindustrial society, connected with a craving for wealth and individualism.

3.2 Mari ethical values as evidenced in paroemias

As we noted, *ethics* occupied the first place in the list of priorities. The word 'ethics' encompasses several meanings. In our paper we viewed it as a system of moral rules governing the conduct of a person or the members of any social group or profession.

As is known any nation's ethics might be formulated in two different ways: 1) either as a laudation of ethic virtues or 2) criticism of its vices and moral flaws. In the present research the accent was made on the latter.

We again addressed the proverbs and sayings containing the concept *ethics* for the analysis of the most intolerable vices and moral flaws condemned in the Mari society. As in many dictionaries, we understand the term 'vice', 'ethical vice' as 'an evil conduct', 'an evil habit' and 'a defect of character or behavior'.

Analyzing *ethics* we may say that there are many human vices (moral sins) but in each society they are not all intolerable equally. Clearly, that the more often the vice is mentioned the more intolerant it is for the ethnic group.

Mari proverbs and sayings criticize *violence* ('A fighting rooster loses its head', 'Freedom for a man is dearer than anything else'); *laziness* ('Rain drops will fall on the lazy person three days earlier'; 'Wind closes the doors after the lazy'); *greed* ('Rich man's greed is bottomless'; 'A greedy man will die for a copeck and kill for two'); *lies* ('A liar is afraid of the truth'; 'Lie only once and you will be labeled a liar for the rest of the life'); *stupidity* ('A head without brains is like a lamp without light', 'An empty head is like a lamp without light'); *cowardice* ('A cowardly sparrow frightens others', 'A fur-coat made of hare skins tears off at every dog's barking'); *theft* ('One steals—forty are suspected'; 'Help a thief and ruin the honest man'); *envy* ('Envy is like a coal: even if you do not burn yourself you'll become dark/black'); *heavy drinking* ('Vodka is not mother's milk'; 'If you look into the vodka bottle you will see a wolf at the top, a bear in the middle, and a swine at the bottom'); *ingratitude* (Kindness is at the tree-top, grateful words are at the Tamaraikino (non-existing) village'; 'Fulfilling the promise is like a buttered porridge, forgetting it is a nine-tail whip) (Grachyova 2001; Ibatov 1960; Kitikov 2004).

The difference in criticism of *violence*, *laziness* and *greed* is very slight. Tolerance to violence was reflected in many paroemias: 'If you are constantly beaten even a dog gets accustomed to it'. *Violence* is justified by the ethnic group only when people struggle for their freedom and independence. This fact can be explained by the history of the Mari people. Nevertheless the majority of proverbs and sayings warn that use of force will lead to no good: 'Life of an aggressive person ends up in other people's hands'. *Laziness* is mainly associated with doing nothing. Condemnation of *greed* speaks of the necessity of sharing with others in a closely-knit community to survive under severe climatic conditions.

The results of the componential and contextual types of analysis carried out with the already described algorithm applied to the investigation of values are shown in the table (Table 2) and the diagram (Fig. 2).

On the basis of this table one more diagram was built.

The x-axis shows ethical vices and moral flaws enumerated in the table and labeled with the corresponding numbers in the diagram. The y-axis shows the average probability distribution. The arrow with the number one indicates the enumerated vices; the arrow with the number two designates a cumulative curve which shows the result of the approximation of probabilities distribution. The diagram does not show the dramatic differences in the criticism of the enumerated vices: the cumulative curve has a rather smoothly descending character of the sum of the normal (Gaussian) and even distribution.

3.3 Mari students' system of ethical values

To see and understand changes going on in the Mari society and compare the difference between the traditional values and the values of a new economic and ideological situation in the country we interviewed Mari students of the Mari State University whose mother tongue was Mari. There were 212 people, male and female, during five years of investigation (2002–2006).

Table 2. Averaged probability distribution of ethical vices and moral flaws criticized in Mari paroemias.

№	Vice	Probability
1	Violence	0,192
2	Laziness	0,175
3	Greed	0,169
4	Lies	0,119
5	Stupidity	0,107
6	Cowardice	0,085
7	Theft	0,062
8	Envy	0,034
9	Heavy drinking	0,028
10	Ingratitude	0,028

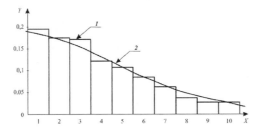

Figure 2. The diagram of probabilities distribution of the criticism of ethical vices in Mari paroemias.

Table 3. Averaged probability distribution of ethical vices and moral flaws mentioned in Mari students' questionnaire answers.

№	Vice	Probability	№	Vice	Probability
1	Violence	0.155	10	Lies	0.048
2	Theft	0.122	11	Greed	0.045
3	Heavy drinking	0.073	12	Cowardice	0.044
4	Envy	0.059	13	Jealousy	0.043
5	Hypocrisy	0.056	14	Stupidity	0.036
6	Ingratitude	0.055	15	Stubbornness	0.033
7	Disrespect of elders	0.054	16	Atheism	0.03
8	Haste	0.052	17	Untidiness	0.036
9	Laziness	0.05	18	Dislike of forest	0.018

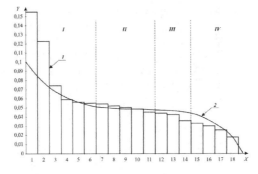

Figure 3. The averaged diagram of probabilities distribution of criticism of ethical vices by Mari students. Roman figures show the groups of vices: I – intolerable; II – secondary; III – insignificant; IV- nearly permissible.

As *ethics* was singled out as the leading value on the material of paroemias we decided to analyze students' attitude towards this concept.

Mari students' answers to the questionnaire on *ethic values* showed that the system of *ethnic ethics* has undergone certain changes if we compare these results with the outcomes of the proverbs analysis. The most disapproved moral flaws were: *violence, theft, heavy drinking, envy, hypocrisy*.

The last group includes weaknesses, tolerable by the Mari young people: *stupidity, stubbornness, atheism, untidiness, dislike of forest*.

On the x-axis ethical vices and moral flaws enumerated in the table (Table 3) are situated. They are labeled with the corresponding numbers in the diagram (Fig. 3). The y-axis shows the averaged probability distribution of ethical vices and moral flaws. The arrow with the number one indicates the revealed moral flaws; the arrow with the number two points at a cumulative curve which shows the result of the approximation of probabilities distribution. It is the cumulative curve that highlights the trends in the dynamics of changes.

Comparison of this diagram with the diagram of probabilities distribution of the criticism of ethical vices in Mari proverbs and sayings allows us to make the conclusions on the dynamics of Mari ethnic ethical system of values (in the form of criticism of its vices).

4 CONCLUSION

It is known that different nations' values are not the same. This difference may lead to cultural misunderstanding and intolerance when peoples come into contact. Therefore knowledge of other peoples' value systems is of a high priority for intercultural communication.

The Mari ethnic value system as evidenced in paroemias embraced a wide range of priorities, singling out eight factors as most important. Their enumeration and summary showed a list of quite trivial values. But they were basic for the clan survival when climatic circumstances were not favorable, consumption was limited and one lived within a big extended family.

It is very well known that there exist universal moral standards, principles and norms but the importance, ranking and systemic ties among them are different. At present due to globalization processes and changes in the life of our own country traditional value systems are undergoing certain changes. It is vividly seen in the axiological paradigms of the younger generation. The present research has shown three trends and results in the shift of certain moral values.

The first trend in the dynamics of changing system of values lies in the fact that the differentiation in criticism of vices has become more prominent. It is vividly shown by diagrams (cf. Fig. 2 and Fig. 3). The degree of the differentiation among the vices increased due to the positive dynamics of the approximating function. If the normal (Gaussian) distribution was typical of the analyzed Mari proverbs and sayings then the diagram drawn on the basis of students' answers to a questionnaire was better approximated by a cubic parabola. This clearly demonstrated a stronger differentiation in criticism of vices.

The second trend consists in the fact that modern Mari youth disapproves of *violence* most of all and is quite tolerant of *laziness* and *stupidity* in comparison with Mari paroemias. More criticism got such vices as *theft, heavy drinking*, and *envy*. *Laziness, greed, stupidity* were less criticized.

The third trend is seen in the difference of meaning in a linear correlation coefficient. A linear correlation coefficient, showing the correspondence between the traditional Mari ethics and the youth's answers, constituted 0.48, which reflected the significant transformation of Mari people's life during the last 30 years due to economic, political and ideological reasons. Traditional Mari ethnic system of values especially in the criticism of its ethical vices is indirectly considered by a younger generation as not corresponding to the present times and the demands of the contemporary life.

The outcomes of the study can be useful for school teachers, university lecturers, tutors to understand the youth's attitude to traditional ethnic value systems. Besides, the results will be able to help educators in their everyday work with young people. Some more sociological investigations into the problem can confirm the present results or show other tendencies in relation towards the traditional values of Mari ethnic social community.

REFERENCES

Glukhov, V.A. & Glukhova, N.N. 2008. Mari Ethnic Value System. Ünnepi írások Bereczki Gábor tiszteletére. Urálisztikai tanulmányok 19. Budapest: 221–228.
Glukhov, V.A. & Glukhova, N.N. 2012. Systemic Reconstruction of Mari Ethnic Identity. The Finno-Ugric Contribution to International Research on Folklore, Myth and Cultural Identity: Fifth International Symposium on Finno-Ugric Languages in Groningen. Maastricht [The Netherelands]: 45–62.
Glukhov, V.A. & Glukhova, N.N. 2013. A System of Techniques and Stratagems for Outlining a Traditional Ethnic Identity. Approaching Methodology. Second revised edition with an introduction by Ulrika Wolf-Knuts. Edited by Frog & Pauliina Latvala With Helen F. Leslie. Helsinki:399–412.
Glukhova, N. et al. 2015. Traditional Values of Eastern Finno-Ugrians as Evidenced in Proverbs and Sayings. Review of European Studies. Canadian Center of Science and Education. Vol. 7. No. 8: 156–170.
Grachyova, F.T. 2001. Hill Mari proverbs, sayings, riddles, weather signs. Yoshkar-Ola: Mari State University.
Ibatov, S.I. 1960. Mari proverbs, sayings, riddles. Yoshkar-Ola: Mari Publishing House.
Jung, Carl Gustav, 1981. The Development of Personality. The Collected works of C.G. Jung. Volume 17. Editors: Sir Herbert Read, Michael Fordham, M.D., M.R.C.P., Gerhard Adler, Ph.D. Translated by R.F.G. Hull. Routledge and Kegan Paul. London and Henley: 407–487.
Kitikov, A.E. 2004. Proverbs and sayings of Finno-Ugrians. Yoshkar-Ola: Mari Publishing House.
Shabykov, V.I. 2000. Itogi sotsiologicheskogo issledovaniya. Mariitsy: problem sotsial'nogo i natsional'no-kul'turnogo razvitiya. Yoshar-Ola: 171–206 (Shabykov B.I. The Results of sociological research. The Maris: problems of social and national-cultural development. Yoshkar-Ola; 171–206).
Shkalina, G.E. 2012. Ethnic ethical aspects in Mari traditional outlook. Bulletin of Kazan State University of Culture and Arts. 3: 48–53.

On the importance of comparative phonetic analysis for bilingual language learning

I.G. Ivanova, R.A. Egoshina, N.G. Bazhenova & N.V. Rusinova
Department of Romance and Germanic Philology, Faculty of Foreign Languages, Mari State University, Yoshkar-Ola, Russia

ABSTRACT: Under the condition of natural Mari-Russian bilingualism, Mari students studying French at the faculty of foreign languages of MARSU need to learn the articulation base of this foreign language for the purpose of its further teaching. In the conditions of the same artificial bilingualism the comparative phonetic analysis is of great importance. Comparison of two language systems is important for the solution of questions of universality and uniqueness of the language phenomena. During the research of the problems connected with bilingualism it is necessary to consider processes and factors which promote the emergence of an interference.

It is considered that the less there is typological distance between native and studied languages, in other words, the higher the degree of their similarity and the smaller the distinction, the more there is the possibility of the emergence of an interference. If this distance is large, that is the languages are genetically unrelated as, for example, French and Mari, then there will be less mistakes and cases of automatic hyphenation.

The real image of phonetic violations in the speech of bilinguals can differ considerably from predicted ones. At the same time, experimental and phonetic researches of the speech of Mari students' studying French receive great significance. Data from the comparative analysis help to find and explain the reasons of certain distortions.

Keywords: bilingualism, interference, comparative analysis, phoneme system, prosodeme system, multi-structured languages, experimental and phonetic methods

1 INTRODUCTION

Language contacts inevitably lead to bilingualism. The main criterion for determining bilingualism is the degree of proficiency in languages. An example is the Russian-Mari bilingualism in the Republic of Mari El in Russia. Under the conditions of natural Mari-Russian bilingualism, Mari students who study French at the Faculty of Foreign Languages need to learn the articulatory base of the French language as the first foreign language for the purpose of further teaching it. At the same time, the teaching of correct French pronunciation is hampered by the influence of the native (Mari) language. In this regard, all the studies carried out in this field acquire special significance in conditions of bilingualism in the Republic of Mari El.

The relevance of comparative studies of the phonetic systems of the French and Mari languages is due to the bilingual situation in the study of a foreign (French) language by the speakers of the Mari language. In recent years, the question of comparing the phonetic systems of typologically unrelated languages, for example, French and Mari, in the conditions of artificial bilingualism, has been increasingly raised.

Until now, bilingualism was not considered on the basis of functional linguistics but was reduced to a description of the relationships between structures and individual structural elements of languages. In recent decades, in linguistics, functionality has been interpreted as communicational (Weinreich, 1979). In Russian linguistics, a functional (communicative) approach to the consideration of bi-linguism is reflected in the works of V.A. Avrorin (1975), Z. Zorina (2000), and others.

In the opinion of many researchers (E. Bryzgunova, 1984), M. Gordina (1997), T. Nikolaeva (1984) complete and fluent possession of a non-native language leads to the action of an intermediate system becoming a kind of independent language, which differs from the natural language by the presence of two signatory systems.

Studying the correspondence between different signifiers in Mari and French (French speech of Mari language speakers), researchers find that the correspondences outlined here represent a third system, a language where the same signifier is compared with two other signifiers. This creates a third, intermediate system, which includes signifiers, common to the two languages, as well as general rules for the transition from them to the corresponding signifiers (Lyubimova, 2012).

In the linguistic literature there is no single interpretation of this phenomenon. F.S. Akhmetzyanova calls the object of violations "the zone of interference" of contacting languages, meaning only intonational interference (Akhmetzyanova, 2010). In our opinion, the phenomenon of interference is more complex and multifaceted.

Real phonetic violations in the speech of bilingualists can significantly differ from the predicted violations. Comparative analysis helps to uncover and explain the causes of a specific distortion. Therefore, at the segmental (phonological) level, the comparison should be based on an analysis of the acoustic characteristics of sounds and their differential features; and the suprasegmental (prosodic) level is focused on the description and comparison of the melodic contours of phrases of the French and Mari languages.

Both Russian and international science notes that interference zones can present various violations. With the deepening of knowledge and the development of sustainable skills, the boundaries of interference are compressed. In the zone of inferential phenomena, errors of various kinds and levels are generated: at the segmental and suprasegmental levels (Lyubimova, 2012).

Bilingualism is defined as the free command of two languages and their alternate use in the communication process. Artificial bilingualism is a type of bilingualism, in which the second language is not acquired in a native way, but is purposefully learned. The authors of the article put forward the hypothesis that with artificial bilingualism of students of the Mari State University studying French, one-sided interference is observed due to the imposition of a non-native language system on the native language system.

With respect to the sound system, one can speak of phonetic conditioning of phonological processes. It can be of two kinds. First, a non-native speaker can not hear the difference between the sounds of native and foreign languages. In this case, possible errors are determined by the perceptive properties of the speaker. Secondly, with adequate perceptual evaluation, the correct implementation of the sounds of a foreign language may not be possible due to the influence of the properties of the artificational base of the native language. In this case, by their nature, the errors are caused by motor skills (Orazaeva & Ivanova, 2016).

When researching problems related to bilingualism, it is necessary first of all to take into account those psychological processes that contribute to the appearance of interference. Analogy, when the ways of denoting and constructing utterances in a foreign language are intuitively assumed to be identical to the native language and the speaker unconsciously strives to assimilate them, plays a significant role.

Artificial forms of language contact such as bilingualism resulting from the formation of a secondary language system occurs in specifically created conditions outside of a natural language environment attracted the attention of the authors. The aim of the study is to obtain objective data on the differences in the phonological and prosodic systems of the Mari and French languages under conditions of artificial bilingualism by means of a comparative method.

2 MATERIALS AND METHODS

The study of a bilingualist's speech requires the use of complementary methods and techniques, as well as a comprehensive analysis of extensive theoretical and experimental materials that provide a comprehensive approach to the issue under study and the reliability of the data obtained.

In modern linguistics, experimental and phonetic studies devoted to the comparison of phonological and intonational systems of various languages are becoming increasingly widespread. Comparison is important, first of all, to address issues related to the problem of the universality of linguistic phenomena. Undoubtedly, each language has a certain set of categories that are common to most languages. At the same time, the analysis of phonological and prosodic systems of languages with different structures is important in the context of the influence of the native language (Mari) on the foreign language (French).

Experimental phonetic methods represent a variety of techniques and methodologies of working with the material under study. Methods of experimental phonetics arose in the middle of the 19th century. Thanks to two factors: the development of technical and linguistic thought.

As is known, V. Bogoroditsky was the first to apply some physiological instruments in the analysis of articulations. P. Russlo created a special laboratory in Sorbonne, which later became widely known and became a school for many foreign and domestic phoneticists.

L.V. Shcherba, the founder of the Leningrad (Petersburg) phonological school, created in 1909 a laboratory of experimental phonetics at Petersburg University. Moscow Phonological School (MFSh), represented by such linguists as R.I. Avanesov, P.S. Kuznetsov and A.A. Reformatsky were the first to propose the necessity of applying the morphological criterion in determining the phonemic composition of the language.

Experimental phonetic methods have received wide development, mainly due to modern computer technology. Acoustic methods of investigation (from spectrographic analysis to speech modeling) represent sound in the form of a curve or spectrum. Somatic methods are the investigation of articulations made by the human speech apparatus.

The research is carried out on the material of sound recordings of native French speakers and the French speech of the Mari students. In comparative analysis, experimental acoustic phonetic methods are used: oscillographic, histographic, intonographic. Methods of instrumental analysis are supplemented by auditory, statistical and theoretical linguistic methods, which reveal similarities and differences in the system of vowels and consonant phonemes and the intonation of the French and Mari languages.

In particular, in carrying out this study, the authors used an intonographic method to compare the melodies of various communicative types of Mari and French phrases.

3 RESULTS AND DISCUSSION

With artificial bilingualism, one-sided interference occurs in the speech of the bilingualist. With respect to the sound system, one can speak of the phonetic conditioning of processes. When studying problems related to bilingualism, it is first of all necessary to take into account those psychological processes that contribute to the appearance of interference.

Analogy, when the ways of designating and constructing utterances in a foreign language are intuitively assumed to be identical to the native language, and the speaker unconsciously strives to assimilate them, plays a significant role. The depth and volume of interference depends on subjective and objective factors. Subjective factors are determined by the individual language abilities of the speaker, his or her language competence.

The objective factors include degree of genetic affinity of the contacting languages, individual systemic and structural properties determining specifics of the language examined.

It is believed that the less the typological distance between native and learned languages, that is, the greater the degree of their similarity and less difference, the greater the probability of interference. Genetically unrelated languages, such as French and Mari, have fewer cases of automatic transfer, and therefore errors. The Mari and French languages are genetically different. However, experience of teaching shows that the Mari language, while having an articulatory base different from the French language, at the same time has vowel sounds that are similar to French phonetic characteristics. This phenomenon has a strong negative interfering influence on the French pronunciation of Mari students (Zorina, 2000).

The vowel phonemes of the Mari and French languages are represented in the following summary tables. Comparison of vowels allows us to reveal their common and differential features.

The system of vowel phonemes of the Mari and French languages is characterized primarily by the quantitative difference. In the Mari language there are 8 vowels, while the phonetic system of the French language includes 15 vowel sounds (Tables 1 and 2).

The comparative analysis of the phonemes of the Mari and French languages allows us to distinguish their similar and different features. The similar features of vowels of the compared languages are:

– Front character of pronunciation;
– differentiation of vowels according to the degree of ascent (upper, middle, lower);
– differentiation of vowels according to the degree of lip movement: labialized/non-labialized vowels (Tables 1 and 2).

Along with the similar features found in the system of vowel phonemes of the French and Mari languages, there are also significant differences. As can be seen from Tables 1 and 2, the vowels of the French language form four groups according to the ascent, and the Mari language—three groups.

Despite the fact that in both the French and the Mari languages the vowels are differentiated according to the ascent and the number, there are specific differences between the vocal systems of the languages. The French language has phonemes [e–ɛ], differing in degree of openness. In the Mari language, only one neutral phoneme [ə] is represented, which is characterized by less openness and greater advancement than the French sound [ɛ], and accordingly, greater openness and less advancement than the French [e]. The opposite nature of the French phonemes [ɔ – o] in terms of the degree of openness corresponds in the Mari language to the neutral [o], occupying an intermediate position between the French [o] and [ɔ].

Both languages show the polarization of the phonemes of the front and rear rows. In the Mari language of The French opposite phonemes [a – ɑ], which, however, is unstable, correspond to the Mari neutral sound [a]. The main difference between the French and Mari vowel systems is the presence of nasal sounds in French that are absent in the Mari language.

Despite the absence of nasal vowels the phenomenon of nasalization is also present in the Mari language. Under the influence of nasal consonants m, n, n', ŋ, the Mari vowels are

Table 1. Vowel phonemes of the Mari language.

Frontal			Back labial
Labial	Non-labial	Mixed non-labial	
ӱ	и		у
ӧ	э	ы	о
	а		

Table 2. Vowel phonemes of the French language.

Voyelles								
	Antérieures				Postérieures			
	Non-arrondies		Arrondies		Non-arrondies		Arrondies	
Degrés d'aperture	Orales	Nasales	Orales	Nasales	Orales	Nasales	Orales	Nasales
1	i		y				u	
2	e		ø				o	
3	ɛ	ɛ̃	œ (ə)	œ̃			ɔ	ɔ̃
4	a				ɑ	ɑ̃		

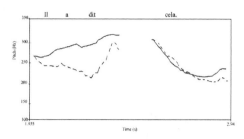

Figure 1. _____ model (French speaker); - - - - - - sample (Mari speaker). Il a dit cela. / He said this.

subjected to nasalization, the degree of which depends on the position of the vowel in relation to the nasal consonant.

In this case, the most typical deviations from the normative characteristics in the articulation of French vowels are:

- the replacement of French ovalized vowels with the sounds of the native Mari language similar in their phonetic characteristics;
- non-normative realization of the unglazed vowels of the front row with insufficiently vigorous stretching of the lips;
- incomplete nasalization of nasal French vowels, formed mainly by pronouncing the native stop consonant [ŋ];
- replacement of more open French vowels [a, ɛ, œ] with sounds of a native language similar in their characteristics, but more closed [a, e, ö] (Ivanova & Egoshina, 2015).

The experimental procedure included an instrumental analysis of the recorded research material in the Mari language and its comparison with the reference samples in French.

As you know, French narrative emotionally unpainted phrases are pronounced with an ascending-descending melody: a significant drop in tone is realized on a pre-syllable syllable. A slight increase in tone on the stressed syllable is an effective way of linking phrases in the context (Ivanova & Egoshina, 2015). Instrumental analysis demonstrates some discrepancies between the model speaker and the recording by a Mari student: a descending-ascending-descending tone without an improvement on the last stressed syllable.

Violation of the bilingual melodic system of the French language leads to the appearance of interference errors, which creates a foreign accent. The likening of melodic units of a foreign language to units of the native language is involuntary and is explained by the inadequate antithesis of the two linguistic systems in the consciousness of bilingualism.

As the examples illustrate, the experimental phonetic methods ensure the reliability of the research while observing certain methodological principles—relevance, objectivity, versatility and systemic nature.

These methods, as the researchers suggest, will help confirm or disprove the hypothesis that the two compared languages, for all their differences, can formally have identical characteristics of both phonemic and prosodic units. However, this similarity can not always serve as a support in mastering French pronunciation due to semantic differences in the phonemic and intonational structures of the French and Mari languages.

Materials can be explored from the point of view of different modern phonological concepts that are in mutually complementary relations, which contributes to the construction of an integrated complementary concept, oriented to solving applied problems.

4 CONCLUSION

Thus, the problem of interaction between languages in the modern world has evolved from a purely thoretical one into a problem of linguistic practice. The theory of inference, defined as the process and the result of the interpenetration of the systems of contacting languages, arose on the basis of the study of situations of natural language contact, in which the formation and functioning of bilingualism took place as a result of everyday interlingual communication.

Objective data on the differences in the systems of contacting languages, in particular French and Mari, explain the linguistic causes of many typical features of emphasis on the segmental and suprasegmental levels on the basis of experimental phonetic methods. For Mari students who study French, it is the comparative phonetic analysis of phoneme systems and prosodem that is important in order to determine the degree of influence of the articulatory base of the native Mari language on the acquired French language.

So, the phonetic interference in the psycholinguistic aspect appears as a distortion of the pronunciation norms of the system of the foreign language being studied (French), arising from the interaction in the minds of the speaker phonetic systems and the pronunciation norms of the two languages: the native, for example Mari, and the foreign, in our case French by means of transfer of auditory and pronunciation skills.

Moreover, comparative phonetic analysis in the study of interference presents objective data on differences in the systems of contacting languages and explains the linguistic causes of many typical features of the accent. Such an analysis takes on special significance in the teaching of Mari students to French pronunciation in conditions of Russian-Mari bilingualism.

REFERENCES

Avrorin, V.A. 1975. Problems of studying the functional side of language (to the question about the subject of sociolinguistics). Leningrad: Nauka.

Akhmetzyanova, F.S. 2010. Definition of interference zones as a result of the comparison of the means of expression causes in the compared languages. Bulletin of the Bashkir University, 2, vol. 15, 411–415.

Bryzgunova, E.A. 1984. Emotional-stylistic differences of Russian of spoken language. Moscow: Publishing house of Moscow state University.

Weinreich, W. 1979. Language contacts. Current status and problems of research. Kiev: Vyscha shkola.

Gordina, M.V. 1997. Phonetics of the French language. S.-Peterburg: Publishing house of S.-Peterburg state University.

Zorina, Z.G. 2000. Implementation of the phonological system in speech of a bilingualist. In Tartu (ed.), Congressus Nonus Internationalis Fenno-Ugristarum (p.285–286), Tartu, 7–13 August 2000. Pars II. Tartu.

Ivanova, I.G. & Egoshina, R.A. 2015. On the problem of intonational interference of typologically unrelated languages (on the example of French and Mari languages). Philological Sciences. The theory and practice, 12 (54), 80–84.

Lyubimova, N.A. 2012. Comparative study of the phonetics of the Russian language: linguistic and methodological aspects. Russian language abroad, 14–19.

Nikolaeva, T.M. 1984. Functional syntax and semantics. Intonation. Diachronic linguistics. Phrasal prosody: Parallels and divergence. Actual problems of intonation. Moscow: Publishing house of PFUR, 3–16.

Orazaeva A.A. & Ivanova I.G. 2016. Comparative analysis of the phonemic composition of the Mari and French languages. Actual problems of Roman and Germanic Philology and teaching of European languages in schools and universities. Yoshkar-Ola: Mari state University, 38–46.

Resource of a public debt in forecasting the development of the economy

M.V. Kazakovtseva & E.I. Tsaregorodtsev
Mari State University, Yoshkar-Ola, Russia

ABSTRACT: In recent years, questions about managing a public debt and exact forecasting of its level have become one of the crucial subjects in the economy and politics of the Russian Federation. The development and deployment of managing the regional finance of modern methods in the sphere of debt policy against the background of economic and financial crisis is one of the most important purposes of the program of reforming the system of regional finance.

1 INTRODUCTION

Unbalanced consolidated budgets of subjects of the Russian Federation is one of the topical issues of the public and municipal finances.

Financial and economic crisis is a serious threat for the system of public finances of territorial subjects of the Russian Federation. The collapse of the credit market and an economic decline have led to considerable fall of the income of regional budgets recently. The need of financial support for the vital sectors of economy and carrying out the stimulating economic policy do not result in significant cutting down of budget outlays. Regional budgets are not able to provide financings of all volume of powers now. As a result, practically all regions of the Russian Federation face a problem of considerable budgetary deficiency and sharp growth of a debt load. In fact, it would be better if regions coped with a deficiency problem by own forces, for example, due to attracting investors, increase in tax base and optimization of expenses. However, in a developed economy, it is almost impossible; therefore, using debt as a tool is inevitable (Ajupov A.A., Kazakovtseva M.V. 2014).

The current situation in the world financial markets remains unstable, as experts have predicted possible repetition of the crisis phenomena. Such state of the economy demands from public authorities of territorial subjects of the Russian Federation of carrying out extremely weighed and responsible budgetary and debt policy.

The gain of the budgetary services provided to the population and economic entities with simultaneous maintenance of a debt load at economically safe level has to become the purpose of effective debt policy at the level of the territorial subjects of the Russian Federation. The resource of a public debt can be considered:

- from the viewpoint of a source of growth of the public expenditures and the consumer demand connected with this growth and
- from the viewpoint of limit and optimum structure of a public debt for ensuring financial stability of the region.

This work is devoted to the latter aspect of a resource of a public debt.

2 RESEARCH MATERIAL

The main objective of implementing an effective debt policy by territorial subjects of the Russian Federation is to maintain the volume of a central government debt within the optimal size at the operation level. In other words, it is the compliance of the size of a central government debt to sources of its repayment during any period of time while considering all possible risks. In this case, it would be possible to take into account economically safe level of a debt load on the budget.

For the last 5 years, debts of subjects and municipalities have increased by 74% and 47%, respectively. These data confirm decrease of regional budgets in own means. The debt of territorial subjects of the Russia Federation as of January 1, 2017 was 2.35 trillion rub, and the debt of municipalities was 0.36 trillion rub. This constitutes about 3.2% of GDP of the Russian economy.

However, for comparing own income of regions, debt load of regions is essential – it makes about one-third of their own income. It is considerable, let and not critical size.

Furthermore, 42% of this debt is made by a nonmarket debt of regions of the Russian Federation (its annual percent 0.1%, on commercial credits, is higher than 10%). It is clear that at such percent, the share of expenses on debt servicing on average in budgets of subjects is very small – on average 2.3%.

The debt of regions has increased considerably in the last 10 years. During the previous crisis (2008–2009), it increased from 15% to own income of regions of 25%. The current crisis increased it to 32.8% in 2016 (Mikhaylov A. 2017).

In recent years, the policy of gradual replacement of commercial credits has been pursued by budgetary. The share of commercial credits in national debt of subjects decreased from 66% in 2013 to 54% in 2016.

The process of replacing commercial credits pursued by budgetary is a temporary crisis response measure of the Ministry of Finance. In the federal budget for 2017, under budgetary crediting, only 200 billion rubles against were reserved against 338 billion rubles for 2016.

The share of securities in a debt of regions remains invariable the last 2 years – 19% (2013–26%). In the beginning of 2017, 44 regions (more than a half) had no debt in securities. According to the Ministry of Finance, in 2016, attraction of the bank credits was planned by 74 territorial subjects of the federation, and issue of bonds by only 27 (actually carried out in its 22 regions).

Regions do not consider fully the preference of use of bonded mechanisms of attraction of financial resources before the credits of banks. The use of state securities as a source of

Table 1. Dynamics of a central government debt of territorial subjects of the Russian Federation (RF) and a municipal debt (one billion rubles).

	2012	2013	2014	2015	2016
Central government debt of territorial subjects of the RF	1351	1737	2089	2319	2350
Municipal debt	245	289	313	342	360

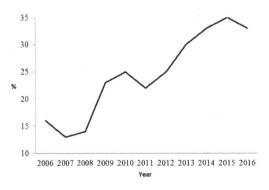

Figure 1. Debt of Russian regions in relation to their own income (%).

long-term financing of budget deficit has the following several advantages not characteristic of the credits of commercial banks: a possibility of attraction of "longer" money, lack of dependence on one creditor, loans with repayment in parts during the circulation period, but not at a time in date of repayment that also allows to form more flexible hours of repayment of obligations and is the additional instrument of regulation of volume of expenses on debt servicing (Kazakovtseva M.V. 2015). However, this tool is available first of all to regions with large volume of budget and good credit history; therefore, it is impossible to claim that the region attracting commercial loans pursues less effective debt policy than the region that has issued securities. Many subsidized subjects of the Russian Federation have no opportunity to place the securities as there are no buyers of such securities. It should be noted that demand for regional bonds in the market is insufficient nowadays.

For optimization of structure of loans, it is necessary to raise long-term borrowed funds (more than 5 years) for financing of investments into public infrastructure, long term and medium term (more than 1 year and up to 5 years), on refinancing of the available debt.

From the viewpoint of increase in efficiency of expenses, short-term loans are not allowed to be used for financing of long-term investment projects. It is possible to carry out attraction of short-term loans (for less than 1 year) only for maintenance of the current liquidity. The optimum level of short-term borrowed funds, according to recommendations of the Ministry of Finance, should not exceed 15% of volume of loans. Thus, in case of excess of volumes of short-term loans of the specified threshold in 15% of the total amount of loans, excess volumes can be carried to inefficient expenses (Kazakovtseva M.V. 2013).

The decision on the use of any tools of debt policy or cut in expenditure has to be carried out proceeding from priorities of regional policy and opportunities of the regional budget. Furthermore, an opportunity to attract financial resources and cost of loans is influenced by the following factors:

– current situation in the financial markets;
– credit history of the subject of the Russian Federation; and
– scale of alleged loans.

In an unstable financial situation (during crisis), loans will be more expensive; at the same volume of loans, expenses of subjects of the Russian Federation will be higher than those in the period of relative stability (Kadnikova T.G. 2014).

If a subject of the Russian Federation already declared once a default or was on the verge of a de-fault, loans for him/her will be more expensive than for the subject with a good credit history even if the subject carries out responsible policy of debt management at present.

Another important factor influencing the cost of loans is the total amount of means, which the subject of the Russian Federation plans to attract. Potential creditors prefer to deal with large volumes; therefore, loans will be more expensive to the region requiring to raise the small volume of funds, and securities it, perhaps, would not be able just to place (there will be no subjects interested in buying them).

There is no uniform indicator measuring effective management of a central government debt as, carrying out this activity, the state has direct impact on various areas of public life now. There are no techniques of an assessment of positive influence of a central government

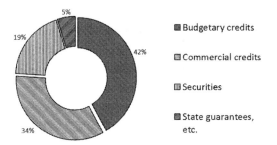

Figure 2. Structure of a central government debt of regions as of January 01, 2017.

debt on the budget and monetary circulation of the country, investment process, degree of trust of the population to financial activity of the state, and so on. It is possible to distinguish quantitative and qualitative criteria from criteria for evaluation of efficiency of debt policy of the territorial subjects of the Russian Federation. Indicators of an assessment of a debt load and stability of the territorial subject of the Russian Federation, indicators of service of a central government debt of the subject, and indicators of a structural assessment of a central government debt of the subject belong to quantitative criteria. It is possible to refer the following to qualitative criteria: observance by the subject of requirements of the budget legislation, carrying out monitoring of a central government debt of the subject, existence of a method of calculation of a debt load on the budget of the subject, introduction by the subject of own restrictions for the debt size, openness and availability of data on a condition of a central government debt of the subject, and existence of credit rating of the territorial subject of the Russian Federation (Kazakovtseva M. V. 2016).

The Budget Code of the Russian Federation defined two criteria of debt stability of subjects:

– a debt in relation to own income;
– expenses on debt servicing to the general expenses of the budget.

By the second criterion at regions, everything is alright because of a high share of inter-budgetary credits in a debt with almost 0%. But here, by the first criterion, there are obvious problems. More than a third of the Russian regions come under the "low debt stability" and a tenth are behind the line of debt risk.

Officially, the Ministry of Finance of the Russian Federation—no more than 50% – recommends to maintain the debt in relation to own income (to high-subsidized regions—no more than 25%), and a share of expenses on debt servicing in the general expenses—no more than 5%.

Significant increase in the debt load of territorial subjects of the Russian Federation requires the formation of an effective strategy of management of debt obligations for the formation of a reasonable budgetary policy as one of the components of long-term budget planning.

Effective management of a central government debt of territorial subjects of the Russian Federation should be aimed at balancing the regional budget, preventing unreasonable growth of a debt load on the budget, reducing volumes of budgetary deficiency, and, finally, reducing the size of the central government debt.

The main objective of implementation of an effective debt policy by territorial subjects of the Russian Federation is maintenance of volume of a public debt within the optimum sizes

Table 2. Regions with debt levels higher than the levels of own income.

№	Region	Debt in relation to own income (%)	Debt (billion rubles)	Share of budgetary credits (%)	Share of commercial credits (%)	Share of state securities (%)
1	Republic of Mordovia	176.0	40.2	56.0	15.0	24.0
2	Republic of Khakassia	145.5	22.9	13.0	37.0	50.0
3	Kostroma region	143.3	21.3	40.0	53.0	8.0
4	Astrakhan region	135.4	29.5	49.0	51.0	0.0
5	Republic of Karelia	115.1	22.6	52.0	29.0	17.0
6	Smolensk region	113.0	33.0	59.0	36.0	5.0
7	Mari El Republic	100.9	13.5	29.0	68.0	3.0
8	Jewish Autonomous Region	100.6	4.9	39.0	61.0	0.0
	Total		188.0			
	Share in a total debt of territorial subjects of the Russian Federation			8%		

at the operated level, that is, compliance of the size of a public debt to sources of their repayment during any period of time, considering set of all possible risks.

For the purpose of effective management and exact forecasting of the level of public debt and expenses on service by regional authorities, implementation of the following budgetary and program events is expedient (Kadnikova T.G. 2014):

1. Carrying out inventory of the existing debt obligations, including regarding compliance to standards of the Budgetary Code of the Russian Federation.
2. Flexible response to the changing conditions of the domestic financial market and use of optimum sources and forms of loans.
3. Control of a condition of receivables and payables.
4. Mobilization of new credit resources only for financing of priority projects and programs on condition of their effective use.
5. Improvement of quality of a debt due to depreciation on service.
6. Monitoring of course of execution of obligations by the principal for the provided state guarantee.
7. Ensuring the advancing decrease in growth rates of a public debt in relation to growth rates of the tax and nontax income of the budget.
8. Formation of proportions of a debt portfolio of the region in favor of the nonmarket credits.
9. Maintenance of the assigned credit score with the prospect of increase.

Audit of the efficiency of debt policy of the territorial subjects of the Russian Federation is necessary for defining the degree of effective management of a public debt of the territorial subjects of the Russian Federation and effectiveness of use of borrowed funds in the regional budget at realization of the priority directions of social and economic development. This estimated tool will allow revealing the existing risks, shortcomings, and discrepancies; designating the positive and negative moments in the existing regional legislation in the sphere of management of the regional budget, a public debt of the territorial subject of the Russian Federation; offering the possible solutions of the available problems promoting increase in efficiency (improvement) in the debt policy of regions.

3 CONCLUSIONS

The regional debt policy is a specific factor of the change in the financial stability level of the budget, which is connected with the intense use of the sources of budget deficit financing, with the features of the subject's emission policy. The average level of debt of regions is not threatening to the country. Today, even debt leaders are very far from a possibility of a default. Nevertheless, the Ministry of Finance has taken measures to constrain the deficiencies of regional budgets. In 2016, the increase of regional debt had practically stopped, and its structure for regions had improved. In other words, the share of budgetary credits had increased.

Following the results of 2017, experts expect increase in the volume of a public debt by 5–7% and the tax and nontax incomes of regional budgets within 10%. In that case, the debt load will be 32–33%, and in comparison to the result of 2016, it will not change significantly.

However at deterioration in a macroeconomic situation, reduction of prices of oil and growth of federal budget deficit instability of a federal situation can immediately be thrown at a regional level. Therefore, the creation of a transparent and effective debt policy at the regional level plays an important role in exactly predicting the development of the economy of a region.

ACKNOWLEDGMENT

The authors thank the Russian Science Foundation for their support by the grant RSF-16-18-10017 "The program complex for forecasting of economic development of the region".

REFERENCES

Ajupov A.A., Kazakovtseva M.V. 2014. Management of financial stability of the non-tax income of regional budgets.
Procedia – Social and Behavioral Sciences. T. 131. C. 187.
Ermakova E.A. 2014. Methodical approaches to an assessment of efficiency of debt policy of the territorial subject of the Russian Federation, Finance and Credit. No. 28 (604). Page 32–39.
Kadnikova T.G. 2014. Separate aspects of debt policy of the Republic of Karelia, Studia Humanitatis Borealis. No. 1. Page 102–109.
Kazakovtseva M.V. 2015. Effective management of a central government debt of the territorial subject of the Russian Federation, Bulletin of the Mari state university. Series: Agricultural sciences. Economic sciences.. No. 3 (3). Page 60–64.
Kazakovtseva M.V. 2016. Improvement of a control system of a central government debt, Urgent problems of economy of modern Russia. No. 3. Page 134–139.
Kazakovtseva M.V. 2013. Methodology of an assessment of efficiency of budget outlays, Economy and Business. No. 11 (40). Page 145–152.
Kazakovtseva M.V. 2013. Methodical tools of an assessment of efficiency of the budgetary income and budget outlays, Innovative development of economy. No. 2 (14). Page 43–47.
Kazakovtseva M.V., Gumarova F.Z., Tsaregorodtsev E.I. 2015. Forming of Competitive Advantages of Regional Agrarian and Industrial Complex as Mechanism of Ensuring Economic Safety, Mediterranean Journal of Social Sciences. T. 6. No. 3. Page 213–220.
Mikhaylov A. 2017. A Debt hole for governors http://www.profile.ru/economics/item/115341-dolgovaya-yama-dlya-gubernatorov (accessed 11.06.2017).

Conceptual theories of forming leadership qualities of a competitive university student

S.Y. Lavrentiev, D.A. Krylov & S.G. Korotkov
Mari State University, Yoshkar-Ola, Russia

ABSTRACT: The urgency of the investigated problem is due to the socio-economic development concept of Russian Federation for the period until 2020, it is noted that socially active, competitive professionals with creative thinking, leadership qualities, taking responsible decisions, are required for a modern society dynamically developing along an innovative path, calculating their possible scenarios in a situation of free, independent choice. The article gives an overview and analysis of approaches to the successful formation of professional and personal leadership qualities of a competitive graduate of a modern university. An interdisciplinary study of developments in the field of social anthropology, sociophilosophy, pedagogical management, management psychology, generalization of the achievements of modern psychological and pedagogical studies of this phenomenon is given, a critical analysis of the eclectic of social and philosophical views and an acute shortage of conceptual unity to the problems of leadership by science and practice are given. It is shown that the most famous was the behavioral approach, based on the basic provisions of which were formulated the factors that contribute to increasing the effectiveness of the formation of competitive leadership behavior of a student at a modern university. A leadership is defined as the mutual influence of the leader and his followers, having a positive impact on each other, working together to achieve effective changes and achieve planned, results-oriented goals. Thus, the formation of leadership qualities of a competitive university graduate can be considered in the context of interdisciplinary integration of behavioral, transactional, motivational leadership theories.

Keywords: Leadership, competition, theories, interdisciplinary approach

1 INTRODUCTION

In the socio-economic development concept of Russian Federation for the period until 2020, it is noted that socially active, competitive professionals with creative thinking, leadership qualities, taking responsible decisions, are required for a modern society dynamically developing along an innovative path, calculating their possible scenarios in a situation of free, independent choice. Modernization of the entire socio-economic, cultural and educational system enhances the need for professional training of creative, enterprising specialists.

Considering the modern technological changes and the growth of competition promoted by the processes of markets globalization, information and digital innovations, the concept of the Federal target program for the development of education for 2016–2020 proposes a broad application of the personality-oriented model of education that will promote the competitiveness of the future specialist. The formation of its leadership qualities, the human capital of educational institutions and ultimately the economy and the state.

The study of such a complex, multifaceted phenomenon as leadership requires an interdisciplinary systematic approach and a synergetic interpenetration of the latest theoretical and empirical innovations in various fields of scientific thought—biophilosophy, social philosophy and social anthropology, pedagogical management, organizational behavior and management psychology, etc.

2 LITERATURE REVIEW

It would be wrong to argue that the problem of leadership has not received due attention to date. A number of interesting scientific publications related to this problem appeared in the twentieth and early twenty-first centuries in Europe and the USA (R. Stogdill, 1957; F. Fiedler, 1995; J.D. Mayer, 1999 etc.). In Russia, until the seventies and eighties of the last century, this topic was bypassed by many psychologists and sociologists for a number of reasons, or was addressed in its separate aspects. The situation has changed in the past three or four decades (T.V. Bendas, 2009; O.V. Evtikhov, 2007; A.L. Zhuravlev, 1983; R.L. Krichevsky, 1998; E.V. Kudryashova, 1996; B.D. Parygin, 1971 et al.).

It is noteworthy, however, that the issues of leadership were analyzed most often in relation to such branches of activity as politics, economy, business, public practice in its various modifications, sports and some others. At the same time, theoretical and methodological problems, related to the formation of a competitive student leadership qualities of a modern university, have so far been little studied.

According to the philosophical dictionary, where leadership is interpreted (from the English as a leader, manager), as one integrative mechanism of collective activity, when a leader is the individual or group of people that is part of a social group, through the management of the activities of the entire group, host and supporting his actions.

In the Russian pedagogical encyclopedia, the concept of leadership is explained as rendering the dominant influence of the group member, that is, the leader, on each member of the group.

Traditionally, Russian researchers viewed the leader as a member of the group, which is nominated for the first positions, regulates interpersonal relations and organizes group activities. To the main feature of leadership qualities, researchers tend to attribute the ability to influence people united by a common goal. Thus, the term "leadership" was interpreted only as a group phenomenon. The leader was perceived not as part of the organization, but as an element of intra-group relations in this organization. Often, the contradictory nature of claims to leadership and the degree of other group members readiness to take his leading role was emphasized.

3 RESULTS

The problem of forming the desired leadership qualities is at the center of attention of many scientific and practical discussions. It is the object of numerous studies by scientists around the world and now. In the works of ancient philosophers from Plato to Plutarch, leading figures became outstanding leaders of that time. Medieval thinkers, studying the qualities of outstanding representatives of that time, noted that leadership derives from individual characteristics of the human person. Philosophical representations of that time often reduced the essence of leadership to one of the most significant characteristics or properties.

4 DISCUSSIONS

In the XIX century, many researchers addressed the study of the phenomenon of leadership, among them – F. Galton, S. Rhodes, etc. One of the founders of the concept of the psychology of individual differences, the English anthropologist F. Galton, combining the existing notions of hereditary conditioning of individual psychological differentiation between people, argued that some leadership qualities can be transmitted by inheritance (F. Galton, 2001). An outstanding South African politician and successful businessman, of British descent, Cecil John Rhodes, on the contrary, convinced that the formation of leadership qualities is possible through appropriate education.

In the 20th century, due to the development of a variety of psychological, pedagogical and sociological diagnostic tools, the first attempts to empirically substantiate theoretical studies of the essence of leadership behavior appeared. Nevertheless, despite many theoretical and empirical interpretations of the essence of leadership, there is no consensus among researchers, practitioners, and to date, a solution to the problem of effectively forming the leadership qualities of a competitive personality. In the psychological-pedagogical, management literature, there are several theories that differ in the genesis of leadership: the great man theory; personal qualities; behavioral; probabilistic; influence; relationships.

An analysis of modern research allows us to highlight conceptual theories that have gained popularity in the field of studying the essence of leadership. It is advisable to designate the first theory as personalistic (B. Bass, 1990; F. Woods, S. Klouchek, A. Meneghetti, O. Teed, etc.). The line of original research in the vein of personalistic theory was built from the earliest leadership theories, which were developed in the early twentieth century, focusing on the study of individual leadership qualities. In the studies of supporters of the situational approach, empirical evidence is evidenced that qualitatively different circumstances can "demand" qualitatively different leaders. In specific situations of group activity, members of the group who have the most personally sought-after professional qualities who become leaders in the given conditions are brought to the forefront. Accordingly, in order to identify the qualities of organizational leaders, it is necessary to study the characteristic features that have a dominant effect on the results of their professional activities, as well as the subjective views of the group's participants on the most demanded leadership qualities in situations of command intra-group interaction.

In the mainstream of the problem we are examining, theories dealing with issues of leadership behavior and interaction (J. MacGregor Burns, 1978; K.X. Blanchard, 1977; V. Vroom, 1998; F. Yetton, R. Likert, 1967; C. Lewin, 1939; R. Tannebaum, 1973; P. Hersey, 1977 et al.). The first research developments in the framework of behavioral theory were based on the concept of behaviorism, respectively, the central object of research in this field was directly targeted activity, and the quality of leadership was considered "leadership style". In further studies (Warren Blank, 1995 et al.), the attention of the authors was focused not so much on the behavioral style of leadership, but rather on the features of the interaction of leaders and their followers.

The emphasis on the perception by the followers of the motives that guide the leader, his behavior and actions, as well as on expectations from these actions, is investigated by B. Calder, T. Parsons, J. Pfeffer, F. Heider, E. Hollander and others in the channel of the transactional the theory, based on attributive analysis of leadership.

Attributing the intentions of the group members through the evaluation of interpersonal perception was reflected in the study of the essence of the discussed phenomenon of leadership, which became the core of the created model of so-called "credit" (trust, merit, reputation, respect, etc.) E. Hollander. According to the developed model, the "credits" of the ability to influence members of the group are earned by the leader during a certain period of group life, and is determined by how much the followers understand his competence in significant situations and to what extent his behavior corresponds to the norms existing in the group.

Differentiation in the views of representatives of social and humanitarian thought, manifested itself in the development of a motivation-value theory. A. Maslow (2003), T. Mitchell (1974), G. Feyrholm, R. House (1974), K. Hodgkinson, S. Evans and others, oriented the system of coordinates of scientific search on the study of worldviews, values, motives of the leader and his followers' activity, development of models, creation of principles interaction and interaction between the elements of the system. Theoretical and empirical interdisciplinary representations of leadership behavior were integrated in the motivational theory, where a correlation was found between the effectiveness of the leader and his ability to influence the motivation of followers.

In the framework of this study, attention should be paid to the basic provisions of the value theory of leadership developed by S. and T. Kuchmarsky. According to the first statement, the leader has a significant influence on the development of the values of both the organization as a whole and its individual representatives. Theoretical and empirical calculations of the second indicate that in the process of activity, value-oriented leadership needs to be constantly trained (Evtikhov, 2007).

Expanding the author's view on the theory of leadership, scientists formulated indicators of value leadership in the organization. Susan and Thomas Kuchmarski identified the priority values of leadership phenomenon: the establishment of positive interpersonal relationships; knowledge of the personal goals of each subject of the leadership process; formation of a sense of belonging in the group and involvement in the society; the success of solving conflicts of an interpersonal nature; training work on developing leadership behavior of followers; creation of conditions for the realization of the personal and professional potential of other individuals; communicative polylogical interpersonal interaction; connection of internal culture with external representation; demonstration of enthusiasm and support for diversification. In the theory presented, leadership is formalized as a permanent dynamically developing process, manifested in the "leader-follower" relationship, and the role of the leader is in the internalization of the above-mentioned value principles, teaching them followers and training with them.

5 CONCLUSION

Considering the above-described theories, leadership is defined as the mutual influence of the leader and his followers, having a positive impact on each other, working together to achieve effective changes and achieve planned, results-oriented goals. Thus, the formation of leadership qualities of a competitive university graduate can be considered in the context of interdisciplinary integration of behavioral, transactional, motivational leadership theories.

ACKNOWLEDGEMENT

The reported study was funded by RFBR and Government of the Mari El region according to the research project № 16-16-12005.

REFERENCES

Bass B. Stogdill's handbook of Leadership. N.Y., 1981. Retrieved from http://discoverthought.com/Leadership/References_files/Bass%20leadership%201990.pdf.
Bendas T. Gender Gender Studies of Leadership // Questions of Psychology. 2000. № 1. PP. 87–95.
Blank W. The Nine Natural Laws of Leadership. New York: AMACOM, 1995. Retrieved from http://www.celtagora.com/the-nine-natural-laws-of-leadership.pdf.
Evtikhov O.V. Training of Leadership: Monograph. – SPb.: Speech, 2007. – 256 p.
Fiedler F.E., Potter E.H. Dynamics of Leadership effectiveness // Small group and social interaction. V.I., London, 1983.
Galton F. Inquiries into human faculty and its development// First electronic edition, 2001. Retrieved from http://galton.org/books/human-faculty/text/galton-1883-human-faculty-v4.pdf.
Heider F. The Psychology of Interpersonal Relations. N.Y., 1958.
Hersey P., Blanchard K.H. Management of organization behavior. N.J., Prentice Hall, 1977. - 412 p.
Hollander E.P. Conformity, Status and Idiosyncrasy Credit // Psychological Review. 1958. № 65.
House R.J., Mitchell T.R. Path Goal Theory of Leadership // Journal of Contemporary Business. – 1974. – №3. – pp. 81–97.
Ivanenko N.A., Akhmetov L.G., Lavrentiev S.Y., Kartashova E.P., Lezhnina L.V., Tzaregorodtzeva K.A & Khairullina E.R. Features of modeling the formation of teaching staff competitiveness. Review of European Studies. Vol. 7. No 3; 2015, pp. 37–42. Retrieved from http://dx.doi.org/10.5539/res.v7n3p37. DOI:10.5539/res. v7n3p37.
Ivanenko N.A., Lavrentiev S.Y, Khrisanova E.G., Tenyukova G.G., Kuznetsova L.V., Yakovlev S.P. & Shvetsov N.M. Basic Principles for Forming Teaching Staffs' Competitiveness in Vocational Training Institutions. Review of European Studies; Vol. 7, No. 5; 2015, pp. 118–123. Retrieved from http://dx.doi.org/10.5539/res.v7n5p118. DOI:10.5539/res. v7n5p118.
Komelina V.A., Mirzagalyamova Z.N., Gabbasova L.B., Rod Y.S., Slobodyan M.L., Esipova S.E., Lavrentiev S.Y., Kharisova G.M. Features of Students' Economic Competence Formation.

International Review of Management and Marketing/ Vol. 6(1), 2016, pp. 53–57. Retrieved from URL: http://www.econjournals.com/index.php/irmm/article/view/1736/pdf.

Krichevsky R.L. If you are a leader. M.: Case, 1998. 400 p.

Kudryashova E.V. Leadership as a subject of social and philosophical analysis. PhD.M., 1996. 359 p.

Lavrentiev, S.Y. Socio - economic and pedagogical preconditions for the formation of competitiveness of the future specialist in the educational process of the modern university/S.Y. Lavrentiev, D.A. Krylov//Modern problems of science and education. – 2015. – No. 5; Retrieved from URL: www.science-education.ru/128–22145.

Lavrentiev S.Y., Komelina V.A. Genesis and evolution of the concept of "competitiveness" in the history of human thought//Bulletin of the Mari State University. 2016. No. 4 (24). pp. 22–27. Retrieved from URL: http://vestnik.marsu.ru/view/journal/download.html?id=1280.

Lavrentiev S.Y., Shabalina O.L., Krylov D.A., Korotkov S.G., Svetlova V.A., Rybakov A.V. and Chupryakov I.S., 2016. Future Specialists' Competitiveness Development: Pedagogical and Social-Economical Aspects. The Social Sciences, 11: 1855–1860. DOI: 10.3923/sscience.2016.1855.1860. Retrieved from URL: http://medwelljournals.com/abstract/?doi=sscience.2016.1855.1860.

Lavrentiev S.Y, Krylov D.A., Komelina V.A., Arefieva S.A. Theoretical Approaches to the Content and Structure of Competitiveness of Future Teacher. Review of European Studies. Vol. 7. No 8; 2015, pp. 233–239. Retrieved from http://www.ccsenet.org/journal/index.php/res/issue/view/1324. DOI: 10.5539/res.v7n8p233.

Likert R. The Human Organization. McGrow-Hill, 1967. – 288 p.

Mayer, J.D. Emotional Intelligence: Popular or scientific psychology?//APA Monitor. September, 1999. 30, p. 50. Emotional Intelligence Information; John D. Mayer. – Los Angeles, 2008; http://www.unh.edu/emotional_intelligence/ EI%20 Assets/Claims/EI1999MayerAPAMonitor.pdf.

McGregor Burns J., Leadership. N.Y., Harper & Row, 1978. – 498 p.

Maslow, A. Some Educational Implications of the Humanistic Psychology//Harvard Education Review. 2003. – V. 38.4 – P. 688.

Orekhovskaya N.A., Lavrentiev S.Y., Khairullina E.R., Yevgrafova O.G., Sakhipova Z.M., Strakhova I.V, Khlebnikova N.V., Vishnevskaya M.N. Management of young professionals in the labor market. International Review of Management and Marketing/ Vol 6, No 2, 2016, pp. 254–269. Retrieved from URL: http://www.econjournals.com/index.php/irmm/article/view/2098/pdf.

Parygin B.D. Foundations of the Socio-Psychological Theory. M.: Thought, 1971. 351 p.

Pfeffer J. The ambiguity of leadership // Academy of Management J. 1977. N 2. Retrieved from https://cnx.org/resources/57a9239e98cfe3a054423cb2d81f6023f4b8c8ad/Pfeffer.pdf.

Philosophical Encyclopaedic Dictionary. – Moscow: Soviet Encyclopedia. Ch. Edition: LF Il'ichev, P.N. Fedoseev, S.M. Kovalev, V.G. Panov. 1983.

Russian Pedagogical Encyclopedia. – M: "The Great Russian Encyclopedia". Ed. V.G. Panova. 1993.

Shabalina O.L., Svetlova., Krylov D.A., Lavrentiev S.Y. The Model of Formation of Readiness of the Future Pedagogues to Self-Education by Means of Interactive Technologies. Mediterranean Journal of Social Sciences. Vol. 6. No 3. S.7. June 2015, pp. 133–137. Retrieved from URL: http://www.mcser.org/journal/index.ph...e/view/167. DOI: 10.5901/mjss. 2015. v 6 n 3 s7 p 133.

Stogdill R.M. Handbook of Leadership. A survey of theory and research. N.Y., 1974.

Tannenbaum R., Schmidt W.H. How to Choose a Leadership. Harvard Business Review. – 1973. – 36 (May–June). – Pp. 95–101.

The concept of long-term socio-economic development of the Russian Federation for the period until 2020. Order of the Government of the Russian Federation of November 17, 2008 No. 1662-r. Retrieved from URL: http://www.consultant.ru/document/cons_doc_LAW_82134.

Zhuravlev A.L. Socio-psychological problems of management/Applied-Problems of social psychology: Collection of articles. Moscow: Nauka, 1983. Pp. 173–189.

Pragmatic aspect of implicit addressing means in different spheres of communication

T.A. Mitrofanova, T.V. Kolesova, M.S. Romanova & E.V. Romanova
Faculty of Foreign Languages, Pedagogical Institute, Mari State University, Yoshkar-Ola, Russia

ABSTRACT: In political and scientific communication fields an addresser's intention to prove his point of view, to influence the addressee's opinion and his evaluation of the problem are realized more effectively while using implicit addressing means. The linguopragmatic method gave a new impulse to research implicitness. Linguistic means can be hidden by word's meaning, their syntactic links in utterance. Addressing phenomenon is expressed in semantic, syntactic and pragmatic levels of a language. Thus we can speak about its super-pragmatic nature.

Keywords: addressing, addresser, addressee, implicitness, linguistic pragmatics

1 INTRODUCTION

At present, in the age of information technology, when a person has to handle a large amount of information, studies of implicit ways of conveying the pragmatic meaning of an utterance acquire particular urgency. The purpose of this article is the consideration of means of implicit transmission of the addressing in such spheres of human communication as politics, science.

Implicit addressing means in the political and scientific spheres of communication are considered in this study from the point of view of linguistic pragmatics. The focus of this approach is a person with his socio-individual characteristics, which influence the choice of linguistic means of addressing to provide a pragmatic impact. In the political and scientific spheres of interaction, pragmatic influence through implicit means of addressing plays an important role in achieving pragmatic goals, because the addresser must convince the addressee to accept his point of view.

2 STATEMENT OF PROBLEM

The twentieth century was marked by the personification of personality. Each person is viewed as a complex personality, versatile in a psychological sense, requiring a differentiated approach, so the person became the central figure of the language, which determined the anthropocentrism of modern linguistics. The origins of the anthropocentric approach lie in the sphere of linguistic pragmatics (pragmalinguistics) (Maslova, 2007), which explores the relationship of signs to their interpreters, that is, to those who use language systems, who creates, accepts and understands the sign (Arutyunova, 1985). Language is explored as an instrument of action and influence in different conditions as well as for different purposes, therefore linguistic pragmatics appeals primarily to the addressee as a criterion in the choice of linguistic means. This is particularly evident in the construction of political and scientific discourse, since the addresser implements in them two basic communicative and pragmatic intentions: firstly, to influence an addressee's consciousness, his opinion and evaluation by language means; secondly, encourage him to commit certain actions, somehow change, direct his behavior (Chernyavskaya, 2006).

3 CONCEPTUAL FRAMEWORK

3.1 *Research instrument*

A communicative-pragmatic approach occupies a central place in modern linguistics. This approach is a natural consequence of the development of research in the field of language, which evolved from the system-structural, i.e. the attention was paid to the study of the language "in itself and for itself" in isolation from external factors, to the communicative-pragmatic, when the language is considered in its functioning and use, pragmatic factors such as the communicative situation, the characteristics of communicants, the intentions of the addresser, the reaction of the addressee, the channel (oral/written) of information transfer are taken into account.

The change of the scientific paradigm makes it possible to look at linguistic phenomena from a new position and reveals those properties that were not taken into account before, remained out of sight (Mitrofanova, 2009). A number of linguistic units, coming from a linear system-structural plane into a pragmatic volume space, reveal additional properties (Akhmanova, 1966) that convey the pragmatic value of the addressing and find both an explicit and an implicit implementation.

4 RESEARCH MATERIAL

The study of the phenomenon of implicit addressing within such types of institutional discourse as political and scientific is of special interest in connection with the active expansion of mass information communication into the everyday life of people. Because of the transparency of the boundaries of discourse, the characteristics of different types of discourse are often superimposed on each other.

In this study, linguistic analysis was conducted on the basis of public speeches by German politicians and a scientific article in a specialized journal. The intention of the addresser to defend his point of view, to influence the opinion and evaluation of the addressee, in our view, is most effectively achieved by using implicit means of the addressing, which also represent the axiological sphere (Kaptsov, 2008) of communication participants.

5 FINDINGS

The study of the phenomenon of addressing (the address) the statements was made by both domestic and foreign linguists (M. M. Bakhtin, A. M. Pesh-kovsky, A. G. Rudnev, V. E. Goldin, A.V. Polonsky, S. L. Nistratova, N. P. Volvak, E. G. Zheludkova, E. Benveniste, R. O. Jacobson, H. Brinkmann, J. Erben, H. Glinz, A. Kohtz and others). Some linguists view addressing as a functional semantic category with a grammatical center—a grammatical category of addressability, the semantic dominant of which is the addressee as the second subject of the communicative speech act (pragmasemantic addressee) (Polonsky, 1999).

Others consider the address as one of the essential characteristics of the utterance, which is created by actualizing the addressee's role by the speaker (Volvak, 2004). Still others talk about the category of addressing and understand it as an addressing to another (Zheludkova, 2004). Different approaches in the interpretation of the direction of the utterance, the absence of a unified terminological base, and the study of mainly explicit means of addressing indicate a lack of elaboration of the problem of implicit methods of transmitting addressing in various spheres of human communication and the need to clarify the linguistic essence of this phenomenon.

The addressing (the address) of the utterance is widely considered. This notion includes not only means of addressing, but also all the language means associated with such pragmatic characteristics of utterance as the two-way communication act, emotionality and expressiveness, i.e. with the characteristics connected with the recipient of the message (Nistratova, 1985).

The implicit problem is revealed when we are thinking what a person is saying and what he wants to say, what he means, what he hints at. Researchers drew attention to the fact that a number of abstract grammatical meanings are expressed indirectly. Hidden grammar was and is the subject of searches of many linguists of the past (V. Humboldt, A.A. Potebnya, A.A. Shakhmatov, L.V. Shcherba, O. Espersen etc.) and the present (V. L. Medynskaya, I. I. Revzin, V. A. Zvegintsev, N. S. Vlasova, E. I. Shendels and others). The opinions of linguists agree that the relationship between form and content in grammar is inconsistent not only in the sense that several grammatical forms can correspond to one grammatical function and, conversely, several functions may correspond to one grammatical form, but also in that deeper sense, that not every grammatical category gets a direct and immediate expression in the grammatical forms of a given language. Many grammatical categories appear, from this point of view, hidden in the meanings of words and the syntactic connections of words in the sentence (Katznelson, 1972), for example, the category of addressing (A.V. Polonsky, N.P. Volvak, E.G. Zheludkova, V. E. Chernyavskaya et al.).

6 DISCUSSION AND INTERPRETATION

The development of pragmatics gave a new impulse to the study of implicitity, because scientists came to the general opinion that the content of the message (utterance) is formed as a result of additional efforts of the listener and is not reduced to a simple identification of linguistic signs. We adhere to the point of view of linguists who understand the term "implicit" as "hidden, intended, not evident" (O.S. Akhmanova, E.I. Shendels) (Shendels, 2006). Researchers note that such phenomena as ellipse, zero modes of expression (zero morpheme, zero article), pronouns, "background knowledge" (G. Brinkman) can not be attributed to the implicitity, because they presuppose the immediate completion of the missing elements of the utterance from the immediate context or situation, and implicitness presupposes an indirect, implied expression of any meaning (E.I. Shendels).

In the speech of the agents of politics, German Federal President Horst Köhler (2004–2010) and Federal Chancellor Angela Merkel, the orientation toward the addressee is manifested in the use of a wide range of implicit means of addressing: utterances of different communicative orientation, inclusive pronominal-verb forms of the 1st person plural, the prepositionless dative case, modal verbs, passive constructions, subordinate clauses of conditional type. These means acquire real addressing only in a concrete utterance, which in modern linguistics is not considered outside the categories of the author and the addressee, addressing is recognized as his communicative, functional-semantic foundation (Volvak, 2004).

It should be noted the high degree of addressing of speech acts, which presuppose an immediate verbal response of the addressee or calculated for the subsequent perlocutive effect. The question as a type of utterance combines several subjective perspectives. On the one hand, it is put by the author, expressing subjectively refracted information, i.e. it is a mean of maintaining addressability. On the other hand, the question presupposes is the existence of an opinion that is not identical to the author's, and strengthens the address of speech.

Researchers distinguish three main groups of questions: rhetorical questions, dialogical unity, question-answer complexes (Chernyavskaya, 2006). By their logical component of rhetorical questions they are directed to the textual content, and by the expressive component—to the reader in order to activate his position. Rhetorical questions can have the function of exclamation especially when following one another (Romanova 2015).

In his speech at the annual meeting of the founding union of German science, the Federal President Horst Köhler (2004–2010) addressed to the scientific community and noted that ideas and innovations in science are important for the welfare of Germany, and noted that Germany's main resource is a man. The Federal President asks the question, whether conditions are created for the development of creativity in science. Following each other rhetorical questions with the included personal pronoun of the 1st person plural *wir* 'we' contain a reproach, addressed to the scientific community, e.g.:

1. Wir wissen das natürlich alles. Aber handeln *wir* danach? Sorgen *wir* unermüdlich für die Bedingungen, die Kreativität zulassen oder gar ermutigen und erleichtern? (Speech by Federal President Horst Köhler).

Of course *we* know everything. But are *we* acting after that? Are we tirelessly concerned about the conditions that allow or even encourage and facilitate creativity? (Speech by Federal President Horst Köhler).

The parceled structure of the response turn of the question—dialogic unity additionally emphasizes its correlation—logical, syntactic—with the preceding question (Fadeeva, 2016). An immediate and univocal answer to the question reinforces the author's element (Chernyavskaya, 2006), e.g.:

2. Aber wie verläuft eigentlich der Weg von der Feststellung eines Problems zu seiner Lösung? Wie werden aus einer guten Idee viele gute Produkte? Wie entstehen Innovationen, die uns das Leben leichter machen, die es bereichern?

Ob in der Kunst oder in Wissenschaft und Forschung: immer kommt es vor allem auf eine Fähigkeit an: auf Kreativität. (See ibid.).

But how does the path actually run from a ascertaining of the problem to its solution? How does a good idea become many good products? How do innovations arise that make life easier for us and enrich it?

Whether it is in art or science and research, it always depends on an ability: on creativity. (See ibid.).

The question-answer complexes realize in the institutional discourse the idea of dialogueness of the speech and form a complex combination of the authorization and addressing vectors. They distinguish between question-answer complexes, in which the question or statement is perceived as initiated by the addressee, but is pronounced by the author of the speech. H. Köhler responds with an indirect question-statement addressed to entrepreneurs (*Ihr als Unternehmer*), which are foregrounded by the conjunction *als*, the personal pronoun of the 2nd person plural *Ihr*, e.g.:

3. Und wenn ich manchmal den Einwand höre: *Hierzulande finden wir ja längst nicht mehr genug hochqualifizierte, gute Mitarbeiter*, dann gebe ich darauf die Frage zurück: Könnt *Ihr als Unternehmer* nicht sehr viel dazu beitragen, dass sich daran etwas ändert, indem *Ihr* die Schulen und Hochschulen unterstützt ... (See ibid.).

And when I sometimes hear the objection: *Here we do not find for a long time enough highqualified, good employees*, then I return the question: Can not *you, as entrepreneurs*, help to make a difference to this by supporting the schools and colleges ... (See ibid.)

Such question-answer complexes are a special method to put forward the additional, contextually important information.

The dominant function of the speech of the Federal Chancellor of Germany A. Merkel, dedicated to the solemn event of the fund "Erinnerung, Verantwortung und Zukunft" 'Memory, responsibility and future', is an expression of gratitude. This pragmatic vector and the addressing implied by the speech act of gratitude (gratitude is always addressed to someone), are realized with statements containing:

a. performative verb *danken* 'to thank' (Romanova & Smirnova 2015), e.g.:

4. Ich *danke Ihnen*. (Speech by Federal Chancellor Angela Merkel).

I *thank you*. (Speech by Federal Chancellor Angela Merkel).

b. A structure representing the modality of necessity and possibility, as well as the evaluative attitude of the Bundeschancellor. The structure *haben zu verdanken* 'have to thank' is used in relation to persons with high social status, e.g.:

5. Dass dies gelingen konnte, *haben* wir vornehmlich dem amerikanischen und dem deutschen Verhandlungsführer *zu verdanken*: Stuart Eizenstat und Ihnen, Graf Lambsdorff. Sie sind die Väter dieser historischen Einigung. (See ibid.).

That this succeeded, we have to thank mainly the American and German negotiation leaders: Stuart Eizenstat and you, the Earl of Lambsdorff. You are the fathers of this historical union. (See ibid.).

Using a separate nominal address in the postposition after the pronoun in the polite form *Ihnen, Graf Lambsdorff* 'you, the Earl of Lambsdorff' performs not only a phatic function, but also contains an evaluation component, the attitude of the Chancellor.

c. The common noun *Dank* 'thank' with the verb in the indicative mood.

Summing up, the Bundeschancellor mentions once again all those to whom she expresses her gratitude, uses the indefinite pronoun *alle* 'all', the subordinate clause attributive *die dazu beigetragen haben* 'who have contributed to the fact', the possessive pronouns with nouns *meinen Respekt, meinen aufrichtigen Dank* 'my respect and my sincere thanks' contain the evaluation, e.g.:

6. Kurzum: *Allen, die dazu beigetragen haben*, dass dieses weltweit beachtete politische Zeichen zustande kommen konnte, spreche ich *meinen Respekt* und *meinen aufrichtigen Dank* aus. (See ibid.).

In short, I would like to express *my respect and my sincere thanks* to *all* those *who have contributed to the fact* that this worldwide respected political symbol could be achieved. (See ibid.).

In a strong position, at the end of speech, the addressing to listeners is strengthened.

The structural-semantic addressee allocated by some linguists, i.e. object-addressee, is transmitted by a specialized form of the prepositionless dative case—dative of the indirect object, or dative addressee (Polonsky, 1999). "The person's name in the dative case with the predicate of speech acquires the semantic-syntactic function of the addressee of speech" (Katznelson, 1972): *Ihnen* 'you' (4); *dem amerikanischen und dem deutschen Verhandlungsführer* 'the American and German negotiation leaders', *Stuart Eizenstat und Ihnen, Graf Lambsdorff* 'Stuart Eizenstat and you, the Earl of Lambsdorff' (5); *Allen, die dazu beigetragen haben* 'all those who have contributed to the fact' (6).

Implicit addressing and impulse for action within the utterance are found by some modal verbs (E. I. Shendels, Duden etc.). The utterances with the verbs *müssen* 'must, need to' and *sollen* 'must' are considered as typical causative sentences addressed to the addressee. The modal verb *dürfen* 'may' in combination with the negation of *nicht* 'not' marks the "external" negation and expresses a definitive prohibition (Duden, 2006). The modal verb *wollen* 'will, want' in the passive sentence passes the value of necessity, but unlike *sollen* or *müssen* in combination with its basic meaning, it shifts responsibility for the action to the subject of the sentence (metaphorical transfer (Duden, 2006)). The Federal President calls on the state and citizens (inclusive personal pronoun of 1st person plural *wir* 'we') to realize the value of an independent, cognitive learning that should start not in school, but in the parents' home and kindergarten. To convey the addressing and the greatest impact, to emphasize the importance of this thought, the modal verbs *müssen, dürfen, wollen* are used in the indicative mood, e.g.:

7. Wir *müssen* uns vielmehr den Wert des spielerischen, des selbständigen, des entdeckenden Lernens noch stärker *bewusst machen*, das übrigens *nicht* erst in der Schule *beginnen darf*, sondern schon im Elternhaus und in der Kita *angeregt sein will*. (Speech by Federal President Horst Köhler).

Indeed, we *must* become even more aware of the value of the playful, the independent, the discovering learning, which, by the way, *can* not begin at school, but *wants* to be stimulated in the parents' house and in the day-care center. (Speech by Federal President Horst Köhler).

The adverbial clause with the conjunction *wenn* 'if' is of a contact-establishing nature, because it explicates a dialogical situation in the form of a condition (Kostrova, 1990). The explicit structure of *wenn ..., dann ...* 'if ..., then ...' contains an implicit addressing to citizens, to their national identity, to their axiological basis (If we want ..., then ...). The addressing of the utterance of this type reinforces the repeated inclusive personal pronoun of the 1st person plural *wir* 'we', e.g.:

8. Und *wir* wissen natürlich, dass *wir* im internationalen Wettbewerb nur *dann* bestehen können, *wenn wir* mindestens so viel besser sind, wie *wir* teurer sind. (See ibid.).

And *we* know, of course, that *we* can only exist in international competition *if we* are at least as much better as *we* are more expensive. (See ibid.).

In a scientific article that presents scientific discourse, addressing is realized through a system of formal and content-related means and methods to attract the reader's attention and to actualize in the textual structure the "image of the reader" as a communication partner. Considering in his article "Macht Erfahrung klug?" 'Does experience make wise?' the problems of subjective experience, its social significance and scientific generalization, Morus Markard uses the means of implicit transmission of the addressing to the reader which have a great influence potential: various types of interrogative sentences, modal verbs, prepositional object, metadiscursive inserts and others.

The scientific article is characterized by the use of question-answer complexes in which the author foregrounds the problem, marks the transition from awareness of the problem to its formulation and solution. In the introduction, the main theme of the article is allocated: What kind of experience and how to use, and what is generally understood by the concept *klug* 'wise'. The chain of questions, which represents several variants of the definition of *klug* 'wise', contains elements of the parcellation introduced with the conjunction *oder* 'or'. In the next passage, the conjunction *oder* 'or' is inclusive; i.e. these meanings of *klug* are components of this concept (Duden, 2006). The conjunction *oder* emphasizes dialogueness and implicates the addressing of the utterance to the addressee, who is invited to make a choice. The author thereby intends to involve the reader in the activity, in co-creation, e.g.:

9. Bedeutet "klug" bspw. das gewitzte Zurechtkommen im Rahmen eines geschmeidigen Opportunismus, der jene Rundung sich zuzulegen weiß, mit der ein Anecken nicht mehr möglich ist?

Oder human inspirierte, raffinierte bis subversive Taktiken im Dschungel institutioneller Menschenverwaltung?

Oder die Analyse des Verhältnisses fachlicher und politischer Aspekte psychologischer Arbeit, etwa bei den Begründungen für die Verlängerung der Aufenthaltsmöglichkeit aus Bosnien geflohener, "traumatisierter" Frauen (vgl. Rafailovic 2006)?

Offensichtlich hängen die beiden Aspekte—was man unter "klug" versteht, und wie man welche Erfahrungen aufschlüsselt—zusammen (Markard, 2007).

Does "wise" mean, for example, the clever cunning, within the framework of a smooth opportunism, which can round off the rounding with which a hitting the corner is no longer possible?

Or human-inspired, refined to subversive tactics in the jungle of institutional human administration?

Or the analysis of the relationship between professional and political aspects of psychological work, such as the justification for the extension of the possibility of sojourn of the from Bosnia escaped "traumatized" women (See Rafailovic 2006)?

Obviously, the two aspects—what one understands as "wise", and how to break down what experiences—are linked up. (Markard, 2007).

Modal verbs *müssen* 'must, need to' and *sollen* 'must' explicate in scientific text the meaning of motivation for action, the addressing to the addressee-specialist, whom is recommended precisely such a formulation of the problem and such solution of it. Moreover, it should be noted, the motivation with *müssen* is more categorical than the motivation with *sollen* (Duden, 2006), e.g.:

10. Und in diesem Sinne *müssen* theoretische Aussagen über Handlungen Aussagen über Prämissen-Gründe-Zusammenhänge sein. (See ibid.).

And in this sense, theoretical statements about actions must be statements about premises-reasons-connections. (See ibid.).

Metadiscursive inserts are a specific means of representation of the textual addressee. The text, as a special way of translation of cognitive, scientific and other meaningful experience of the author, transmitted to the addressee being structured in accordance with the peculiarities of the recipient's perception. The author, taking into account the model of the addressee, his social and psychological appearance and the functional tasks facing him, forms specific, impressive means of activating perceptions, attention and understanding, that is, specific signs of memory of the addressee, or text punctuation marks as communication signals aimed at the addressee, corresponding to the content of the text, with its functional-stylistic and genre features, etc. (Polonsky, 1999).

Introductory sentences referring the reader to the above mentioned information, a prepositional attributive and other metadiscursive inserts intensify the attention of readers, recall the main topic of the article, prepare them for further information perception and structure the statement, e.g.:

11. ... ob es um – *wie schon beispielhaft angeführt* – Erfahrungen in beruflicher Praxis oder im Alltag geht ... (See ibid.).

... whether it is a case of experience in professional practice or everyday life, *as already mentioned* ... (See ibid.).

7 CONCLUSION

Hence, the differentiation and integration of science are constitutive characteristics of modern scientific knowledge, oriented to a man, to its manifestation in different spheres of activity. In this connection, the role of a single method of describing and interpreting a number of phenomena is growing. An integrating factor in the interpretation of implicit language means of expression of the phenomenon of addressing in political and scientific discourses is the cognitive-discursive approach and the consideration of the linguopragmatic aspects of the utterance. Regarding the specifics of institutional communication, in our opinion, of particular interest are indirect methods of influencing the addressee, which represent the author's intentions indirectly, implicitly. Due to attraction of extralinguistic factors, such linguistic means as utterances of different communicative orientation, inclusive pronoun-verb forms of the 1st person plural, modal verbs, modal passive constructions, conditional clauses, indirect dative object, prepositional object, the definite article, metadiscursive insertions, etc., reveal the ability to implicit the pragmatic significance of addressing in political and scientific communication.

Thus, the study of implicit ways of representing the linguopragmatic phenomenon of addressing in the political and scientific spheres of communication, explicated at different levels of the language: semantic, syntactic, pragmatic, makes it possible to raise the question of the super-pragmatic nature of the phenomenon studied, which were mentioned in Stepanov's (Stepanov, 2007) and Revzina's (Revzina, 2005) works.

REFERENCES

Akhmanova, O.S. 1966. Dictionary of linguistic terms. Moscow: Soviet Encyclopedia, p. 506.
Arutyunova, N.D., Paducheva, E.V. 1985. Origins, problems and categories of pragmatics. New in foreign linguistics (16): 3–42.
Chernyavskaya, V.E. 2006. Interpretation of scientific text. Moscow: Flinta: Nauka.
Chernyavskaya, V.E. 2006. The discourse of power and the power of discourse: problems of speech influence. Moscow: Flinta: Nauka.
Duden, 2006. The grammar. Volume 4. Mannheim: Dudenver-lag.
Fadeeva, L.V. 2016. Stylistic advices of violence of the sentence stricture in modern German language. Language, culture, comminication. Orel: Orel State University. 242–246.
Kaptsov, A.V. 2008. Diagnostics of personal values and the structure of the axiological sphere. [Online] Available: https://psibook.com/articles/diagnostika-lichnostnyh-tsennostey-i-struktury-aksiologicheskoy-sfery.html.
Katznelson, S.D. 1972. Typology of language and speech thinking. Leningrad: Nauka.
Kazantseva, Y.M. 2004. Grammatical categories and the change of linguistic paradigms. Vestnik MGLU (482): 5–87.
Kostrova, O.A. 1990. Pragmatic structure of motivational phrase in speech communication. Communicative-pragmatic functions of language units: interuniversity collection: 58–65.
Markard, M. 2007. Makes experience wise? Subject-scientific considerations on the relationship between subjective experience and scientific generalization. Journal of Psychology (3). [Online] Available: http://www.journal-fuer-psychologie.de/jfp-3-2007-4.html (April 14, 2008).
Maslova, A.Yu. 2007. Introduction to pragmalinguistics. Moscow: Flinta: Nauka.
Mitrofanova, T.A. 2009. Addressing as a pragmalinguistic phenomenon in the institutional discourse (on the basis of the German language): Dis. ... cand. philol. sciences. Moscow: Moscow State Linguistic University.
Mitrofanova, T.A. 2016. Dialogical nature of the institutional dicourse. Actual problems of Romano-Germanic philology and teaching of European languages at school and university. Yoshkar-Ola: Mari State University, 118–124.

Mitrofanova, T.A. 2017. Syntactic means of expression of the addressing in the institutional discourse. Innovative technologies in the teaching of foreign languages: from theory to practice. Yoshkar-Ola: Mari State University, 37–42.

Nistratova, S.L. 1985. Syntactic means of expression of addressing in oral scientific speech. Scientific literature. Language, style, genres. Academy of Sciences of the USSR: 81–100.

Polonsky, A.V. 1999. Categorical and functional essence of addressability. Moscow: Russkij dvor.

Revzina, O.G. 2005. Discourse and discourse formations. Criticism and Semiotics (8): 66–78.

Romanova, M.S. 2015. Exclamations structure in German online discourse. Languages, culture, ethnoses. Formation of language picture of the world: philological and methodical aspects. Yoshkar-Ola: Mari State University, 45–49.

Romanova, M.S. Smirnova V.A. 2015. Grateful expressions in internet communication. Languages, culture, ethnoses. Formation of language picture of the world: philological and methodical aspects. Yoshkar-Ola: Mari State University, 95–99.

Shendels, E.I. 2006. Selected Works: To the 90th Anniversary of the Birth. Moscow: MGLU.

Speech by Federal Chancellor Angela Merkel on the occasion of the ceremony of the "Remembrance, Responsibility and Future" Foundation. [Online] Available: http://www.bundeskanzlerin.de/nn_5296/Content/EN/Rede/2007/06/2007-06-12-rede-merkel-stiftung.html (January 20, 2009).

Speech by Federal President Horst Köhler at the annual meeting of the Stifterverband für die Deutsche Wissenschaft. [Online] Available: http://www.bundespraesident.de/Reden-und-Interviews (January 20, 2009)

Stepanov, Y.S. 2007. Concepts. Thin film of civilization. Moscow: Yazyki slavyanskih kul'tur.

Volvak, N.P. 2004. The factor of the addressee in the public argumentative discourse. Yuzhno-Sakhalinsk: Sakhalin State University.

Zheludkova, Y.G. 2004. Functional-pragmatic aspect of the addressing category (on the basis of French and Russian languages): Dis. ... cand. philol. sciences. Kemerovo: Kemerovo State University.

The problem of early development of creative abilities via experimentation in productive activities

O.A. Petukhova, S.L. Shalaeva, S.O. Grunina & N.N. Chaldyshkina
Department of Preschool and Social Education, Mari State University, Yoshkar-Ola, Russia

ABSTRACT: This article is devoted to the issues of early show up of children creative abilities through experimentation with visual materials in the conditions of preschool and family. The problem of the early development of the creativity of children has been and remains among the most pressing problems both in psychology and in pedagogy. Various aspects of creative development are being actively studied, starting from the second half of the 19th century. This article reflects both organizational and methodological aspects of pedagogical work with children of middle and senior preschool age, determined during the activity of the experimental site on the basis of the countryside Municipal pre-school educational institution "Morkinsky kindergarten No. 2" of Mari El Republic. The possibilities of using various materials in the visual arts of preschool children are discussed. Conditions for the early development of the creative abilities of children in the process of experimentation with visual material are singled out. Psychological and pedagogical aspects for organizing and conducting classes using experimentation in productive activities are defined. The article analyzes the results of the development of children's creative abilities after the psychological and pedagogical work. The conclusions drawn from the results of the study can be useful for teachers and psychologists working with pre-school children.

Keywords: Creativity, creative abilities, experimentation, search (investigation) activity, productive activities, children of preschool age

1 INTRODUCTION

The problem of giftedness today is very actual in the scientific and everyday spheres of life. Public changes that have occurred in recent decades have clearly marked the problem of the personal identity – its independence, creative activity, ensuring its rights and freedom. At these conditions the process of mental development of the individual from the very birth should be provided with thoughtful psychological and pedagogical support, including educational impact, teaching, developing, stimulating one's own activity. However, all this bunch of influence on the individual has to take into account those social transformations and the change of moral orientations and values that contribute to the emergence of certain contradictions that should be monitored by social institutions (Shalaeva 2015 a, b).

Many scientists from different countries developed conceptual models of giftedness. The first experimental studies of giftedness were carried out within the framework of "associative psychology" (A. Beng, W. Wundt, D. Mil, G. Spencer, T. Tsigsn, etc.). One of the options for solving the problem of studying giftedness can serve as a system proposed by the Russian scientist G.I. Rosslimo. His diagnostic methods, in a shortened version, presupposed the study and measurement of the five main functions: thinking, attention, will, receptivity, memory. Almost in parallel, a different approach called integrative was developed. His supporters viewed giftedness not as a special combination of the levels of development of individual functions, but as an integrative personal characteristic. Thus, the notion of "general giftedness" (V. Stern, E. Clapared, etc.) came into use. Many aspects have not been solved yet, however, one should acknowledge, that many aspects of this problem have already been studied, and that has allowed overcoming one-sidedness both in the diagnosis of giftedness, and in development.

One of the most popular modern concepts of giftedness is the model developed by the American specialist in the field of teaching gifted children, J. Renzouli. According to this concept, giftedness is considered as a combination of three characteristics: intellectual abilities (exceeding the average level), creativity and perseverance (motivation, task-oriented). At the same time, his theoretical model takes into account knowledge (erudition) and a favorable environment (Renzulli 1986).

The concept of J. Renzulli is quite popular and is actively used in the study of applied aspects of the giftedness development. Disclosing the essence of giftedness as a natural phenomenon in quite a detail, the concept rather definitely indicates the directions of pedagogical work on its development. It should be noted that in the title the term "giftedness" is replaced by "potential". This shows that the concept is a universal scheme, applicable not only to work with gifted children, who have a natural gift, but to all children.

Thus, 'creativity' or 'creative abilities' become one of the basic characteristics in the model of general giftedness. Their development, from an early age in productive activities, is the content of our research.

This article is devoted to the issues of the early development of the creative abilities of children as the main characteristics of the giftedness with the use of visual experimentation in kindergarten and family. The purpose of the paper is to describe the experience of the experimental site on the basis of a rural preschool institution. The aim is also to define organizational and methodical aspects of pedagogical work with children of middle and senior preschool age for the successful development of their creative abilities using experimentation in productive activities.

2 STATEMENT OF PROBLEM

One of the most effective ways to develop the creative abilities of children, in our opinion, is to independently seek ways to solve new problems, new ways and means to achieve the goal through experimentation. Problems of search activity and experimentation of children still reasonably attract the attention of researchers in the field of pedagogy and psychology. Teaching methods based on the problematic presentation of the material as well as research and design methods are the most effective ones in developing the creative abilities of children, in formation of cognitive interest in the surrounding world, in the strength and awareness of study. "Tell me – and I'll forget, show – and I'll remember, give it a try – and I'll understand," – says the ancient wisdom. The process of cognition is a creative process, extremely important in the personal development of a preschooler. T.S. Komarova has noted that pre-school children need not only knowledge, but also mastering the methods of activity, skills and experiences (Komarova 2005).

Now there appears a fairly large number of studies devoted to experimenting with living and static nature, that investigate the physical phenomena and capabilities of the human body, experimenting with the subject man-made world. We would like to share the experience of organizing experimentation of children in the visual arts, in particular, in the creation of application and collage artworks. The study of the possibilities of experimenting with visual materials was carried out during the flow on the research project "Experimenting with visual materials as the way of developing the creativity of preschool children", which was implemented on the basis of the Municipal Pre-school Educational Establishment "Morkinsky Kindergarten No. 2" of Mari El Republic from 2008 to 2011. 67 children of middle and senior preschool age, 6 teachers of pre-school educational organization took part in it. The supervisor of studies is the candidate of pedagogical sciences O.A. Petukhova.

The idea of carrying out experimental work on this topic arose on the basis of the Zaporozhets Research Institute for Preschool Education staff multi-year research of the problems of experimenting children in constructive activities (from construction material, designer parts, paper, natural materials, etc.). They have convincingly proved that at first acquaintance with new material it is necessary to use "disinterested" experimentation, that is, when the child is not given specific tasks of a productive nature, allows children to freely and liberally get acquainted with the properties and features of this material and in the future will expand the possibilities for its application. Next is the process of transferring to children the necessary knowledge on the possibilities of using this new material to create new expressive images. The second stage is "utilitarian" experimentation, when children during their experimenting are looking for ways to solve the problem stated by an adult (Paramonova 1999).

However, the studies were devoted to working with constructive material, while the problem of experimentation in visual activity remained open. Therefore, we are formulating the research problem: what are the forms of organizing children's experimentation with visual materials, what is its content and what are the possibilities in selecting materials for children's experimentation.

3 CONCEPTUAL FRAMEWORK

The necessary initial stage in the implementation of the work was the methodical training of educators for experimental activity. It was necessary to help teachers to acquire new pedagogical thinking, readiness to solve complex problems in the development of creativity of preschool children, to improve their pedagogical skills in an unfamiliar area. Pedagogical councils and seminars were held, during which the educators who participated in the experiment became acquainted with the method of developing the creative abilities of children in the process of learning activity, realized the importance of free experimentation in the emancipation of the creative potential of children, with the causes of difficulties arising in the realization of creative ideas in children Applications (Petukhova 2010).

The main goal of the experimental mode kinder-garten methodical work, is to promote the self-development of a teacher which is a subject of managerial, pedagogical and experimental impact, as well as to optimize the process of upbringing and education of preschool children (ie, acquiring the skills of planning, results of their work, make adjustments and identify the most rational forms and methods of organizing children's experimentation). Here we remark that the methodological work is most effective when it is organized as an integral system. It becomes possible only when the entire teaching staff is working on a single research topic.

Not all teachers of this preschool institution immediately began to participate in this experimental work. Their inclusion in the work went gradually. First, the teachers participated in the experiment, they worked with the children of the middle group (children of 4–5 years). In the second year of the study, the teachers of two groups participated: the middle (children of 4–5 years) and the older (children of 5–6 years). Teachers of three age groups participated in the third year: the middle age group (children of 4–5 years), the older (children of 5–6 years) and preparatory (children of 6–7 years). The children of all mentioned age groups were fully involved in the experimental study. The research group of teachers also included research supervisor of the research project, the head of the pre-school educational institution, preschool teacher's methodologist and psychologist.

To achieve high results, we used different collective and individual forms of organization of methodical work. Collective seminars, workshops, methodological associations, small creative groups of teachers (creative micro-groups) are included. Individual forms include individual counseling, mentoring, work on a personal creative topic, individual self-education, etc. Observing the experimental work, we found that a great effect is the work of creative micro-groups. They include teachers who are engaged in experimental work with great interest and high activity. They are looking for new ideas, discussing, applying in work with children and sharing their findings with colleagues in methodological associations and pedagogical councils.

The research material was collected in the "bunch of pedagogical skills" having inside the abstracts of lessons, consultations for parents, scenarios for thematic entertainments, etc. Based on the results of this experimental work, the manual for teachers of pre-school institutions was published (Petukhova, 2013).

4 FINDINGS

Monitoring of research activities of children assumes the systematic study of current and intermediate results, as well as their assessment as problematic or successful, in other words, analysis of individual elements of creative experimentation and development dynamics.

To carry out monitoring and conduct reflexive discussions, we have used the materials of diary entries, the results of diagnostic sections, creative reports of teachers. "Diary of the creative development of the child", the form of which was developed in the process of working on the project, allow to track the smallest achievements of a child, his successes in the manifestation of creativity in the visual activity. Records have to be written after each lesson or self-activity act of a child. They are laconic, which does not take much time from a teacher, but allows a comparative analysis of the creative growth of both the individual child and the group as a whole.

Diagnosis of the child's creative development was carried out at the beginning and the end of the school year in the experimental and control groups. As diagnostic materials, the methods of diagnosing universal creative abilities for children aged 4–5 years were used by V. Kudryavtsev and V. Sinelniov (Kudryavtsev and Sinelnikov 1995):

1. "Sun in the room" method (Founding: realization of the imagination Purpose: revealing the child's ability to transform the "unreal" into "real" in the context of the given situation by eliminating the inconsistency).
2. Method "Folding picture" (The basis: the ability to see the "entire" before "the parts". Purpose: to determine the ability to preserve the integral context of the image in a situation of its destruction.)
3. The method "How to save the hare" (The basis: the suprasituational-transformational nature of creative solutions. Purpose: To assess the ability and turn of the "choice problem" into the "transformation task" when the properties of a familiar subject are transferred to a new situation.)
4. Method "the board" (Founding: child experimentation. Purpose: to evaluate the ability to experiment with transforming objects.

The results of the control diagnostics showed a significant increase in the creative abilities of the children of the experimental group in comparison with the results of the ascertaining stage. We believe that pedagogical work contributed to this, including both the organization of independent experimentation of children with visual materials and the training sessions.

In addition, the creative growth of a child was taken into account directly in the analysis of the products of activity according to such parameters as originality, diversity of intentions, richness of the topic (Grigorieva 1999).

To study the dynamics of the development level of children's creativity in experimental work, we also used indicators characterizing the aesthetic attitude, interests and abilities developed by NA Vetlugina:

- developed creative imagination, on the basis of which the past experience is transformed, the ability to "enter" into the circumstances depicted, in Conditional situations;
- Enthusiasm, passion to activity;
- Sincerity, truthfulness, spontaneity of experience;
- Special artistic abilities that enable them to successfully solve creative tasks;
- The change in the motives of activity, which gives pleasure to its results (characterizes the direction of the imagination);
- The rise up of the need for creativity.

Also in the diagnostics were used indicators characterizing the ways of creative actions:

1. Additions to, changes, variations, transformation of already familiar material, application of known skills into new situations, creation of new variants on the basis of already available experience.
2. Self-searching for solutions to creative tasks.

3. Finding new, original solutions, independence, initiative.
4. Rapid reactions, resourcefulness in action, orientation within new conditions, flexibility.

In addition, the creative growth of a child was taken into account directly in the analysis of the products of activity according to such parameters as originality, diversity of intentions, richness of the topic (Grigorieva 1999). During the time of the study, the level of development of the creative abilities of the children of the experimental group increased significantly, in comparison with the level of development of children who did not take part in the forming part of the experiment. Children who were at a low level of development of creativity at the beginning of experimental work had an average and above average level by the end, most of the children who were at the average level of development of creative abilities at the beginning of the experiment at the end showed a high level. The results of diagnosing children convincingly prove the effectiveness of our work on the organization of experimentation in the visual activity.

5 DISCUSSION AND INTERPRETATION

This study allowed us to identify the following pedagogical conditions in order to adhere to the following aspects of the introduction of creative pedagogy:

– benevolent atmosphere and emotional touch, affectionate intonations, positive attitude to the initiative of children;
– Individual approach to each child;
– interesting content of classes (an entertaining plot, a game exercise, a surprise moment), which is the basis of a shared emotional experience;
– Integration of all means of aesthetic education and types of artistic and creative activity of children (play, dramatized, musical, graphic, artistic and verbal);
– a variety of classes, both in form and content, because it is novelty, unusualness that creates an emotional mood for creative activity.

An important aspect in pedagogical work is the creation of conditions for children's experimentation in the group. Therefore, at the first stage, a transformation of the group's subject-developing environment is necessary, which provides children with freedom of choice and success in implementing their tasks. The subject-developing environment is based on the principles of accessibility, novelty and assumes the personality-oriented interaction model. A zone was created for independent creative experimentation of children ("Corner of experimentation"), where a variety of materials were presented, which children could use at their own pleasure in their visual works. It was a natural and junk material: candy wrappers, pieces of cloth, leaves, dried flowers, buttons, plant seeds, tinsel, "Christmas rain", etc. A mandatory requirement for the material was safety, hypoallergenicity, aesthetic appeal. In addition, interior decoration of the group action and dressing rooms of the pre-school institution included the creative artworks of the educator and children, in which non-traditional pictorial material was used and which were made by various techniques.

One of the conditions for the successful completion of the assignment is the creation of a mood for activity, namely content interesting for children (an entertaining story, a game exercise, a surprise moment), which is the basis of a shared emotional experience. In the classes it is desirable to have a variety, both in form and content, since it is the unusualness, originality that creates the emotional mood for experimentation.

A teacher can offer children a problem situation. For example, "How can we tell in the work that the puppy is shaggy, the kitten is fluffy, the toy is soft?" Children offer their options for using non-traditional techniques to create an expressive image.

Experimenting children becomes more meaningful and goal-oriented in those cases when children already have some experience of using variational technologies and materials. The creative work of children on the basis of processing previously obtained knowledge of the visual possibilities of the material was brighter, more expressive, more original and diverse.

Children should be able to combine different methods of activity into a new one (having mastered the technique of break-out applications, they use it to create clouds, snow lumps, mountains, animals). In the process of experimentation children show the ability to alternative solutions, learn to see new visual functions of objects (contour application with threads, use of puzzle pieces as elements of mosaic work in the application).

6 CONCLUSION

Thus, the offered approach to the introduction of child experimentation into the process of acquisition of productive activities increases the level of creative development of children. It includes the timely diagnosis and monitoring of the development of creative abilities, the creation of a material and technical base for children's experimentation and the provision of game motivation for the activities of children both in kindergarten and in the family.

Further research on the development of the creative abilities of preschool children will be associated with studying the process of experimenting with kinetic sand in playing games, as well as drawing with sand on light tables, since this material has great potential both in motivating children and in using the properties of children themselves.

REFERENCES

Grigorieva, G.G. 1999. Development of the pre-school child in visual activity. Moscow: Academy.
Komarova, T.S. 2005. Children's art creativity. Moscow: Mosaic-Synthesis.
Kudryavtsev, V., Sinelnikov, V. (1995) A pre-school child: a new approach to the diagnosis of creative abilities. Preschool education, 9, 52–59.
Paramonova, L.A. 1999. Children's Creative Design. Moscow: Karapuz Publishing House.
Petukhova O.A. 2010. Experimenting with visual materials in kindergarten and family. Kindergarten from A to Z., 6 (48), 79–86.
Petukhova O.A. (editor). 2013. Experiment, play, create. From the experience of the work of the MdOU "Morkinsky kindergarten No. 2 on the organization of children's experimentation with fine materials". Yoshkar-Ola: Quad.
Renzulli, J.S. The three-ring conception of giftedness: A developmental model for creative productivity// Sternberg R.J., Davidson J.E. (Eds.). Conceptions of Giftedness. New York: Cambridge University Press. 1986. P. 53–92.
Shalaeva, S.L., Shalaeva, A.V. 2015. Axiological analysis of intergenerational relations in a globalized society: myth and reality. Review of European Studies, 7 (8), 246–252.
Shalaeva, S.L. 2015. Intergenerational dimension of the national security of modern Russian society. Social and human sciences in the Far East, 3 (47), 102–106.

A negative assessment of the concept insolvency in French and English paroemias

T.A. Soldatkina, S.L. Yakovleva, G.N. Kazyro & G.L. Sokolova
Mari State University, Yoshkar-Ola, Russia

ABSTRACT: In this study, we assess the verbalization of the concept insolvency reflected in the paroemiological view of the world of the French and English languages. The study of the proverbial view of the world allows us to describe peoples' attitude to insolvency and reconstruct part of the linguistic view of the world which conceptualizes the phenomenon of adversity in different languages. The objective of this study was to perform a comparative cognitive analysis of the French and English paroemias to assess the verbalization of the concept insolvency, with the aim of revealing its universal and specific features. The analysis of the paroemias selected using the continuous sampling method from the significant French and English dictionaries of proverbs and sayings thus allowed a semantic and contextual analysis of lexical units to verbalize the negative assessment of the concept "insolvency" and its logical opposition "wealth". Thus, the concept "insolvency" is objectivized through paroemias, which can be grouped into the following four topics: (1) adversity as a moral and material condition of the person; (2) characteristic features of adversity; (3) category of entities of the subjects characterized as poor; and (4) lexical units, reflecting the people's attitude toward the state of poverty and the miserable's behavior. The comparative analysis of languages show that paroemias that verbalize the state of insolvency contain a negative connotation, which coincides with the traditional moral values of society. Insolvency and wealth are the key universal concepts in the English and French languages. In this study, we reveal that the English nation is generally rational and favors wealth and assets. In comparison, the French mentality displays a negative attitude toward wealth as the material value. We also reveal that the general understanding of the selected concept differs in a number of features in both the English and French cultures.

Keywords: proverb, paroemiological view of the world, concept, assessment, adversity, insolvency, the French language, the English language

1 INTRODUCTION

In this study, we reveal the general classification of paroemiological units in the concept "insolvency" in English and French languages. Paroemiological articles from the English and French dictionaries have been analyzed. The choice of the dictionaries as lexicographic sources is proved by their practical value for the English and French paroemiologies. We treat paroemiological units as universally recognized folk truth, passed from mouth to mouth in the form of brief, complete, rhythmical sentences with didactic sense, which can function independently. Regardless of the implied notion, they contain the experience of generations, state and evaluate people's properties and events, and prescribe a certain pattern of behavior. Classification of paroemiological units that express negative assessment in the concept of insolvency is also addressed in this study. Merriam Webster dictionary defines insolvency as "a state, condition, or instance of serious or continued difficulty or adverse fortune". We treat "insolvency" as a state of misfortune or affliction as well as a stroke of misfortune or a calamitous event. Thus, in this study, the concept of insolvency is assessed using criteria such as the lack of property and tangible assets (money), which generally allows people to be classified as poor.

2 OBJECTIVE OF THE STUDY

Adversity (poverty) and insolvency represent one of the most important aspects of the paroemiological view of the world. The objective of this study is to perform a comparative cognitive analysis of the French and English proverbs and sayings verbalizing the assessment in the concepts of insolvency. We hypothesize that in the English and French languages, paroemias verbalizing the state of insolvency contain a negative connotation, which coincides with the traditional moral values of society.

According to the definition proposed by V. A. Maslova, a concept is a mental nationally colored construct where the plane of content is represented by the whole body of knowledge of wealth as a phenomenon and the plane of expression are all lexical units nominating and describing the object (Maslova, 2004, p. 28).

3 MATERIALS AND METHODS

Continuous sampling method, as well as a complete semantic, contextual, and comparative analysis of lexical means, representing the concepts of adversity and insolvency in the French and English languages are used in this study.

This article belongs to the authors' series of articles devoted to the study of different concepts and assessment in the paroemiological views of the world of the English, Mari, Finnish, and French languages (Yakovleva, Soldatkina 2015, p. 253–262; Yakovleva, Kazyro, Soldatkina 2015, p. 21–33; Yakovleva, Kazyro, Soldatkina 2016, p. 1881–1884; Soldatkina 2016, p. 178–180; Soldatkina 2016, p. 241–247; Soldatkina 2017, p. 64–72).

The Concise Dictionary of Proverbs contains essential and most useful proverbs in English. Now reissued and updated, this invaluable work of reference provides the reader with more than 1000 of the best-known English proverbs from around the world edited by John Simpson and Jennifer Speake. This invaluable work contains the best-known proverbs in the English language from around the world. *The Concise Oxford Dictionary of Proverbs* provides the interesting history of our proverbs. John Simpson is Chief Editor of the Oxford English Dictionary and a Fellow of Kellogg College, Oxford. He is the Co-editor, with John Ayto, of the Oxford Dictionary of Modern Slang. Jennifer Speake is a freelance writer. She is the editor of the Oxford Dictionary of Foreign Words and Phrases and the Oxford Dictionary of Idioms.

The French-Russian Phraseological Dictionary by Ya.I. Retsker is the most complete of all the existing French–Russian phraseological dictionaries. It has been compiled on the basis of the French–Russian phraseological dictionary edited by Y.I. Retsker. The dictionary now has updated and renewed expressions, quotations, and examples. It is meant for specialists working with the French language, translators, researchers, teachers, students, and all those who are interested in the French language, as well as for foreign specialists who study the Russian language.

4 DISCUSSION AND RESULTS

The attitude to others' property is reflected in the following thematic classification of paroemiological units expressing a negative assessment in the compared languages.

1. The threat of property loss due to a third party's interference or a nomadic way of life. Thus, in the English language: *a rolling stone gathers no moss*; in the French language: *pierre qui roule n'amasse pas (de) mousse*. French paroemias predict great problems for those who own property as they constantly need to fight for it: *qui terre a guerre a—those who possess land should be ready for war*.

2. The necessity to prove property's appeal, for example, in French: *bonne est la poule qu'un autre nourrit—the chicken that is fed by others is always good.*
3. Dramatically negative attitude of representatives of the regarded cultures to embezzlement, for example, in the English language: *ill-gotten gains never*; for example, in French: *le bien mal ne profite jamais—you won't make profit of what is ill-gotten.*
4. The negative consequences in case of embezzlement, for example, in the English language: the camel going to seek horns lost his ears; in French: *l'avarice perd tout en voulant trop gagner—if you are greedy you will lose everything.*

Thus, none of the representatives of the cultures under investigation profits of embezzlement. In paroemias, it is strongly recommended not to pursue someone else's property, because it is obviously possible to lose one's own.

Considering the concept of insolvency, it should be noted that all material things have their value and are sold for money. Thus, the economic formula "commodity-money" is applicable in linguistics. Moreover, people who are constantly in need for money, and therefore can hardly make two ends meet are classified as the destitute. The English and the French are confident that the complete lack of livelihood is the worst burden. Paroemiological units that express negative assessment in the logical opposition of the concepts "insolvency/wealth" are addressed in this study as well. Moral-value component of the concept "adversity", belonging to a social category, allows us to define structural and hierarchical relationships that are formed according to the principle of axiological marking by means of framing analysis.

The English and French are unanimous in the opinion that adversity cannot be included in the category of human vices. This idea is verbalized in the English paroemia of the French origin: *poverty is no skin (or a vice)*. While the French make a remark: *la pauvreté n'est pas vice (mais c'est une espèce de ladrerie, chacun la fuit)*. The thought that poverty is not a disgrace is justified in the following English proverb: *poverty is not a shame, but the being ashamed of it is*. However, the English suffer from pricks of shame for unfair profits more strongly than being ashamed of the precarious financial situation: *a clean fast is better than a dirty breakfast*. Poverty is seen as fatal: *a beggar's purse is bottomless*. The French believe that who was born poor does not get rid of the scrip: *au gueux (или au pauvre) la besace*.

The state of the extreme poverty is reflected in all linguistic cultures. The English say: *a beggar can never be bankrupt*. The Russian equivalent states that as the beggar has nothing to lose he is not afraid of robbery. Thus, the French believe that the poor got used to be hungry: *assez jeûne qui pauvrement vit*. According to the Russian proverb: *you even can't take a shirt off a beggar*.

Financial stability predetermines either one has relatives and friends or not, for example, in French: *pauvreté n'a point de parenté—no relatives for the poor; les malheureux n'ont point de parents/d'amis—miserable have neither friends nor relatives*. More pragmatic British believe that poverty kills even such feelings as love: *when poverty comes in at the door, love flies out of the window*.

The poor are responsible for everything, so those who cannot pay off the jail will most likely be punished, as in English: *the poor man pays for all*. In French: *le gibet n'est fait que pour les malheureux—gibbets are made for the miserable only*.

The data collected for this study demonstrated that the ethics of the French is more robust. The French believe that it is better to panhandle than to rob: *il vaut mieux tendre la main que le cou—it is better go begging than go to gibbets*. The English proverb runs: *adversity (misery or poverty) makes strange bedfellows*. Adversity is not only the cause for poor health (in English: *bare walls make giddy housewives*), but it also launches the instincts of self-protection when a person is supposed to look for the way out of the given dilemma. Thus, in English: *needs must when the devil drives*; in French: *la faim chasse le loup hors de bois—hunger makes wolf hunt outside the forest*. The worst aspect of poverty is that it breeds strife and breaks the law.

Thus, the concept "insolvency" is objectivized through the paroemias, which can be grouped under the following four topics: (1) adversity as a moral and material condition of the person; (2) characteristic features of adversity; (3) category of entities of the subjects characterized as the poor; and (4) lexical units, reflecting the peoples' attitude to the state of poverty and the miserable's behavior. In the compared languages, paroemias verbalizing the state of insolvency contain a negative connotation, which coincides with the traditional moral values of society. The concepts of "insolvency" and "wealth" are key universal concepts in the English and French languages.

The English and French are unanimous in their opinions that fortune, which gives the person an opportunity to enrich, is rather capricious, so in English: *fortune is fickle;* in French: *fortune et vent varient souvent—wind and fortune change often.*

In the paroemias, those who enriched all over sudden are rebuked, as in English: *honours change manners;* in French: *il n'est orgueil que du pauvre enrichi – a person who has become rich puts on side his being rich.* It is believed that wealth cannot be earned in an honest way, so in English: *muck and money go together;* in French: *ce qui abonde ne nuit pas.* The group of paroemias selected during this study proves that wealth is not the key to happiness, so in English: *great fortune is a great slavery; nightingales will not sing in a cage*; in French: *la (plus) belle cage ne nourrit pas l'oiseau; des chaînes d'or sont toujours des chaîne – golden chain is always a chain.*

The paroemias under study allow us to conclude that paroemiological units with negative assessment of the inheritance as a component of wealth predominate in the French language. The French believe that *only a fool would make his doctor as the heir: c'est folie de faire de son médecin son héritier.* Moreover, no one can impose an inheritance against the will: *n'est héritier qui ne veut.* The pragmatic features of the paroemias with the concept "wealth" are revealed through the fact of unwise reliance on someone else's property. Thus, in English: *he goes long barefoot that waits for dead man's shoes.* The same in French: *il ne faut pas compter sur les souliers d'un mort pour se mettre en route; si ce n'était le si et le mais nous serions tous riches à jamais – we would have been rich long ago if the right conditions ever go.*

5 CONCLUSIONS

In this study, we revealed the similarities and differences in the perception of the anthropocentric environment typical of the cultures under study. The comparative analysis of languages showed that paroemias that verbalize the state of insolvency contain a negative connotation, which coincides with the traditional moral values of society. Thus, none of the representatives of the cultures under study profits of embezzlement. It is strongly recommended in paroemias not to pursue others' property, as it is obviously possible to lose one's own.

As for the concept "insolvency", it is objectivized through the paroemias, which can be grouped into the following four topics: (1) adversity as a moral and material condition of the person; (2) characteristic features of adversity; (3) category of entities of the subjects characterized as poor; and (4) lexical units, reflecting the public's attitude to the state of poverty and the miserable's behavior.

We also revealed that the English nation is generally rational, as well as favors wealth and assets. In contrast, the French mentality displays a negative attitude toward wealth as a material value. It was also shown that the general understanding of the selected concept differs in a number of features in both the English and French cultures.

REFERENCES

Gnezdilova, V.G. French-Russian Dictionary of proverbs and sayings. 3d Edition/V.G. Gnezdilova. – Moscow: Mirta-Print, 2010. – 72 p.

Kunin, A.V. The English-Russian Phraseological Dictionary/A.V. *K*unin. – Moscow: The Russian Language, 1984. – 944 p.

Margulis, A. Russian-English Dictionary of Proverbs and Sayings/A. Margulis. – McFarland & Company, Inc., Publishers, 2000. – 494 p.

Maslova, V.A. Kognitivnaya lingvistika: uchebnoye posobiye / V.A. Maslova.– Mn.: TetraSistems, 2004. – 256 s.

Mitina, I.E. English Proverbs and Sayings and Their Russian Analogues/I.E. Mitina. – St.Petersbourg: KARO, 2006. – 336 p.

Soldatkina, T.A. Leksicheskie sposoby vyrazheniya ocenki vo frazeologii i paremiologii angliyskogo yazyka // Filologicheskie nauki. Voprosy teorii i praktiki. 2016. № 3 (57). – CH.1. – S. 178–180.

Soldatkina, T.A. Leksiko-sintaksicheskie sposoby vyrazheniya ocenki v paremiyah angliyskogo yazyka Vestnik CHGU. Gumanitarnye nauki. 2016. № 2. S. 241–247. (rezhim dostupa: http://www.chuvsu.ru/index.php?option=com_k2&view=item&id=1235).

Soldatkina, T.A. Metaforicheskie modeli frazeologizmov s ocenochnym komponentom v mariyskom i francuzskom yazykah //Finno-ugorskiy mir. 2017. № 1. S. 64–72.

The Concise Oxford Dictionary of Proverbs (CODP). – Oxford, New-York: Oxford University Press, 1996. – 316 p.

The Dictionary of Proverbs. French-Russian And Russian-French/by I.E. Shvedchenko. – Moscow: Rus. L. Media; Dropha, 2009. – 165 p.

The French-Russian Phraseological Dictionary / by Ya.I. Retsker. – Moscow: Publishing house for foreign and national dictionaries, 1963. – 1112 p.

Yakovleva, S. The Frame Structure of the Concept FAMILY in the Paroemiological View of the World of the Finnish Language/S. Yakovleva, M. Pershina T. Soldatkina, E. Fliginskikh//Review of European Studies, Vol. 7, No. 8, 2015, pp. 253–262.

Yakovleva, S. Concept "Family" and its Metaphorical Models in the Mari and Finnish Paroemiae / S. Yakovleva, G. Kazyro, M. Pershina, T. Soldatkina E. Fliginskikh//Mediterranean Journal of Social Sciences. – MCSER Publishing, Rome-Italy, Vol. 6, No 3. S7 (2015). P. 21–33.

Yakovleva, S. Representation of the Top Level "Nuclear Family" in the American Paroemiological Fund / Yakovleva S., Kolesova T., Kazyro G., Soldatkina T., Zhaksylykov A.//The Social Sciences Year: 2016 | Volume: 11 | Issue:8 | Page No.: 1881–1884.

Yakovleva, S. Verbalization of the Concept Wealth in the Mari and American Proverbs and Sayings: Universal and Specific Features/Svetlana L. Yakovleva, Tatyana A. Soldatkina, Ekaterina E. Fliginskikh, Galina N. Kazyro, Marina A. Pershina, K.Y. Badyina//Sociont 2017, 4th International Conference on Education, Social Sciences and Humanities Dubai, UAE.

Training of the Russian Federation's national guard officers in information security

R.V. Streltsov
Department of Designs Armored Techniques, Perm Military Institute of Russian troops of the National Guard, Perm, Russia

T.A. Lavina
Department of Computer Technology, Chuvash State University, Russia

L.L. Bosova
Faculty of Theory and Methods of Teaching Computer Science, Moscow State University of Education, Moscow, Russia

E.I. Tsaregorodtsev
Faculty of Economics and Finance, Mari State University, Yoshkar-Ola, Russia

ABSTRACT: The aim of this study was to develop scientific methodological approaches for training Russian troops in information security. We deal with problems related to the fundamentals of information security of the troops of the Russian Federation's National Guard. A literature analysis was conducted to understand the importance of training troops in information security in the world community. The results of the initial pedagogical experiment on the level of training that officers receive in information security are presented. The development of graduates' military professional competencies is ensured by obtaining a body of expertise, knowledge, and skills to guide units in a modern combined-arms battle in an armed conflict using Information and Communication Technologies (ITC).

1 INTRODUCTION

In the Russian Federation, information security is understood as the state of protection of individual, society, and the state from internal and external information threats, which ensures the implementation of constitutional rights and freedoms of a person and citizen, sovereignty, territorial integrity, sustainable social and economic development, defense, and state security of the Russian Federation. Ensuring information security is implementing interrelated legal, organizational, operational search, intelligence, counterintelligence, scientific and technical, information and analytical, personnel, economic, and other measures to forecast, detect, deter, prevent, repel informational threats, and eliminate the consequences of their manifestation.

At present, information security is the most urgent concern. If military actions on land or water occur locally and pass, as a rule, in certain time intervals, then cyberwar will continuously occur. In this connection, the problem organizing regular training of troops in information security is important.

2 RESEARCH FRAMEWORK

Thus, the Robert E. Crosslera study "Future Directions for Behavioral Information Security Re-search" addresses the current problems related to information security, as well as the prospects for their development. These areas include examining the behavior of deviant persons, approaches to understanding hackers, and improving compliance with information security.

In his work, Yu-Ping Ou Yang offers an in-formation security risk assessment model that could improve information security for companies and organizations.

In his article "A Situation Awareness Model for Information Security Risk Management", Jeb Webb offers the ISRM (SA-ISRM) situation model in addition to the information security risk management process. This model eliminates the identified shortcomings by collecting, analyzing, and reporting risk information.

The research "From Information Security to Cyber Security" by Rossouw von Solms declares that often cyber security is used interchangeably with the term "information security". This study argues that despite the significant overlap between cyber security and information security, these two concepts are not entirely analogous. In addition, it argues that cyber security extends beyond traditional information security, which includes not only the protection of information resources but also other assets, including the person himself.

For all the aforementioned reasons, the need for organizing regular officer training in information security is crucial.

This study is aimed at organizing regular training for National Guard officers in information security. To achieve the basics of information security, it is necessary to master first of all the conceptual apparatus, which contributes to the formation of primary ideas about the tasks of the object under study. The use of ICT tools by troops is also an integral part of training in information security. The continuity of training is due to the dynamic development of the capabilities of ICT tools.

3 RESULTS

To achieve training of troops in information security, on the basis of guidance documents, we formulated minimum requirements to the level of information security, which must correspond to information systems used in the troops of the National Guard of the Russian Federation.

To these requirements, we refer the following:

- information security should be directed to the implementation of tasks assigned to the troops;
- processing, storage, and transmission of data are aimed at assessing risks and threats that may affect the operation of troops;
- have adequate resources to ensure the requirements for information security;
- monitoring and evaluation of information security regulators;
- opportunities to address vulnerabilities on in-formation security;
- security of personnel;
- practical training troops for performing combat missions related to information security; and
- development and modification of plans for implementation of activities related to information security, as well as plans for responding to various emergency situations, backups, if necessary, and so on.

On the basis of these requirements, we identified the levels of development of troops training in information security (low, medium, and high), as well as indicators of their structural components (motivational, cognitive, and activity). The definition of the structural components of the levels of development of troops training in information security allowed us to specify their indicators of troops training in this area.

At the low level, the motivational component is characterized by a lack of need to study problems related to information security, and there is a "static" interest in the development of information and communication technologies aimed at ensuring information security.

The cognitive component means a low level of knowledge on information security and basic knowledge of computer and office software. The activity component is determined by the lack of practical computer knowledge and skills.

At the middle level, the motivational component is interpreted by the development of knowledge on information security as required. The cognitive component means knowledge of applied software and computer operating principles. The activity component is understood as the ability to apply practical knowledge when working with a personal computer aimed at information security.

At the high level, the motivational component is characterized by the priority values for studying problems related to information security, a socially significant focus in the development of in-formation and communication technologies. The cognitive component is determined by advanced knowledge on information security. The activity component means professional use of operating systems and application of theoretical knowledge to achieve the objectives (solve the set tasks) on problems related to information security.

The criterion for determining the levels of training in information security is the formation of their structural components: motivational, cognitive, and activity. Indicators of levels of training on information security show the formation of these levels and determine the technology of regular troops training in this direction.

In the initial stage of the study, troops' knowledge meeting the minimum information security requirements was checked.

We used questionnaires (to check the motivational component) and tests (to check the cognitive component) to verify knowledge on information security proficiency, as well as troops carried out a comprehensive project, during which it was planned to solve a quasi-professional task on information security. The degree of formation was determined on the basis of expert judgment. Experts were educators of a military high school. The study introduced the following quantitative indicators: 1 point – low level; 2 points – middle level; and 3 points – high level. The total score for all points of the formation of components varies from 19 to 57 points. The choice of intervals for determining the level boundaries for all points in this study was determined using the method of V. P. Bespalko. The quantitative characteristic of the boundaries of formation levels for all components is in Table 1.

The summary characteristics of the motivational, cognitive, and activity components are calculated (in percentage) using the following formula:

$$\frac{\sum_{i=n}^{n} {}_nK_i}{n*192}*100\%$$

where k is the number of answers for each question, i is the question number, n is the number of questions, and 192 represents the total number of respondents. Consolidated, average, absolute, and relative indicators of the formation of the motivational, cognitive, and activity components by levels are presented in Table 2.

Table 1. Quantitative characteristic of the boundaries of formation levels.

Level	Low	Middle	High
Points	19–29	30–40	41–57

Table 2. Consolidated, average, absolute, and relative indicators of the formation of the motivational, cognitive, and activity components by levels.

Components	Low Abs.	Low Rel.	Middle Abs.	Middle Rel.	High Abs.	High Rel.
Motivational	18	9.4%	144	75.0%	30	15.6%
Cognitive	8	4.2%	145	75.5%	39	20.3%
Activity	0	0.0%	167	87.0%	25	13.0%

The results show that the majority (>75%) of the troops of military higher educational institutions have an average level of knowledge on information security, which does not meet the requirements for training officers.

4 CONCLUSION

Determining the level of training received by Russian troops in information security allows us to identify the training components (data in Table 2 showing that special attention should be paid to the activity component). Therefore, this should be emphasized while organizing the learning process. After analyzing the study results, we conclude that training in information security perhaps plays a decisive role in the training of officers in the modern army. The results of our study will help organize the training of officers in information security, regardless of the training profile.

REFERENCES

Klepatskya I.N. 2015. The implementation of innovative pedagogical technologies in the teaching of reading in a foreign language of cadets of a military higher educational establishment. Pedagogical skills and pedagogical technologies: materials of the 5th International scientific-practical conference. Cheboksary: CNS "Interaktiv plus": 83–85.

Kok, Ayse. 2014. Education and information technologies. A conceptual design model for CBT development: A NATO case study 19 (1): 193–207.

Lavina T.A. 2005. Information and communication training in system of continuous pedagogical education. Pedagogical Informatics 2: 41–50.

Robert E. Crosslera, Allen C. Johnstonb, Paul Benjamin Lowryc, Qing Hud, Merrill Warkentina, Richard Baskerville. 2013. Computers & Security Future directions for behavioral information security research 32: 90–101.

Robert I.V. 2004. The interpretation of the words and phrases of the conceptual apparatus of Informatization of Education. Computer science and education 6: 63–70.

Rossouw von Solms. From information security to cyber security School of ICT. Nelson Mandela Metropolitan University, South Africa.

Streltsov R.V. 2014. About relevance of training of officers in information and communication technologies: Science and business: ways of development. Moscow 4(34): 202.

Webb Jeb 2014. A situation awareness model for information security risk management. Computers & Security 44: Pages 1–15.

Yang Y.O. 2013. A VIKOR technique based on DEMATEL and ANP for information security risk control assessment Information Sciences 232: 482–500.

Models of the learning styles within the adaptive system of mathematical training of students

V.I. Toktarova
Mari State University, Yoshkar-Ola, Russia

ABSTRACT: This article considers the issues connected with pedagogical design of the adaptive system of mathematical training of students within the electronic information and educational environment of the higher educational institution. It provides basic principles and patterns for building adaptive systems based on the style characteristics of a student. The authors review basic models of teaching styles (Gregorc's Mind Style Model, The VARK Model, Felder–Silverman Teaching Style Model), and suggest the criteria to create the adaptive mechanisms for building individual educational paths on their basis.

1 INTRODUCTION

The adaptive learning is an overall, multifaceted process based on the principles of differentiation and individualization. These days the return of researchers to the ideas of adaptive learning has become possible and necessary. On the one hand, the reason is the objective needs of present-day society to train specialists ready to meet professional challenges in a dynamically changing environment. On the other hand, it is due to the emergence of a new interdisciplinary theoretical and methodological basis for research which takes into account the achievements in the field of teacher training and psychology, mathematics and information science, management theory and artificial intelligence.

The basic principle for implementing the adaptive learning is management of the learning process, which means that the adaptive learning technologies are oriented toward activating students' self-study mechanisms connected with their individual characteristics and abilities through continuous management of the educational process.

The problem of developing and implementing the adaptive learning technologies is determined by the need to optimize and adapt the principles of traditional didactics to the challenges of the modern information society and solve the following tasks: improving the content and the methods of education taking into account universal application of information technologies; developing the electronic information and educational environment (IEE) providing access to training materials and freedom of choosing methods, forms of learning, and evaluation of results (Toktarova, 2015).

Personal orientation of mathematical training of HEI students contributes to the rethinking of didactic techniques used in higher school. The analysis of the content of the third-generation federal state educational standards, basic educational programs and curricula has allowed us to distinguish the following contradictions between:

- the need of the modern society in qualified specialists with high level of mathematical training and the level of mathematical training of graduates which does not correspond to these needs;
- the increase of the role of mathematics in modern science and practice and the decrease of the quantity of academic hours to study the mathematical subjects;
- the existing potential of electronic information and educational environment and its insufficient use mathematical training of HEI students.

Thus, the purpose of the work is to describe the way of designing the adaptive system of mathematical training for students in IEE conditions, aimed at improving the quality of education.

2 ADAPTIVE SYSTEM OF MATHEMATICAL TRAINING OF STUDENTS

According to the Concept of developing mathematical education in the Russian Federation (Concept, 2013), study of mathematics plays a systemically important role in education developing human cognitive abilities, including logical thinking, and influencing the teaching of other subjects. Qualitative mathematical education is necessary for everyone to have a successful life in the present-day society.

The content of modern mathematical education shall be directed to the development of student's professional and personal skills; educational technologies shall be based on the mechanisms of adaptation to students' individual characteristics; the learning strategy shall take into account student's internal motivation and the targets of training and shall be based on nonlinear technology (Fedorova and Toktarova, 2016). Intensification of student's comprehensive development can be achieved through knowledge-intensive and practice-oriented updating of the content based on interdisciplinary and adaptation in accordance with society needs and student's personality.

Building of the mathematics teaching process for students in the adaptive system provides an opportunity to build mental activities of every student in accordance with their individual characteristics and preset requirements, which is achieved by clear definition of the purpose and stages of activity, methods and tools, content of education and educational technologies.

Patterns of building the adaptive systems of mathematical training of HEI students with the purpose to promote personal and professional development of future specialists include the following (Toktarova, 2017):

- fundamental character of mathematical training of HEI students and necessity to increase practice orientated content of education;
- the content of modern educational technologies of mathematical training shall include substantive personal and profession oriented integration;
- quality of mathematical training of students may be assessed by graduates' competitiveness and HEI ranking.

Thus, the adaptive system of mathematical training of HEI students is an open dynamic system which:

- consists of different objects of education;
- is oriented to meet the social, professional, and personal needs of students in mathematics, skills, abilities, and competencies;
- is characterized by adapting to students' individual characteristics and talents aimed at improving the quality of mathematical training.

3 MODELS OF THE LEARNING STYLES

Since the main purpose of the mathematical training of students of higher school is to train a specialist with fundamental knowledge and competence, a fully developed personality capable of being able to develop and educate him/herself, the design of the adaptive system shall take into account the style issues of learning/teaching. The analysis of scientific and theoretical literature has presented numerous classifications depending on cognitive style, intellectual characteristics, emotional peculiarities, temperament, and other characteristics peculiar to the individual. The complexity of the category under study explains the lack of a common interpretation and the presence of different approaches to the understanding of the essence of "learning style" concept. Nevertheless, the analysis of the variety of "learning style" concept helps to highlight several fundamental aspects:

- learning style, on the one hand, is seen as a system of ways for the student to perceive and process educational information that promotes the formation of individual sustainable properties and peculiarities of educational task solution, on the other hand, as a means of interacting of a student with educational environment;

- defining the individual learning style includes not only the diagnosis of the level of intellectual development and personal abilities and characteristics of a student, but also the identification of the preferred learning methods;
- learning style is a kind of learning strategy that contains and describes specific response of the student to the requirements of the learning situation;
- learning styles shall be flexible and variable depending on the type of learning activity and the form of training.

In conditions of present-day realities, when addressing the design and development of the adaptive system of mathematical training of students within the electronic information and educational environment of HEI, it is essential to identify, take into account, and develop students' individual abilities, their thinking and perception style, and the achievement of a high level of knowledge and education.

The introduction of the adaptation algorithm into IEE can be achieved with the help of software-based methods by building an individual educational path for every student. The educational path defines educational structure and learning content, correlation between its individual elements, including theoretical, practical, control, and reference ones, as well as the ways of managing the learning process. To carry out qualitative building of individual educational paths, it is necessary to define a number of criteria which influence the building of an educational path, for example, a learning style. It helps to select individual student learning technologies for a student, as well as self-study technologies. There are several models of styles, and the most wide spread are the following:

The VARK Model. According to the model of styles developed by Neil Fleming, the learning process is based on the individual and psychological characteristics of personal cognitive structure, its aptitude to use the ways of interaction of the student with training information (Fleming, 1995). The classification of students is based on the channels of educational information perception: Visual (studying of training materials is based on the visual world, visual memory), Aural (oral speech, sounds are perceived better), Read/Write (written information, text is perceived better), Kinesthetic (students study training material using perceptive principle in practice).

Gregorc Learning Style Model. According to this model, learning is defined as a behaviour that ensures knowledge of students' abilities, their learning style. A. Gregorc (2017) distinguishes four primary learning styles which explain how an individual thinks and studies best: Concrete Random, Concrete Sequential, Abstract Random, Abstract Sequential.

Felder–Silverman Teaching Style Model. Richard Felder and Linda Silverman (1988) formed a rather popular model of learning styles built on students' preferences from the point of view of accepting and processing information trainees and based on four factors with two opposite values: Visual and Auditory; Active and Reflective; Sensory and Intuitive; Sequential and Global.

Having studied and analysed the features of all models, it was found out that each model includes a combination of factors influencing the adaptation of educational environment, some of which may intersect. Taking into account different factors depends on the objectives and tasks of the models.

4 CONCLUSION

Being based on the learning principles depending on the models of styles, let us specify a number of fundamental criteria for building the adaptive mechanisms: type of presenting the training material (textual description, video lessons (video lectures, video practices, webinars), audio lectures (audio dictionaries, audio reference books); level of difficulty (beginner, intermediate, advance); volume of training material (study in brief, detailed study); strategy of presenting the training material (fragments of small volume, full presentation of training material); teaching techniques (providing guidance and instructions, creating case studies, games, and simulations, building a teaching plan, organizing communications with experts, record-keeping, etc.); forms of organising learning activity (theoretical training, laboratory and practical works, preparation

Table 1. Taking into account style features when designing the adaptive system of mathematical training of students within IEE conditions.

Name of criterion	Gregorc learning style model	The VARK model	Felder–Silverman Teaching style model
Type of presenting training material	+	+	+
Level of difficulty	+	+	+
Volume of training material	+	+	+
Strategy of presenting training material	+	–	+
Forms of organising learning activity	+	+	–
Teaching techniques	+	–	–
Rate of training	+	–	+

for pass/fail exam, exam, comprehensive course study); rate of training (accelerated, normal, slow). The following table shows the criteria for the models of learning styles (Table 1).

The necessity to use multiple models of styles is conditioned by the fact that each model is not able to adapt the learning process in conditions of the information and education environment as much as possible; individualization in this case is only partial. Such adaptive system of mathematical training of students was designed and introduced into the training of students of the major "Applied Mathematics and Informatics" (reference).

The results of the monitoring activities and the diagnosis of students' preparation level in mathematical disciplines have confirmed the effective implementation of the adaptive system proposed in the article. Upon the completion of the pilot training, the students were suggested to participate in the questionnaire, the results of which showed the comfort of learning in the adaptive environment and the need to introduce it into the educational process; 78.6% stated that they had a complete match in their learning preferences according to the academic scenario, 92.9% wanted to continue studies based on individual paths. The most important criteria mentioned by the students were the form of organising the process (42.9%) and the type of presenting the training material (31.0%).

The prospective directions for the development of the ideas of this study are to examine the efficiency and quality of the adaptive education in higher educational institutions, to design and develop personalized educational environments for lifelong learning.

ACKNOWLEDGEMENT

The present work is supported by the Ministry of Education and Science of the Russian Federation (N 27.8640.2017/8.9).

REFERENCES

Concept of Development of Mathematics Education in the Russian Federation dated December 24 2013, No 2506-p.

Fedorova, S.N. and Toktarova, V.I. (2016). Mathematical Background of Students at the Present Stage of Society Development: Importance, Model, Quality. Proceedings of ADVED 2016 2nd International Conference on Advances in Education and Social Sciences, pp. 489–492.

Felder, R. & Silverman, L. (1988). Learning and Teaching Styles in Engineering Education. Engineering Education, 78, 7, 674–681.

Fleming, N.D. (1995). I'm different; not dumb. Modes of presentation (VARK) in the tertiary classroom. Research and Development in Higher Education, Proceedings of the 1995 Annual Conference of the Higher Education and Research Development Society of Australasia (HERDSA), 18, 308–313.

Gregorc, A.F. (2017). Mind Styles & Gregorc Style Delineator. [Online] Available: http://gregorc.com.

Toktarova, V.I. (2015). Pedagogical management of learning activities of students in the electronic educational environment of the university: a differentiated approach. International Education Studies, 8, 5, 205–212.

Toktarova, V.I. (2017). Mathematical Competence Among HEI Students: Cluster Approach. Proceedings of INTCESS 2017 – 4th International Conference on Education and Social Sciences, 603–607.

The problem of realizing administrative responsibility for an offence provided in Article 19.21 of the administrative code of Russian Federation

A.V. Vissarov, A.M. Gavrilov, V.V. Timofeev & A.A. Yarygin
Mari State University, Yoshkar-Ola, Russia

ABSTRACT: In this study, we analyze the legal norms of the Administrative Code of the Russian Federation, in order to regulate the implementation of administrative responsibility for violating the state registration of rights to real property and transactions therewith. At the same time, one of our objectives is to identify and study the definition of the state regulation, established order of registration, and obligation of the bodies that carry it out. We used a system of general scientific and special methods of legal knowledge, such as analysis, synthesis, and comparative and formal legal approaches. The general scientific method of analysis was used to evaluate the content of the legal rules governing the responsibility for violating an order of registration of property. The formal legal method allowed suggestions to be made to supplement the existing array of legal acts with practical recommendations. We also reviewed the most important articles of the Administrative Code and other legal acts, which describe the state registration of rights to real property, subjects of registration, their rights, and obligations. We revealed the design features of a disposition of the above-stated norm and also the inaccuracies made by the legislator. To resolve the legal gap, we finally suggested a change in the Administrative Code, which helps to escape law enforcement mistakes.

Keywords: Administrative Code, real property, state registration, legal acts, administrative responsibility

1 INTRODUCTION

Implementation of administrative responsibility for violation by the owner, lessee, or other user of the established procedure of state registration of rights to real property and transactions therewith due to the Article 19.21 of the Administrative Code raises several questions and requires substantial analysis and evaluation. According to the authors, the problem was due to the design of the dispositions of respective parts of the aforementioned article, as well as inaccuracies made by the legislator.

In particular, according to part 1 of Article 19.21 of the Administrative Code of the Russian Federation, the legislator establishes the liability of the owner, tenant or other user of the established procedure of state registration of rights to real property and transactions therewith [1].

The subject for the state registration is the right of ownership and other real rights to real property and transactions therewith in accordance with the Articles 130, 131, 132, and 164 of the Civil Code of RF.

Restrictions (encumbrances) of rights to real property arising on the basis of a contract or act of public authority or act of the local government are subject to state registration in cases stipulated by law [6].

According to Article 131 of the Civil Code, the right of ownership and other real rights to real things, restrictions of these rights, and their emergence, transfer, and termination are subject to state registration in the uniform state register of the bodies performing the state registration of rights to real estate and transactions therewith.

In part 2 of Article 131 of the Civil Code of RF stipulated that in cases stipulated by law, together with state registration, the special registration or accounting of separate kinds of real estate may be effected. Parts 3 and 4 of this article established the obligation of authorities performing registration. Next, in section 6 of Article 131 of the Civil Code, it is noted that the order of state registration of rights to real property and grounds of refusal of registration of these rights shall be established in accordance with this Code and the law on registration of rights to real property [5].

In the Federal law of July 21, 1997 No. 122-FZ "On state registration of rights to real property and transactions therewith", the state registration of rights to real property and transactions therewith are established in the 3rd Chapter. However, the majority of provisions of the above-stated law have become invalid on January 1, 2017.

Part 6 of Article 1 of the new Federal law of July 13, 2015 № 218-FZ "About the state registration of the real property" establishes the general obligation of registration of rights to real property and transactions therewith in the unified state register of rights. Analysis of the provisions of the above act leads to the conclusion that it does not contain specific requirements on the procedure of registration of rights, and is addressed to owners, tenants or other users of rights to real property and transactions therewith.

Thus, the current legislation establishes the order of registration procedures, which are addressed to the authorized state body and official. Any regulations about the sequence, timing, and other requirements to fulfill the statutory obligation for the registration of rights to real property and transactions therewith for the owners, tenants, or other users of the rights are not established by the legislator.

Noteworthy is the fact that the textual expression of the provisions of part 1 of Article 19.21 of the Administrative Code provides for responsibility for violation of established order of registration, but not for nonfulfillment of duties of the check that is obviously not the same. As established by the Federal law of July 13, 2015 № 218-FZ "About the state registration of the real property", this order does not cover or covers only the indirect action of the owner, tenant, or other user as the subjects provided for in part 1 of Article 19.21 of the Administrative Code of responsibility.

2 RESULTS AND DISCUSSION

Cases in point are urgent in modern legal practice. It confirms also the existence of scientific articles devoted to the considered subject.

Analyzing the legal norms of the state registration of rights to real estate and transactions therewith, Fedotov V. V. carries it to the state service. He notes that the state service, as well as the state function, really treats administrative procedures that are expressed in actions of the executive authorities having a certain sequence [7]. The similar view is reflected in other scientific publications [2, 3, 4, 10, 11]. Thus, the term "established order of registration" is connected with administrative procedures, which powers on commission that only public authorities and officials have.

This fact drew the attention of Kolokoloff N.A. [8]. He notes that according to the Federal law of 21 July 1997 N 122-FZ "On state registration of rights to real property and transactions therewith" the deadline for registration is only for officers, the owner (lessee or other user) cannot be held liable under Article 19.21 of the administrative code of RF if it does not have an impact on such registration in a given term.

However, we can see a different view of evaluation of these circumstances. Therefore, Mescheryakova T. R. believes that because the initiative to register must come from the owner, lessor's failure within a reasonable period from filing to registration is illegal [9].

We believe that such an evaluation approach to the understanding of the term "the check" creates a situation of legal uncertainty, the possibility of impermissible administrative discretion, and the rule of law.

Therefore, we believe that on the basis of the establishment by the legislator of the formal-legal characteristics, liability under part 1 of Article 19.21 of the Administrative Code of RF may occur only in cases where there is a legal conditionality of performance of duties of the state registration of ownership rights associated with specific circumstances or time.

The fact that, in the cases stipulated by the current legislation, the obligation of state registration of rights to real property is absent should also be taken into account. There may be situations when the person is not able to exercise his/her right for state registration of rights to real property due to the objective reasons not dependent on it. It therefore seems that the involvement in these cases of the persons to administrative liability under part 1 of Article 19.21 of the Administrative Code of RF will violate the fundamental principles of law.

However, we believe that in certain cases, for example, in force in force, legal acts establishing the performance of legal obligations under the registration rights prosecution of a person under part 1 of Article 19.21 of the Administrative Code will be valid. Therefore, there is the obligation of registration within the time and in the manner prescribed by the court, in the evasion of one party from registering the transfer of ownership rights under the contract. The established order of registration rights may be contained in the relevant regulations. Such acts include the ones that are the bases for the state registration of presence, occurrence, termination, transition, and restriction (encumbrance) of rights to real property and transactions specified in part 2 of Article 14 of the Federal law of July 13, 2015 № 218-FZ "About the state registration of the real property" [6].

Difficulty in qualifying offense under Article 19.21 of the Administrative Code is the definition of a specific point for calculating the beginning of the wrongful action (inaction). This is due to the absence of the statutory period for compulsory registration. It appears that in determining the time, which is the sign of the objective side of the offence, in this case, we must follow the rules of Chapter 11 of the Civil Code of the Russian Federation, if other rules are not established in the normative act that defines the term.

3 CONCLUSION

We should also take into account the existence of relevant established rules of registration rights, conditioned by a certain date, term, and period of time containing, for example, the decision of the court or otherwise provided by the legislation legal act. When the legal act is no indication of concrete term of fulfillment of registration of rights to real property and transactions therewith, there is no way to determine a starting point for administrative offences and, accordingly, the possibility of bringing the person to administrative responsibility.

In relation to part 2 of Article 19.21, the Administrative Code of RF only refers to civil contracts tenancy concluded for a minimum term of years or tenancy housing social use. In this case, it is assumed that the period of offence in case of failure by the landlord of the dwelling, a tenant of the premises applying for state registration of limitations (encumbrances) of the right of ownership of the premises, begins upon the expiration of the contract deadline.

When the parties are not set in the contract within that period, there will be a collision. In this situation, we believe that application of the rule in the right time to carry out registration procedures does not completely solve the problems because another question arises: which party should be responsible? In the disposition of the article, the legislator considers landlord's and employer's premises as a subject of responsibility. Thus, the current version of Article 19.21 of the Administrative Code creates legal uncertainty, which in turn creates the preconditions for erroneous interpretation, enforcement, and breaches of the law. We believe that existing conflicts may be resolved by modifications and refinements of the content of the article disposition on the basis of specific goals and objectives of state administration in the field of registration of rights to real property and transactions therewith.

REFERENCES

[1] Administrative Code of the Russian Federation from 30/12/2001 N 195-FZ//the official Internet portal of legal information http://www.pravo.gov.ru – 28.03.2017.
[2] Alkina G.I., GerbaV.A. Essence of the state services//TOGA Bulletin. 2009. N 3. P. 129.
[3] Bartsits I.N. The concept "public service" in the context of the Federal law N 210-FZ "About the organization of providing the state and municipal services" and out of it//the State and the right. 2013. N 10. p. 40.
[4] Bessonova E.V. Administrative barriers on the way of development of small business in Russia. M.: Business, 2010. p. 36.
[5] Civil Code of the Russian Federation (part one) from 30.11.1994 N 51-FZ//the official Internet portal of legal information http://www.pravo.gov.ru – 28.03.2017.
[6] Federal law of July 13, 2015 № 218-FZ "About the state registration of the real property "// The official Internet portal of legal information http://www.pravo.gov.ru – 28.03.2017.
[7] Fedotov V.V. Problems of differentiation of the state services and state functions in Russia at the present stage//the Administrative and municipal law. 2015. N 11. p. 1162.
[8] Kolokolnikoff N.A. Problems of application of Administrative Code//The magistrate. 2006. N 5. p. 26.
[9] Meshcheryakova T.R. Term as sign of the objective side of structure of an offense//Administrative law and process. - M.: Lawyer, 2012, No. 1. Page 46.
[10] Nozdrachev A.F., Tereshchenko L.K. The concept of legislative regulation of standards of the state services//Administrative reform in Russia: A scientific and practical grant/Under the editorship of Yu. A. Tikhomirov. M.: Norm, 2007. p. 286.
[11] Talapina E., Tikhomirov Y. Public functions in economy//the Right and economy. 2002. N 6. p. 3.

Grant-supported collaboration between specialists and parents of children with special psychophysical needs

O.P. Zabolotskikh
Special Pedagogy and Psychology Department, Mary State University, Yoshkar-Ola, Russia

I.A. Zagainov, M.L. Blinova & O.V. Shishkina
Interregional Open Social Institute, Yoshkar-Ola, Russia

ABSTRACT: In this study, we deal with grant-supported collaboration between specialists of different institutions and parents of children with special psychophysical needs. We focus on the most effective ways and forms of this collaboration to help a child achieve educational goals; treat their mental, physical, and speech delay; and encourage their social adaptation. We describe the practical problems commonly faced by the members of families with children having special psychophysical needs, such as integration into the sociocultural surface. Existing experience in the development of personal artistic capabilities in a multicultural learning environment is implemented in the program "Road Map of a Child with Special Psycho-Physical Needs". The article studies a complex impact on the child care system under different social conditions, in order to determine the most favorable conditions of collaboration. The main purpose of a specialist is to provide adequate consultative care while taking into consideration a child's individuality as well as the patterns of their family environment. This approach makes it possible to help socialize both the child and their family. The analysis is based on practical observation materials provided by the autonomous nonprofit organization of higher education "Interregional Open Social Institute" (IOSI), and the Center of Speech Pathology, the Mari El Republic.

Keywords: family with disabled child and its functional patterns; social rehabilitation for families; harmonization of parent–child relations; Center of Speech Pathology; Autonomous Non-profit Organization of Higher Education "Interregional Open Social Institute"; artistic culture; practitioner-family collaboration; artistic culture of a disabled child, individual cultural experience; cultural values; aesthetic conscience; artistic perception; art studios

1 INTRODUCTION

According to the Federal Standards and Requirements to the Educational Institutions, the concept of correctional activities targets to create a system of complex care for children with special psychophysical needs, in order to achieve their educational goals, assist in the improvement of their speech skills, as well as mental and physical development; and secure their adaptation in the modern society. This problem requires a high level of cooperation between specialists and practitioners in areas such as speech therapy, psychology and pedagogics, music, arts, painting and other practical skills, and physical trainings. [1] Basic principles of such activities for specialists have been described by Lev S. Vigotsky, who considered these best applicable for psychologists. Vigotsky's ideas were improved by other Russian psychologists and disability specialists (defectologists). However, those principles still lack specifics, for example, they neither determine psychological methodology nor have general patterns of practical use.

A group of specialists from the Interregional Open Social Institute (IOSI) dealing with the problem of artistic development of children with mental and physical delay created a special project "Road Map of a Child with Special Psycho-Physical Needs". Implementation of the project required a variety of events and activities to be conducted with a common purpose to achieve efficient communication between children with special psychophysical needs and normal children in a specially created environment. The purpose was achieved by cooperation of parents of children with disabilities with volunteer workers trained by specialists from IOSI and SPC and with other children as well. These volunteers had been trained to communicate with children with special psychophysical needs most effectively. The cooperation was realized in the form of the so-called "episodic integration", for example, children with disabilities were made to communicate with normal children of the same age, during their studies, some theatrical activities and other events, all those activities being supervised by specialists of the team.

The mission of a psychologist working in an educational, social, or medical institution is to determine and evaluate the ability of a child to grasp the necessary educational program (preschool, school, etc.). This mission determinates qualification requirements for a psychologist.

The problem of effective advisory guidance for families with children suffering from mental and physical delays is of importance in consultative support, which sometimes becomes the most important of all. A psychologist is often a person who can express a child's needs and interests and describe those to his/her parents. This is the reason for a psychologist to be responsible for coordination of the whole guidance process. We consider this fact to be a modern challenge. Thus, a psychologist faces the necessity of reforming his/her views on both methodological and practical problems of his/her activities. This challenge was dealt with in a series of articles published on the basics of the grant-supported program "Road Map of a Child with Special Psycho-Physical Needs". The key principles of the program were also reflected in a monograph on the subject.

One of the most important principles of the guidance process is the complex approach to a child. This requires a high level of cooperation between specialists of different areas, for example, pedagogics, psychology, speech skills, and medicine. Every specialist evaluates the state of a child with his/her own methods, the result of observation being a contribution into the bigger picture of integrated examination of the child's individual situation. Detailed patterns describing the peculiarities of approach for every specialist mentioned have already been presented elsewhere [4].

The next principle, which is closely related to the complex approach, is the principle of stereoscopic view, which has been discovered and proved during familial advisory services routine. Basically, it implies the necessity of evaluating a child's condition from more than one point of view. To illustrate this principle, it can be effective to analyze the condition of a child using a combination of different approaches, such as neuropsychological and pathopsychological evaluation, with the simultaneous use of psychodynamics and other correctional methods, all of which in close connection with the parents of the child in question. However, this principle should not be underestimated, as it often occurs impossible to diagnose a patient or adequately estimate a child's abilities in terms of pedagogics, or prepare a pattern of development, without the child's condition being analyzed with different methods and by different specialists.

Another principle worth mentioning is the principle of integrity. The child's individual condition, although consisting of some different processes (cognitive, psychic, emotional, etc.), usually manifests itself with complex traits and actions, which depend on the child's physical and emotional states, motivation, and so on. Thus, the diagnostic scenario, as well as the patterns of the child's development and correction, should be based on a thorough analysis of the complex condition, using the most effective diagnostic method. Cognitive abilities, as well as emotional area, should be evaluated by the same principle.

Advisory process should ensure the best interests and well-being of a child. This principle implies the necessity of making family aware of the best possible conditions under which the child could achieve best results in education and development. The right diagnosis, even if explained to the parents, is not enough. A specialist should also give his/her best to help parents understand the utmost importance of a child being educated in accordance with his/her psychic, physical, and cognitive abilities. Consequently, a specialist has to apply a gentle, nonintrusive approach when the diagnosis is explained, as well as provide an all-time psychological support to the parents. These two principles should not be neglected, too [3].

One of advantages of the project "Road Map of a Child with Special Psycho-Physical Needs" is that it provides children an opportunity to attend an art class or an art studio. Children with special psychophysical needs, as well as their parents, can try a variety of activities. Furthermore, with means of distant learning technologies provided by IOSI, it appeared possible to communicate with families of children suffering from mental and physical delays outside the Mari El Republic, to provide them support and advisory care. E-learning technologies also gave opportunity to organize a course for parents to give them some clues on basic principles applied to children with special psychophysical needs, in terms of pedagogics, psychology, medicine, and social studies.

2 CONCLUSION

The problem of learning difficulties in modern neuropsychology is seen as a problem of interaction between the brain, physiological, psychological, and social levels, because the beginning of the development of the PEF lies in the social environment, and not inside brain structures.

In this case, the undeniable fact of interaction and mutual influence of HMF and their brain substrate obeys the principles of dynamic chromogenic system localization of VPF, the maturation of the brain bottom-up, hierarchy and geterogennosti localization, and the gradual lateralization and localization of the VPF system.

Neuropsychological causes of difficulties in training and education can be very different. Neuropsychological diagnostics currently allows us to differentiate symptoms of disorders of the development of the APF and is of not only theoretical but also practical importance. The leading method is "syndrome analysis" developed by A. R. Luria, based on a system of psychological analysis, to determine the condition of the structure or other function, as well as the relationship and interaction of some of the APF to others.

Unfortunately, not always parents, teachers, psychologists, and doctors can assume the real reason for learning difficulties and education of the child with disabilities and take effective action for its localization, provide timely assistance in the development of mental processes, and maintain a level of self-esteem. This happens because of their lack of competency in pediatric neuropsychology caused by the reduced criticality because of the lack of specialist clinical psychology and special educational institutions. Very often the families of these children experience financial difficulties and are unable to teach the child away from the place of residence.

Therefore, learning and growing in the soil of functional disorders represent a socioeconomic problem.

IOSI has every opportunity to organize distance learning courses and provide communication, because modern technologies are available for its members. It is impossible for a parent having a child with special psychophysical needs to attend lectures on a daily basis. For such problems, the IOSI offers a special e-learning program, which gives parents an opportunity to choose the most convenient time and place for a lecture.

Existing experience in the development of personal artistic capabilities in a multicultural learning environment is implemented in the program "Road Map of a Child with Special Psycho-Physical Needs". The project itself can be proof of the fact that a child with special psychophysical needs may – and should – be a person who is of much importance in order to encourage a more tolerant approach to such children in the modern society. The program also gives those children an opportunity for cultural and artistic development.

Collaboration of specialist is necessary from the very first stage, that is, initial examination of a child with disabilities. Initial examination targets at analysis of individual features of a patient's speech, physical traits, and motor skills. It also examines a child's cognitive processes, activities, and individual characteristics. During the course of this grant-supported research, we (psychologists) describe peculiar traits of development, common for children with special psychophysical needs. In close cooperation, specialists choose most appropriate training materials, games, and exercises, which may improve the development of a child's visual and auditory sense, speech skills, attention and memory, intellect, imagination, motor skills, and calisthenics [6].

Collaboration is also important during visual demonstrations, trips, and excursions. A music teacher, an art teacher, a speech skill specialist, and a choreographer does his/her best in cooperation to achieve positive results, such as, to help a child exercise a normal breathing while singing or speaking, make him/her understand how to alter a tempo, timber, or volume of sound while vocalizing; and help a child develop his/her perceptional skills both for speech and music. It is also necessary to cooperate while a child is taught to recognize rhythmical and intonation patterns of speech. Giving children opportunity to listen to different pieces of music and encouraging them to speak of their impressions (which develops lexical baggage) is important as well, along with developing listening, singing, and choreographic skills.

It is obvious that while working with a child with special psychophysical needs, a music teacher does so in a close cooperation with a speech skill trainer. In fact, a music teacher extends the boundaries of a child by giving him/her an opportunity to develop an ear for music, memory, sense of rhythm, breathing skills, all these being a basis for better speech apprehension skills. To achieve this aim, a music teacher has a vast variety of means, such as games and vocal exercises. For an art teacher, it is important to encourage a child to expand his/her motor skills by teaching how to move hands and fingers more subtly and delicately. Thus, cooperation between specialists is not only critical for dealing with speech retardation, but can also have positive effect in helping a child to master sufficient writing skills.

An art teacher should work in coordination with a speech skill trainer, taking into account speech skills of a child and touching upon the same lexical topics at art lessons. It can be done in a form of mastering a certain grammatical construction, for example, using special rhythmic patterns: phonemic or syllable patterns, riddles of some kind, poems, and games of fingers. Children show great results while voice-synchronizing cartoons created during a lesson, as it motivates children to pronounce every line emotionally and with good articulation.

A choreographer teaches a child different forms of expression of musical perception through movements of a body, through dancing. It is vital for a child with special psychophysical needs to have some experience of stage behavior and feel free to express his/her creativity and cultural interests. These forms of activity can be highly motivating for a child and contribute to his/her communicative skills. Cultural life and being part of cultural events helps a child get accustomed to social activities [2].

To encourage imaginational skills of children suffering from mental and physical delay and to introduce an primary stage of occupational guidance, professionals from IOSI organized lessons of tradition handicrafts (basketwork, woodwork, carpentry, lace-tatting, birch bark, and wire-tatting), the craft items being demonstrated at special shows and craft fairs. These activities helped make children more confident and also encouraged with a certain level of harmonization in parent–child relations [5].

The project "Road Map of a Child with Special Psycho-Physical Needs" aims at ensuring a barrier-free environment for children with disabilities and special psychophysical needs, including those with hearing and/or visual impairments and locomotor system diseases. Children with no special needs involved in the project had been given an opportunity to better understand those with disabilities, as well as their problems, needs, behavior, and communicative patterns.

In conclusion, it is necessary to mention that collaboration and close cooperation between specialists of different branches of science and treatment is inevitable for an effective realization of a complex approach to guidance of children with disabilities and special needs. This approach results in a successful socialization of a child, giving him/her a possibility to have a future where effective self-realization and social communication are ensured.

REFERENCES

[1] Cooperation between psycho-pedagogical and health care services on issues relating to the guidance of families with children with special psycho-physical needs/Olga P. Zabolotskikh, Olga V. Shishkina//Kazan Pedagogics Journal. 2016. № 2–1 (115). pages 183–187.
[2] Oksana M. Maslennikova Children's communicative skills development by means of stage and theatrical activities/Oksana M. Maslennikova//Young Scientist. 2014. № 19. pages 581–583. [on-line]. – Direct link: http://moluch.ru/archive/78/13638/ (entry date February 19, 2017).
[3] Nataliya S. Morova Role if a family in a process of social guidance of a child with special psycho-physical needs/Nataliya S. Morova, Olga P. Zabolotskikh//Scientific methodology journal of Russian Academy of Education "Social pedagogics in Russia". 2012. № 3. pages 30–39.
[4] Psycological, medical and pedagogical process of guidance for families with children with special psycho-physical needs: collective monograph/Olga V. Shishkina, Olga P. Zabolotskikh, Nataliya Y. Glazunova. Yoshkar-Ola: IOSI, 2015. 74 pages.
[5] Creativity development of children in the process of realization of the social project "Road map for a child with special psycho-physical needs": guidance materials/compiled by Olga V. Shishkina. – Yoshkar-Ola: "String" Publishing House, 2016. – 223 pages.
[6] Elena V. Yakovleva The impact of visual arts and activities on the process of children's development//Actual pedagogical challenges: materials of the VII International Scientific Conferention (City of Chita, Russia, April 2016). – Chita: Young Scientist. 2016. – pages 67–70.

form
Preparing university educators for tutoring adult learners in distance education

N.A. Biryukova & D.L. Kolomiets
Faculty of General and Vocational Education, Mari State University, Yoshkar-Ola, Russia

V.I. Kazarenkov
Peoples' Friendship University of Russia, Russia

I.M. Sinagatullin
Birsk Branch of Bashkir State University, Birsk, Russia

ABSTRACT: Distance education is one of the effective forms of continuing vocational education and training of adults, which is aimed at meeting their educational needs in lifelong learning. Distance learning introduces new specific requirements to both educators with regard to their competence and the learner. One of the key actors who ensure the effectiveness of the system of distance education is a tutor. Here, a special study has been conducted aimed at studying the readiness of university educators to teach via distance learning, and identifying the attitude of adult learners to distance learning. The study was aimed at identifying students' awareness of the activities of a tutor and studying their opinion on the activities of the tutor. We have also developed a training program aimed at preparing university educators to be tutors. Great importance is given to the readiness of an educator to perform tutoring activities (psychological, subject-specific, and technical).

Keywords: distance education; tutoring; adult student; university educators as tutors

1 INTRODUCTION

The idea of lifelong learning introduced in the 20th century is widely used in all spheres of education currently, including vocational training of adults. Current market situation forces people to keep learning and undergo continuing education courses.

Distance education is one of the effective forms of continuing vocational education and training, aimed at meeting the educational needs of adults in lifelong learning.

Research into distance learning as a new type of education has new challenges, whose solution affects pedagogy, methodology, management and financing, quality assurance, as well as intellectual property rights (Holmberg, 1995). Distance learning introduces new specific requirements to both the educators with regard to their competence and the learner. It is necessary to change the system of priorities, which involves rethinking the methods and the content of teaching, as well as training of educators. It is important to shift the role of an educator to their cooperation with students (Sinagatullin, 2012).

Empirical data show that one of the key actors who ensure the effectiveness of the system of continuing vocational education and training is a tutor—an educator who acts as an advisor, mentor, and organizer of independent activities of students aimed at studying the content of the course as well as personal and professional development and self-development.

Analysis of theoretical and empirical studies carried out worldwide has shown that the term "tutor" is most often used in adult distance education system (M.G. Moore, G. Kirsley, R.G. Williams, G. Gibbs, N. Durbridge, etc.). Moreover, the content of this concept varies depending on specific forms of distance learning. For example, in correspondence instruction, the functions of a tutor are close to the functions of an educator, while in blended learning, the tutor acts as an organizer of students' joint activity aimed at developing their competence.

According to Gibbs and G & Durbridge, N. (1997), a tutor is a mediator between an educator and a learner and an organizer of learners' activities and individual pathways plans. In the system of higher education, built on the basis of a traditional system of education using information and communication technologies, a tutor is an educator's assistant and a mediator between the lecturer (professor) and the student. Tutors in this respect act as educators rather than organizers. In the system of adult distance learning, researchers (Brookfield, 1986, Moore & Kearsley, 1996, Williams, 2000, and others) believe that a tutor is an expert in the field of organizing learning activities and encouraging their self-education, who support adult learners in self-education and the development of their competence.

The analysis of tutors' activities, research into distance learning (Bondarev, 2003, Hayrullina, 2014. Zhuravleva, Cruck B. et al., 2007, etc.), as well as the results of our study (Biryukova et al., 2015) have revealed the difficulties that tutors experience when they start performing their new duties (insufficient understanding of the special role as well as specific tasks of a tutor in the system of distance education, specific principles of interaction with adult learners, etc.). They have also revealed a number of difficulties that tutors face when they perform their main activities from educational process development to assessing student learning outcomes and supporting their further education.

According to Zlotnikova I.Y. (2005), an educator in distance education should be able to:

1. develop a training course (set educational goals and objectives; create a structured, easily perceived, interactive teaching and learning material; and prepare the components of the ICT-learning environment and educational software);
2. organize interaction between learners (organization of teleconferences, videoconferences, discussions on given topics, and joint implementation of tasks);
3. achieve general management of the process of educational and cognitive activity of students (supervision of students' independent work, coordination of the cognitive process, support to a certain educational, and cognitive direction in the learning activity of learners);
4. encourage and conduct discussions on given topics (encouraging students to take part in joint discussions carried out in both "online" and "offline" modes, students' individual responses to the questions sent via e-mail);
5. motivate and stimulate educational and cognitive activity of students (using interactive and active methods of teaching); and
6. monitor students' educational achievements (general supervision of the study process, development of formative and summative assessment tasks, correction of learning outcomes).

In practice, the list of professional duties performed by the educator is more wide. And it differs significantly from traditional forms of teaching. As a result, a large number of educators, even those with more teaching experience, are not able to work in the new conditions of the modern information society (Biryukova & Kolomiets, 2016).

2 MATERIALS AND METHODS

The goals of the research conducted in the Mari State University (hereinafter MarSU) (Yoshkar-Ola, Russia) were to reveal the level of readiness of university educators to teach via distance technologies and to identify attitude of adult students to distance learning. The main research methods were surveys and interviews. The study was carried out via a Web-survey in May 2016.

The research included two steps. First, 35 MarSU educators were interviewed. This part was guided by the following research questions:

a. What is the general attitude of the respondents toward the use of distance learning technologies in adult education?
b. What are the main problems the educators face using distance educational technologies?
c. How do the educators assess their level of skills and competencies in distance education?
d. Are they ready to perform the role of a tutor and what personal qualities are necessary to be an effective tutor?

The second part included interviewing 90 adult students of MarSU, which was guided by the following research questions:

a. Is there a significant impact of distance education technologies on adult students' achievements?
b. What is the level of organizational, advisory, and analytical skills of the tutors?
c. What problems and difficulties do adult learners encounter while studying at MarSU?

3 RESULTS

The results of the survey conducted on educators have shown the following:

Most tutors (75%) have a teaching experience of more than 5 years. All tutors positively evaluate the effectiveness of using distance education technologies. However, only 9% of educators believe that it is reasonable to use distance education technologies when students get first higher education. The majority of the respondents are certain that the use of distance learning is reasonable while delivering continuing vocational education and training courses (54%) or when students get second higher education (37%).

The main problems the respondents indicate with regard to using distance educational technologies are the educators' lack of competence in technical aspects of developing distance learning courses and a considerable amount of time spent on their development. Thus, the greatest difficulty faced by educators is developing a distance learning course.

The main motive underpinning the positive attitude of the respondents toward the use of distance learning technologies is the improvement of methodological knowledge and skills. At the same time, it indicates that the majority of the educators surveyed highly value their personal advisory qualities, and half of those surveyed believe that they have analytical and reflexive skills.

The educators are satisfied with the use of distance education technologies for adult learning: "yes" −66%; "no" −21%; "I find it difficult to answer" −13%. About half of the respondents assess their level of competence in using distance education technologies as average. The respondents are gradually mastering their skills in using this technology; they gather materials that would be of interest to students undergoing distance education courses. At the same time, they note that they cannot use all the variety of available technical and software tools.

Although two-thirds of the respondents are not satisfied with their activities as a tutor in distance education, 88% of the respondents are ready to solve the emerging problems. Furthermore, 91% of the educators note their high level of competence in performing their duties within the traditional system of education, and with regard to distance learning, this percentage is much lower −60%.

The majority of the respondents believe that tutoring is the key factor of improving the quality of education using distance education technologies.

The results of the survey conducted on adult students revealed the following facts.

The majority of the learners (92%) positively assess the effectiveness of using distance education technologies.

In general, most students (87%) are satisfied with the use of distance education technologies. This is explained by the convenience of combining study and work, lower costs, as well as lower time consumption. Dissatisfaction of learners is caused by the following shortcomings in the quality of education related to the human factor: the educator ignores the questions and requests of students; the lack of an individual approach to the training of each student; a lot of time is spent on checking students' assignments; and the students are unaware of the grading criteria.

Table 1. Influence of distance education technologies on student's achievements.

Achievements	Really helped (%)	Partly helped (%)	No influence (%)
To know better fellow students, their professional and personal achievements and challenges	60.7	17.4	8.7
To change attitude to the teacher	52.3	8.7	13.4
To solve problems in the professional field	43.5	21.7	17.4
To develop skills of self-education, research skills	39.1	39.1	8.7
To learn from the experience of other students	34.8	30.4	17.4
To learn new methods and techniques of professional activities	30.4	34.8	17.4
To improve communicative and interactive competence	30.4	21.7	30.4
To increase the motivation of educational activities	26.1	43.5	17.4
To acquire new knowledge in professional field	26.1	39.1	13.4
To identify the focus of professional and personal interests	26.1	34.8	26.1

The respondents consider that the main problem in using distance education technologies is the lack of educators' awareness of technical aspects of using distance learning course and the lack of the relevant competence.

The survey also involved a number of questions about the evaluation of organizational, informational, advisory, analytical, and reflexive skills of the tutors. Most respondents believe that only 40–60% of tutors have the necessary skills.

Controversial opinions were given on the level of the tutor's commitment to successful learning outcomes of students; the opinions of the students were divided almost equally. The answers were: "yes" –52%, "no" –48%. At the same time, according to the results of the survey, the majority of students (88%) consider interaction with a tutor to be an important part of their education.

A proportion of 39% of respondents believe that educators should use interactive learning technologies more widely. A low percentage again indicates the necessity of providing tutors with special training. Furthermore, 91% of students consider distance learning to be an important part of the learning process, and 98% of students positively evaluate the work of the Digital University.

MarSU adult learners have enumerated the following advantages of distance learning:

1. The opportunity to study at a convenient time and in a comfortable rhythm; the opportunity to repeatedly review theoretical material, including lectures, presentations, and electronic textbooks; the opportunity to write a letter with a question to the educator; taking interim and summative assessment tests, which excludes subjectivity of knowledge assessment.
2. Qualified educators, full provision of methodological materials in most disciplines.
3. Accessibility of education: tuition fee, the absence of extra spending forced by educators and the quality of the electronic portal.
4. The opportunity to study any time, at your own speed, wherever you are. The opportunity to combine work and studies, high level of learning outcomes, accessibility of course materials, training in a relaxed atmosphere, convenience for the educator, individual approach to teaching.
5. The opportunity to review materials, save the time on searching for information in some disciplines (where the educators have uploaded course materials on the portal).

The respondents have also noted the following few shortcomings:

1. Tutors do not explain the grades. Distance education does not provide learners with the opportunity to get an immediate answer to their questions.
2. It is impossible to establish fast and efficient communication with some tutors.
3. Communication with the educator is minimized; there is no opportunity of getting preliminary feedback on the completed assignments.
4. Low level of performance of some educators: there are no materials and feedback.

Although there are a few shortcomings in using distance education, its advantages outweigh.

The majority of students (91%) affirmatively answered the question "Do you think that tutoring is an important factor in assuring the quality of distance education?"

Thus, the survey was aimed at identifying students' awareness of the activities of a tutor and studying their opinion on the activities of the tutor. The absence of tutors or their unsatisfactory readiness (professional, psychological, and technical) seems to be the main obstacle in using distance education technologies at university.

4 DISCUSSION

The analysis of the results of the surveys conducted makes it possible to come to the conclusions formulated below. Elaboration on the needs of adult learners in tutor support revealed the areas where learners feel the necessity in it. The authors have developed a teacher training program aimed at preparing them to be tutors. The program has been implemented at the Mari State University. This initiative seems to be well timed, as the survey revealed that two-thirds of educators are not satisfied with their activities as a distance education tutor, with 88% of them willing to improve their skills and showing the desire to use distance learning technologies more effectively.

The survey made it possible to identify the main growth and problematic areas, which should be in the focus of educators' training. The following modules were included in the training program: organization of adult education using distance education technologies; managing adult learning in the system of distance education; current issues in the psychology of adult education; forms of distant interaction; peculiarities in creating and uploading formative, interim, and assessment tasks; distant advising; development of course materials; ICT competence; computer-aided testing of the quality of education; and methods of monitoring student learning and evaluation criteria (Biryukova & Kondratenko, 2014).

Upon successful completion of the program, educators are expected to be ready (subject-related, technical and psychological readiness) for tutor activities. In the framework of our research, we consider readiness in a comprehensive way, being a complex of knowledge, skills, and abilities, as well as consent to carry out tutoring activities. Indeed, a tutor has to be ready as an expert in their field, as a teacher (to have a certain amount of general and subject-specific knowledge and skills, developed algorithms of pedagogical activity), and technically be able to use computer technologies and possess a high level of culture, intellect, values, and so on.

We agree with T.V. Gromova (2007), who argues that readiness for tutoring is a qualitative, systemic, dynamic individual state, being a combination of motivational, cognitive, operational, and reflexive components, filled with qualitative characteristics and indicators.

The *motivational* component includes an educator's value-based attitude to performing the duties of a tutor, satisfaction with their professional activities as well as awareness of the social significance of tutoring.

The *cognitive* component includes subject-specific and pedagogical knowledge of the educator performing the role of a tutor. This is information on the methodology, nature, specifics, and distinctive features of tutor support provision.

The *operational* component includes a complex of skills and abilities, represents tutor's behavior, and provides a set of ways and methods aimed at achieving the goal of the activity.

The *reflexive* component of readiness is a process of cognition and analysis of the phenomena of one's own consciousness and activity. It features the solution of a profession-oriented problem, resulting in a personally colored interpretation of the essence of the problem.

With a significant share of independent work of students, high-quality results of their learning can be achieved provided that a well-developed system of tutoring for adult learners is created. We distinguish between the following types of support: training (gradual, practice-oriented group training); counseling (providing advice on a particular problem, filling the knowledge deficit); modeling (uncovering educators' potential); and supervision (professional counseling by a more experienced expert, excluding formal monitoring and evaluation, creating psychologically comfortable conditions for those involved in professional activity).

We believe that tutoring adult learners in distance education can be carried out in several steps (Kazarenkov & Kazarenkova, 2015).

1. Planning and preparation. This stage involves an examination of the needs of students in help and assistance; identification of the initial level of their knowledge and skills; and preparation of a package of guidance materials.
2. Support itself. It implies extending knowledge, developing students' skills and abilities; counseling on specific issues; providing information and analytical support; organizing information exchange and interaction among students; and foregrounding internal forces and standby capabilities of students.
3. Use of learning outcomes and final analysis. This stage involves the consolidation of acquired knowledge, skills, and abilities; development of the experience in solving professional problems; tracking the progress and changes in students' performance; and final evaluation of the effectiveness of the support.

5 CONCLUSION

To sum up the outcomes of our research into the activity of an educator as a tutor in the system of distance education, the following conclusions can be drawn:

– The problem of interaction between an educator and adult learners is currently insufficiently studied: the conditions for effective interaction between the educator and adult learners have not been thoroughly studied, nor have the models of such interaction been developed for the system of distance CVET.
– The activity of a tutor differs from the activity of an educator in the traditional system of higher education due to a number of features related to communication with students at a distance, including the teaching methods used and the organization of the educational process. In distance education, a tutor supports students.
– Great importance is given to the readiness of an educator to perform tutoring activities (psychological, subject-specific, and technical).
– In the system of distance education, the tutor performs four basic functions (organizational, informational, communicative, developing), each of them requiring certain knowledge and skills.
– A number of requirements are imposed on the tutor with regard to the distinctive features of distance education and the tutor's functions.

Thus, purposeful preparation of educators for tutoring is currently crucial.

REFERENCES

Biryukova N.A. & Kolomiets D.L. 2016. Formation of students' readiness for pedagogical activities based on the analysis of professional difficulties. Contemporary Higher Education: Innovative Aspects. 8. 2. 57–63. [Russian language].

Biryukova N.A. & Kondratenko I.B. 2014. Pedagogical conditions of formation of future teachers' cultural competences in the process of interactive learning. Theory and practice of social development. 9. 62–64. [Russian language].

Biryukova N.A., Yakovleva S.L., Kolesova T.V. et al. 2015. Understanding adult learners as a core principle of effective ESL-educators. Review of European Studies. 7. 8. 147–155.

Bondarev, S.G. 2003. Pedagogical conditions of distance education organization in the process of future teachers' training. [Russian language]. Barnaul.

Brookfield, S.D. 1986. Understanding and facilitating adult learning: A comprehensive analysis of principles and effective practices. San Francisco: Jossey-Bass.

Gibbs, G & Durbridge, N. 1997. Characteristics of open universities tutors. Teaching at a Distance.

Gromova, T.V. 2007. Current aspects of the formation of professional readiness of high school teachers to work in the system of distance learning. [Russian language]. Samara.

Hayrullina A.G. 2014. Readiness of tutors to provide training with the use of distance technologies: results of survey. Bulletin of the South Ural State University. 6. 3. 96–103. [Russian language]. Chelyabinsk.

Holmberg, B. 1995. Theory and Practice of Distance Education. London, Routledge.

Kazarenkov V.I. & Kazarenkova T.B. (2015). Self-education of students in the context of lifelong learning. Lifelong learning. Proceedings of the 13th International Conf. [Russian language]. St. Petersburg. 286–288.

Moore M.G. & Kearsley G. 1996. Distant education: a system view. Wodsworth Pubishing.

Sinagatullin, I.M. 2012. Global education as a cardinal paradigm of the new century. Pedagogics, 3, 14–19. [Russian language].

Williams P.E. 2000. Defining distance education roles and competencies for higher education institutions: a computer-mediated Delphi study.

Zhuravleva, O.B., Cruck B.V. et al. 2007. Managing e-learning in higher education. [Russian language]. Moscow: Hotline-Telecom.

Zlotnikova, I.Y. 2005. Methodological system development of informational and distance training of the subject teachers. [Russian language]. Voronezh.

Theoretical analysis of the rehabilitation study of children with limited health capacities

J.A. Dorogova, O.N. Ustymenko, E.A. Loskutovan & N.V. Yambaeva
Faculty of Physical Culture, Sports and Tourism, Mari State University, Yoshkar-Ola, Russia

ABSTRACT: Currently, the tendency toward an increase in the number of children with limited health capacities can be observed.

The category of such children is extremely heterogeneous, but their common main feature is the violation or retardation of motor activity. In the context of the Russian education process modernization, the problem of creation of optimal conditions for development, upbringing, and education of children with limited health capacities is crucial. The main task of physical rehabilitation of children with limited health capacities is to accelerate recovery processes and prevent or reduce the risk of disability. It is impossible to assure physical and functional development if we limit a natural desire of organism to move. Movement is the main tool of knowledge of the world around that overcomes stereotyped behavior and the tendency to isolate. Motor abilities allow children to act and orientate in the nearest environment and perform activities. Movement develops in children with limited health capacities along with the increase of self-awareness. Motor development of children is achieved with the growth of manual abilities as a result of biological maturation of the body and stimulation as a result of corrective exercises in order to teach them to move. Therefore, health and fitness classes must become a key link for improving the physical fitness of children. One of the means to correct pathologies in children with limited health capacities in their motor sphere is physical therapy lessons. For the development of motor activity, walking training, and rehabilitation of children with limited health capacities, the Ardos simulator was created in Russia, which allows the body to stay in a vertical position for an indefinite period of time. Using a simulator, we can do exercises stimulating the functions of weakened muscles and joints and perform motor actions. Children with musculoskeletal disorders, working with the Ardos simulator, can simultaneously use other simulators or gymnastics equipment.

Keywords: rehabilitation, children with limited health capacities, physical therapy, Ardos simulator

1 INTRODUCTION

Currently, there is an increase in the number of children with limited health capacities due to various reasons. The category of such children is extremely heterogeneous, with the common main feature being violation of motor activity or delay in physical development. In the context of the Russian education process modernization, the problem of creation of optimal conditions for development, upbringing, and education of children with limited health capacities is crucial (Belenovich, 2005).

An increase in the number of infantile cerebral paralysis cases indicates that the problem of pediatric disability requires closer attention. Analysis of this problem showed the following results: currently, children often experience locomotor deficiency, which has a negative impact on their development. Normal development of a child implies a very urgent need for movements. Each movement in the development of a child such as crawling, walking, climbing, and jumping, gives him/her pleasure. If a child cannot move, he/she becomes inert, apathetic, and irritable.

Physical therapy lessons, as a form of adaptive physical education, are designed to meet the complex needs of persons with disabilities in the state of health. This requires special and individual work with children who have impaired motor functions.

In order for children with limited health capacities to grow and learn to walk, the Ardos simulator was created and used in classes on therapeutic physical culture. Its use strengthens the muscles of the locomotor apparatus, facilitates the execution of elementary movements, and improves balance disorders caused by restriction of mobility of joints and disorders of visual-motor coordination. For this purpose, several exercises that give an introduction to the scheme of our own body were suggested.

Most musculoskeletal disorders in children result in decreasing performance capabilities of upper extremities, supporting function of lower extremities, restricted backbone mobility, which dramatically reduce their everyday activities and hinder their social adaptation. Therefore, the objective of physical rehabilitation of children with limited health capacities is to accelerate reconstructive processes and prevent or reduce the risk of disability. It is impossible to assure physical and functional development if we ignore a natural desire of organism to move. Therefore, health and fitness classes must become a key link in improving physical fitness of children.

2 RESEARCH METHODS

The basic method is the theoretical analysis of the scientific-methodical literature on the research problem.

3 RESEARCH RESULTS

One of the most complicated problems is the development of means and methods of motor activity as methods and means of physical rehabilitation of children with disabilities are based on the disease specificity and the initial state of the their body. Rehabilitation complexes of physical exercises recommended in professional literature are carried out by children with disabilities in a lying or sitting position only and are chosen, in most cases, without regard to spare capacities of children doing exercises, which considerably extends the recovery time. Children who cannot keep vertical posture by themselves for various reasons do not practically take recommended exercises.

Currently, there is a serious problem of training children with limited health capacities who experience considerable difficulties with motor and communicative activity due to existent special needs, as well as somatic diseases.

Knowing peculiarities of such children will make it easier for teachers to develop the goal and tasks and apply corrective and educational work methods to practice for the successful introduction into available types of activity and social relations taking into consideration such peculiarities as nature of disability; psychophysiological peculiarities; physical health limitations; children's limited capacities, which do not allow them to participate in play, learning, communication, and labor activity and deprive of normal life in the society; and limited capacities for communication.

Motor activity is crucial for child development. The need for movements constitutes one of the basic physiological peculiarities of a child, being a condition for their normal formation and development. Recently, chronic diseases and various functional abnormalities have been increasingly diagnosed in preschoolers.

Comprehensive physical development of a child is, first of all, timely formation of motor skills and abilities, development of interest in various types of movements available for a child, education of positive moral, and strong-willed personality features. One of the main tasks of physical education is to stimulate positive shifts in the organism and develop motor abilities and skills of physical development aimed at developing and improving the organism, supporting life, and adapting a child in the society (Ayzikov, 2003).

The main goal of physical education lessons is to make children acquire basic motor skills and imitate movements and actions of a teacher or parents. Physical education work, primarily, must be aimed at solving common and corrective tasks. Through physical training lessons, children get an opportunity to acquire the skills of psychoemotional release in a socially acceptable way. They develop functional readiness for learning and interaction with people around them (Vitenzon, 2000).

Children's physical development is greatly influenced by the success of their relationship with adults and peers. Attention should not only be focused on a child's physical defect. Tolerance toward children with limited health capacities is an indispensable quality of teachers and parents.

An integrated approach to training of children with limited health capacities involving specialists and teachers allows children to unleash their potential more extensively, acquire necessary knowledge and skills, experience in communication with peers and adults, and take an opportunity to get ready for future independent life (Gross, 2006).

One of the ways to correct pathologies in children with limited health capacities in their motor sphere is physical therapy lessons, which aim at assuring preservation and improvement of the vitality and potential of teachers and children and allowing them to use acquired abilities and skills independently in extracurricular activity and in real life. When planning and giving lessons, it is necessary to follow main modern requirements for the lessons with various health-saving technologies. As the process of training children with abnormalities is very specific and manifests itself in the lower level of educational material difficulty than in a mainstream school, training at a slow pace, lower density of academic load for students at the lessons, and predominant use of demonstrative training methods are important. Pedagogical methods of corrective work will differ in the fact that they are aimed at stimulating the development of compensatory processes in children with limited capacities and allowing the development of new positive qualities in them. As a result, students develop generalized academic and labor skills, which reflect their level of independence when solving new academic and academic labor tasks (Semyonova, 2002).

One of main factors complicating the development of motor abilities and skills in children with limited health capacities is dysmotilities, which have a negative impact on not only physical development but also personality socialization, development of cognitive and labor activity, and subsequent labor adaptation. Thus, the priority in the work with such children is to use an individual approach taking into account the specificity of mentality and health of each child. When training children with limited health capacities, one of the most important conditions for a teacher is to understand that these children are not handicapped compared to others. Nevertheless, these children need a special individual approach, different from the limits of a standard comprehensive school, and they need to fulfill their potential and create the conditions for development. A key point of this situation is the fact that such children do not adapt to the social rules and conditions but are integrated into life on their own terms, which are accepted and taken into consideration by the society (Gross, 2006).

During a physical therapy lesson, the following special methods are used: step-by-step explanation of the tasks, their sequence; revision of the instruction for the task performance by the student; change of activity; support with audiovisual technical means (application of functional music, audio support of lessons); closeness to children when explaining the task; and individual assessment (permission to redo the task, which the child did not manage to do) (Semyonova, 2002).

Physical therapy lessons allow us to combine physical and mental activities: children must understand the tasks of the lesson, grasp the sense of the teacher's instructions, and analyze using speech and their actions. Encouraging children to answer the questions contributes to consciousness of motor actions and, in addition, to the development of speech and enrichment of vocabulary. Friendly conversation and talk exchange between adults and children create positive emotional background at the lesson.

A child is an acting personality, whose activity is expressed, first of all, in movements. Such work is based on the following: doing exercises outdoors, playing games at breaks, having a fresh air promenade. Application of these technologies is particularly necessary for children with limited health capacities at the physical therapy lessons. Experience shows that it was the right decision to choose this trend for this work, as children with limited health capacities particularly need help of adults, teachers, and tutors to preserve and promote their health. And when we teach children to value, preserve, and promote their health and possess a healthy lifestyle, we can hope that future generation will be healthy and advanced, both spiritually and physically, and will achieve these big goals, which their life determined for them.

A special place in the formation of secondary negative changes in the musculoskeletal system is taken by the absence of the ability to keep vertical posture by children with musculoskeletal disorder, which limits the formation of natural statokinetic reflexes and development of a child's movements at the earliest stages of development. Therefore, it is essential for children with limited health capacities to take up a natural vertical posture, as the ability to keep a vertical position is one of the most important conditions for physical and social development.

At present, there are several rehabilitation methods for children with a diagnosis of infantile cerebral paralysis, but very few of them are based on the use of a vertical position of a child's body applying simulator devices as a base position with the control of the load on the musculoskeletal system to do subsequent active exercises and movements, which certainly hinders the process of physical rehabilitation and considerably decreases its effectiveness.

Motor activity opportunities can expand to a considerable extent thanks to the introduction of methodical techniques and use of simulation devices in a broad scope. Special importance in this respect is attached to the formation of motor activity in children as an ability to perform finely coordinated movements, walking, and other types of body movement in space determined, to a considerable extent, by the ability to keep a certain posture for a long time. The restoration of normal or close to normal locomotor functions depends on how early the development process was started (Gross, 2006).

Development of advanced methods of physical rehabilitation and introduction of simulation devices mostly using active physical exercises will increase the effectiveness of development of not only motor skills but also other qualities necessary for everyday life.

To provide the children with limited health capacities with an opportunity to develop and learn how to work, the Ardos simulator (http://www.ardos-gk.ru) was created based on the Israeli prototype.

It has the following advantages over the prototype (http://www.ardos-gk.ru):

1. Suspension of a child in the area of pelvis is performed in a smooth manner and without skewing back and forth.
2. The Ardos simulator structure is more open, which allows the body to breathe better.
3. The structure of the costume top part support allows us to work and exercise with taller children compared to a foreign-made model.

The simulator was changed in time, and some structural changes were introduced, making its use easier and improving general user and therapeutic characteristics. Recently, the fifth-generation simulators have come into market and already left its Israeli prototype behind by many parameters.

At present, the Russian model, which retained its name "Ardos", provides an opportunity to work with older children. There has been successful experience of the simulator application with 12- to 13-year-old children. The elements of leg position correction, which prevent children's knees from bringing them together, were added. And the costume turned out to be less heavy so that it can be worn in hot weather as well.

It should be noted that in the first lesson, children will be able to show the speed and accuracy of the performed action. This is not a natural-stage process of perception and learning exercises with a simulator, but the desire to return to class, his expectations, motivation, and positive emotions will be.

In the course of medical physical culture, the student will build confidence in his/her own actions, flawless and quick performance of tasks that appears as a sustainable positive attitude, and his/her anticipation and systematic visits, which in turn will increase the level of motor abilities and social adaptation.

The interest on the rehabilitation programs in Russia using Ardos simulator is growing because of the following reasons:

1. The simulator is rather cheap (less than 15,000 rubles with delivery throughout the territory of Russia).
2. It has some advantages over many similar developments.
3. It has positive feedback on its efficiency from specialists and parents.

The Ardos simulator for children with infantile cerebral paralysis unloads the backbone and makes collective walks easier and more enjoyable. It frees parents' hands, allowing them to perform domestic duties together with children involving them in the activity process.

The custom-tailored system of slings holds a child in a vertical position, and paired sandals correct the leg position.

Physical activity of children with the simulator depends on the number of exercises, their difficulty, number of repetitions, pace, and distance traveled. When performing tasks, children deliver their best ability, making verbal encouragement inevitable, and in the case of improper execution of the exercise—correct amendment. Upon completion of the classes, it may be of interest working to the level of their own achievements, the assessment of which must also be correctly and completely satisfying.

4 CONCLUSION

Children with limited health capacities are children having physical defects or mental deficiencies, which hamper successful acquisition of the curriculum by a child. The category of such children is rather diverse: it includes children with disorders of speech, hearing, vision, musculoskeletal disorders, complex intellectual disorders, and disorders of psychological functions (Alabin, 2004).

Physical education is the most important part of the general system of upbringing, education, and treatment of children with abnormalities. That is why, one of the important initial tasks of physical and general education of children is to promote their health. For children with abnormalities, health promotion, as well as development and normalization of movements constitute a unified and seamless process with remedial teaching. Physical education is an integral part of intellectual, moral, and aesthetic education and development of a child. Development of all types of motor skills and adaptive tracking is a basis for the development of all types of children's activity, a prerequisite for the development of oral and written communication, and contributes to the improvement of children's cognitive activity (Velitchenko, 2009).

Musculoskeletal disorder in children is one of the most complicated types of diseases. Motor disturbances in children with limited health capacities occur because of deformation or underdevelopment of functional relationship between different brain regions. Aggravation of innate motor disturbances in the process of a child's growth and development results from enforced muscular akinesia related to a child's limited capacities, as well as imperfect instructional techniques of health and fitness classes. This, in its turn, does not assure, in early ontogenesis, a vertical position of a child's body, which is a basis for harmonious development of the musculoskeletal system, optimal distribution of efforts on the musculoskeletal system links, and the vestibular system development and, eventually, does not develop locomotor capacities in children with limited health capacities in general (Martynov, 1993).

Physical therapy lessons must be given not less than two times a week, lasting 15–20 min, which must include relaxation exercises, respiration exercises, correct posture formation exercises, and exercises developing fine motor skills.

For development, training, and rehabilitation of children with limited health capacities, the Ardos simulator was created in Russia. It allows a user to keep a vertical position for an indefinite period of time. Using a simulator, we can do exercises stimulating the functions of weakened muscles and joints and perform motor actions. Is allows children with locomotor limitations of different degrees to move freely in space in all directions, keeps a vertical position, makes arms and legs free, and secured them against falling. Children with musculoskeletal pathology, working with the Ardos simulator, can simultaneously use other simulators or gymnastics equipment (Samylichev, 1994).

Therefore, conventional means, used during health and fitness classes, do not allow us to develop both motor abilities and physical fitness of children comprehensively with a diagnosis of infantile cerebral paralysis in a broad scope. Application of simulator devices at the lessons improves physical fitness of children with limited health capacities.

Comprehensive development of a child with health limitations is one of the most important tasks of the society at the present stage of development, which requires searching for the most effective ways to achieve this goal. Protection of a child's right for development in accordance with individual capacities becomes a field of activity where interests of parents, healthcare professionals, teachers, and psychologists are closely intertwined.

The top-priority goal of specialists is to develop an optimal individual plan of training a child with health limitations. We should discuss all possible options with parents and collectively decide which form of education best meets a child's needs at each stage.

REFERENCES

Alabin, V.G. Simulators and training devices in physical training and sports: guide. – Minsk: Vysh. shk. [High School], 2004. – 174 p.

Ayzikov. G.S. Physical therapy for children suffering from paralytic diseases. – Moscow, 2003. – 208 p.

Belenovich, V.V. Training in physical education. – Moscow: Fizkultura i sport [Physical training and sports], 2005. – 134 p.

Gross. Yu. A. Application of simulation devices when doing rehabilitation physical exercises of children with musculoskeletal disorders – Moscow: Nauka, 2006. – 28 p.

Martynov, V.L. Analysis of medical social problems of families with disabled children suffering from infantile cerebral paralysis. – Moscow: Zdravookhranenie RF [Healthcare Service of the Russian Federation], 1993. – 14 p.

OOO Ardos Official site http://www.ardos-gk.ru.

Samylichev, A.S. A differentiated approach to students of special needs school when developing motor abilities at physical training lessons. – Moscow: Fizkultura i sport, 1994. – 23 p.

Semyonova, K.A. Clinical picture and rehabilitation therapy of infantile cerebral paralyses. – Moscow: Meditsina, 2002. – 328 p.

Velitchenko, V.K. Physical training for the impaired. – Moscow: Meditsina [Medicine], 2009. – 150 p.

Vitenzon, A.S. Biomechanical mechanism of motor disturbance compensation in walking pathologies. – Moscow: Nauka [Science], 2000. – 41 p.

Analysis of tourism management development of the Bunaken National Marine Park

S.B. Kairupan & R.H.E. Sendouw
University of Manado, Minahasa, Indonesia

ABSTRACT: In this study, we aim to analyze the tourism management of the Bunaken National Marine Park (BNMP), the indicators that may support and impede the tourism development of the BNMP, and the concept of better tourism management of the BNMP. Using the descriptive qualitative approach, the results show that the current mechanism of tourism management has caused responsibility shifting and a sense of injustice in the distribution of income. Factors that support the implementation of tourism development are natural and cultural uniqueness, the regulations of tourism management of the BNMP, and public interest. Factors that inhibit tourism development are natural limitation, limited facilities and infrastructures, low quality of human resources, and the lack of coordination among sectors and regions. Major propositions for better tourism management are implementation of privatization concept in co-management and holistic improvement of a coordination and communication system, budget, as well as community acceptance, and tourism programs.

1 INTRODUCTION

Tourism sector can make a significant contribution to the state or local revenue. Over the past six decades, tourism has experienced continuous expansion and diversification and become one of the largest and fastest growing economic sectors in the world (UNWTO, 2012). Increasing tourist visits will give a positive impact on the economic growth of the communities in both the vicinity of attraction sites and the whole region (Gilaninia, et al. 2012; Suhamdani, 2013; Hilman et al., 2016). Increased economic growth in the area will in turn boost local revenue, which will eventually support the development of various sectors as well as improve the welfare of the community. One of the assets that can be developed is marine tourist attraction, which has a special attraction for domestic and foreign tourists (Kerebungu, 2008). The Bunaken National Marine Park (BNMP) is a flagship marine attraction in Manado. At the events of the World Ocean Conference (WOC), Coral Triangle Initiative (CTI) Summit, and Sail Bunaken in 2009, the BNMP received much attention from the participants.

Foreign exchange from tourism can be realized when the government, private sector, and communities work together to provide the best service for tourists and ensure adequate tourism infrastructure and management policies. The government should set up steering policies that encourage the creation of competitive market mechanisms that involve the private sector and NGOs, so that the country's revenue from tourism sector can be increased. On the contrary, the implementing agency—whether the government or private sector—should win the participation of tourism-related communities in providing the best services for tourists (Gilaninia et al., 2012; Suhamdani, 2013). However, in reality, the people on the main island and the island of Bunaken-Siladen are not fully engaged or less directly involved in tourism activities. As a result, their attitudes and behavior are less responsive to the management and development of tourism. At this time, the general tourism activities in the BNMP are directly handled individually by diving centers or managers based in Manado, who employ people from the city. In other words, labors from Bunaken are not capitalized. In fact, people who live in gateways hold the key to create a "virtuous circle", whereby tourism's contribution to the economy generates incentives to conserve the resources that keep tourists coming, and locals must contribute to sustainable tourism (Stange, J. and David Brown, 2011). The management of the BNMP is not optimal as it is also hampered by tug war between the central, provincial, and municipal governments, as well as private parties (Akpan, 2012; Rares, 2015). At this point, the local government needs to put up policies that will accommodate various management stakeholders of the BNMP by rallying the private sector and communities living around the Bunaken island. The BNMP is managed by the BNMP Management Advisory Board, which is obliged to provide foreign exchange for the city of Manado, and the board cannot work optimally and professionally because its staff members are partly government officials who have other duties to perform. Therefore, the policy needs to address issues concerning effective coordination and communication.

The research aims to describe and analyze the indicators that may support and impede the implementation of the privatization policy of the BNMP as well as analyze the conceptual model of privatization in the management of the BNMP.

2 METHODS

The approach used in this study is descriptive qualitative which aims to find, analyze, and describe a public policy in the management of marine attraction in the BNMP. In this study, the informants were the people engaged in tourism activities of the BNMP, including Head and Secretary of North Sulawesi Tourism office, Head and Secretary of Manado City Tourism Office, two officials of the BNMP Management Advisory Board, an entrepreneur, one technical staff of North Sulawesi Tourism Office, one technical staff of Manado City Tourism Office, and four representatives of the community. The number and variety of informants is meant to capture as much information as possible from various sources in order to specify the findings of the study to be extracted into a proposition.

The necessary data were collected using the techniques of observation, interviews, and documentation. The units of analysis in this study are groups of nautical tourism in the BNMP, officials from institutions associated with tourism activities, and a number of policies related to the marine tourism management of the BNMP.

The process of data analysis was conducted by sorting the data (categorical analysis) on the basis of the information obtained in the field and then reviewing the substance of the information. In this study, the data were analyzed using interactive data analysis techniques proposed by Miles and Huberman (2014).

3 RESULTS AND DISCUSSION

3.1 *Overview of research sites*

The BNMP is located very close to the Manado City, the capital of North Sulawesi Province. It has 6 beach attractions, 5 nature objects, 4 marine attractions, 7 cultural attractions, 2 religious attractions, and 11 artificial attractions.

The BNMP was established as a tourism object by the Governor of North Sulawesi's Decree in 1980 (No. 224/1980) on the location of Bunaken island, Siladen island, and the surrounding islands. Through a ministerial decree dated October 15, 1991 (No. 730/1991), the BNMP is expanded, covering an area of 89,065 ha, including the Bunaken, Siladen, Manado Tua, Mantehage, and Nain islands and some parts of Tongkaina, Wori, Tanjung Pisok, and Wawontulap Arakan coastal regions.

3.2 *Tourism management of the BNMP*

Conditions of the BNMP attractions continue to be neglected, raising the question of who or which agency is responsible. The BNMP Management Advisory Board was established to avoid misunderstandings among the regions sharing the BNMP area. The BNMP

Management Advisory Board has an obligation to coordinate the management of the BNMP. It is a medium for the Provincial Government, (Manado) city and (other) regencies, the BNMP-related agencies, Police Force, community representatives, private sector, and universities to work together in order to strengthen the management of the BNMP for its sustainability. In reality, the presence of the board only leads to the sharing of revenues. At present, the BNMP Management Advisory Board manages admission rates, with the following distribution: 20%, of which 5% is for the central government, 7.5% for the provincial government, 1.85% for the regencies and (Manado) city, 30% for the conservation village (settlement areas), and the rest is for operational funds, while 8 million per month is allocated for every 10 people for hygiene and waste management both on land and at sea. It is clear that the BNMP Management Advisory Board should therefore develop a professional management system.

The following three institutions are responsible for the BNMP: (1) Ministry of Forestry that is responsible for the conservation of the park, as stated in the Minister of Forestry decree No. 730 in 1991. Given that situation, tourism business permit in that area must be obtained from Minister of Forestry. (2) North Sulawesi and Manado City Department of Tourism that are responsible for the marketing of tourist attractions of the park. (3) the BNMP Management Advisory Board that is responsible for regulating the collection and distribution tariff gains of the park attractions according to the Regional Regulation No. 14 in 2000. Moreover, people who inhabit the islands of Bunaken, Manado Tua, and Siladen are headed by the government of Manado City because those islands come under the administrative area of Manado.

Each stakeholder has a different and specific set of duties and responsibilities: the Management Advisory Board addresses tariff management; Manado's Office of Tourism handles marketing and promotion; and the Department of Forestry is responsible for nature conservation, involving NGOs and communities. However, in reality, the attitude of "wait for one another" of each stakeholder is exhibited. From the information collected during the study, it appears that the responsibility is tossed from one institution to another. There is a constraint in coordination and communication. Regardless of a meeting to discuss the management of the BNMP held, almost all stakeholders are absent, or sending a representative person results in meetings with unreliable outcome to produce decision management policies.

Promotion or marketing budget plan is one of the current budget items in funding for tourism development that relies more on funding from local government. According to the Head of Manado City Department of Tourism, North Sulawesi has plenty of potential tourism objects, which are not widely known to the public. It is a great concern for the local government to develop tourism promotion plan with a specific and proper budget allocation.

Consequently, the marine park and its supporting facilities are poorly maintained; the tourist attraction management is full of oversights; and the excursions are not well organized. Ideally, there should be an integrated management system that relies on the synergy of all related components. This reality can be explained by Merton's theory of structural-functional. The fundamental assumption is that every structure in the social system is functional to the other, but could also dysfunction other structures when these structures have negative effects as a result of changes in the social structure or institution. The management system of the BNMP Management Advisory Board of North Sulawesi is changing toward integration, and all relevant agencies have their specific duties and functions. The effect is mutual waiting and shifting of responsibilities, causing the structure to provide negative consequences (dysfunction) to the privatization of the BNMP and the nature resource as well (Huang, Bao, and Lew, 2011; Rares, 2015).

3.3 *Supporting and inhibiting factors of tourism development of the BNMP*

3.3.1 *Government policies*

The policies of Bunaken National Marine Park, which are supporting factors in tourism management, are as follows:

a. Governor SK KDH Level I North Sulawesi number 224 of 1980 which confirms that the park is under the authority of the Government of Manado city, showing that all tourism activities in the premise lie under the supervision of Manado City Government.
b. Ministerial Decree No. 730/KPts-11/1991 dated October 15, 1991, which stipulates that the responsibility is transferred to the Central Government under the supervision of the Forestry Department.
c. Since 1997, according to Ministerial Decree No. 185/SPts-11/1997 dated March 31, 1997, the marine park has been managed by a special board, the BNMP Management Advisory Board.
d. North Sulawesi Provincial Regulation No. 14 in 2000, concerning levies in the BNMP area, outlined in the Governor Rule No. 49 in 2000 and later replaced by Governor Regulation No. 22 in 2007, concerning the implementation of levy tariffs instructions in the BNMP so they would not overlap with the admission rates in the marine park. (The park's entrances are spread across four regions: the city of Manado, Minahasa, South Minahasa, and North Minahasa regencies).

Basically, the objectives of the park management are: (1) to conserve natural resources and to guarantee their existence and the long-term availability of germ-plasma sources; (2) to improve the welfare of local people through the improvement of effectiveness and utilization of natural resources in the area; and (3) to increase local revenues through development and management of tourism in the BNMP.

Particularly relevant to the management and promotion of tourism, the Management Advisory Board is in cooperation with the local government (Department of Tourism) and the related private sectors, such as diving clubs/companies, travel agencies, and hospitality. The board provides an opportunity for the government, in this case the Department of Tourism, to utilize as much tourism potential as possible in order to increase local income and prosperity of the local community. Private parties can also cooperate with the local government to manage the used zone that is in the location of the famous diving-snorkeling spot, Liang. However, they must comply with the conditions set by the Management Advisory Board, mostly covering the issues of preserving the environment and corruption.

However, there is a problem in coordination and communication between stakeholders, the provincial, and city government, as well as with private diving companies because of their different interests.

3.3.2 *Public interest*

People who inhabit the islands are also participating in nautical tourism in the park. They can enjoy the benefits of tourism development in the park, such as new job opportunities due to the development of homestays, local transport services, tour guiding or dive guiding, dive operator, or selling souvenirs. Local community understands the benefits of tourism development. Besides, the preservation of the BNMP natural resources ensures the sustainability of marine resources that have become part of the livelihood of local communities. Local community are very pleased with the number of people who want to work or open a business in their area. They think that if the park is crowded, then they can also have trading activities. Local community can enjoy the benefits of the development of tourism in the BNMP, among others, create new job opportunities, for example, home-stay development, transport services, tour guiding or dive guiding, dive operator, or selling souvenirs. On the contrary, according to some citizens, it is important to limit diving for travelers. They think that the rate of diving now exceeds its limit. The Management Advisory Board conducts community development activities in the island of Bunaken. The council in collaboration with NGOs provided training to local people in making souvenirs. But the problem is that after the training, the participants returned to their original business of selling coconut and banana fritters, instead of applying the skills they obtained from the training. It is because they could hardly get the raw materials. There is a need of better planning of human resource development (Dasaluti, Hubeis, and Wiyono, 2010; Salim, 2015; Hilman, 2016).

3.3.3 Facilities and infrastructure

Since the 1970s, the government has given the opportunity for private corporations, as a form of privatization, to manage tourist sites in the BNMP. Recently, 13 companies have gained access to the BNMP and engaged in diving activities with permission from the government/ governor). There are six types of tours or about 35 destinations to offer in Manado. The Manado City Tourism Office has conducted many promotional activities to increase tourist visits, among others, by publishing guide books and maps of tourist attractions and by carrying out the following events: (1) Festival of Tondano and Bunaken; (2) Festival of Cultural; and (3) Cultural Parade. In addition to the attractions, tourists can also enjoy other entertainment facilities such as pub, billiards, sports centers, theaters, massage parlors, and swimming pools. Lodging in Manado consists of 65 hotels and inns, including 9 five-star hotels, 56 budget hotels, and 2 inns, with a total number of 1905 rooms. The same strategy has been followed by Sri Lanka government in order to attract more tourists. This strategy develops tourism infrastructure by encouraging public–private partnership to increase the number of hotel rooms (Ministry of Economic Development, 2010).

In terms of physical aspects, Bunaken island has the potential to be developed as marine tourism owing to the following physical characteristics:

1. Strategic location and high accessibility.
2. Having a bay where the water is calm throughout the year, with no harming factors such as spinning currents, whales, etc.
3. The marine park is still intact and has a high biodiversity, the structure of coral reefs varies, and so on.

However, there are some limitations too:

1. The climate factor sometimes hampers accessibility to the location.
2. Lack of transportation facilities and infrastructures.
3. Roads need major improvements.

Particularly relevant to the problem of funds and personnel, the Head of Sub-Section Data of North Sulawesi Tourism Office explained that the fund provided by the Government of North Sulawesi is very limited. North Sulawesi Tourism Office could not provide all the necessary facilities to serve tourists (travelers) who visit Bunaken. It is understood that Tourism Office should work on this with the private sector (Akpan, 2012; Alhomod and Shafi, 2013; Primadany, Mardiyono and Riyanto, 2015).

From the interviews conducted for several local communities, it can be reflected that the presence of the dive operators (diving companies) at the site does not benefit the local community. They think that the presence of diving club entrepreneurs in their area (tourism sites of the BNMP) is clearly detrimental to the fishing community in their village, because their presence automatically inhibits fishing around that location. Some people misunderstood the presence of the BNMP Management Advisory Board as the government had taken not only the sea area but also the land (soil) and refunctioned it in the name of nature conservation. This situation will hamper the sustainability of tourism development, and needs to be carefully handled. In addition, it should be examined whether the fame and the large number of visitors can significantly contribute to increasing the Manado City revenue, and especially the welfare of local communities living in Bunaken and Siladen islands.

Positive attitudes were connected with the belief that tourism creates community development, opportunities for earning income, improved services, and a chance for tourism development. A similar case was found in others tourism destinations. The following Several strategies emerged, which can be used to develop an appropriate form of tourism for poor, rural people: encourage cooperatives; encourage the use of local materials and local design; target backpackers and other tourists who are eager to adapt to local conditions; foster local decision-making; integrate tourism with local agriculture; and use tourism revenue for community development. These strategies have created positive attitudes toward tourism and apparently increased the pro-tourism behavior (Scott, 2011; Primadany, Mardiyono, and Riyanto, 2015; Salim, 2015; Hilman, 2016).

4 CONCLUSION

This study concludes that the current mechanism of tourism management has caused responsibility shifting and a sense of injustice in the distribution of income. Factors supporting the implementation of tourism development are natural and cultural uniqueness, the regulations of tourism management of the BNMP, and public interest. Factors inhibiting the tourism development are natural limitation, limited facilities and infrastructures, low quality of human resources, and the lack of coordination among sectors and regions. Minor propositions for a better tourism management of the BNMP are described as follows:

There is a need to fix a tourism management system that relies on the synergy of all stakeholders. The success of coordination and communication lies in implementing integrated tourism management by the National Park Management Advisory Board. Community empowerment is part of a sustainable development of tourism, which serves as a meeting point between the needs of the market (tourists) and the benefits of the local community.

The success of privatizing tourism management depends on the availability of budget, community acceptance, and their attitudes toward tourism programs. The ideal implementation of privatization concept is co-management.

The supporting and inhibiting factors in privatizing tourism management lead to the importance of minimizing risks and increasing the benefits.

ACKNOWLEDGMENTS

This study was supported by the Ministry of Research, Technology and Higher Education of Indonesia.

REFERENCES

Akpan, E. I. 2012. Managing Tourism Sector in Nigeria through Privatization Strategy. Business and Management Research 1(2). 12–124.

Alhomod, S. and Shafi, M. M. 2013. A Study on Implementation of Privacy Policy in Educational Sector Websites in Saudi Arabia. Global Journal of Computer Science and Technology Network, Web and Security 13(1). 22–26.

Dasaluti, T., Hubeis, A. V. S., Wiyono, E. S. 2010. Analisis pengembangan usaha mikro dalam mendukung pemberdayaan perempuan di Pulau Bunaken Kota Manado, Sulawesi Utara. Jurnal Manajemen Pengembangan Usaha Kecil Menengah 5(2). 157–195.

Dinas Pariwisata Provinsi Sulut. 2000. Program Pembangunan Pariwisata dan Kesenian Sulut, Manado. 129p.

Gilaninia, S., Masouleh, S. K., Hasheminasab, M., Masouleh, F. K., Saberhosseini, S. S., Guildehi, R. I., Abkenar, S. S. 2012. Privatization, A Way for Developing Tourism Economy. J. Appl. Environ. Biol. Sci., 2(1). 82–86.

Gouret, F. 2007. Privatization and Output Behavior during the Transition: Methods Matter. Journal of Comparative Economics 35 (2007). 3–34.

Hilman, Y. A. 2016. Regional development of tourism in Ponorogo Regency, East Java. Journal of Indonesian Tourism and Development Studies 4(3). 91–96.

Huang, X, Bao, J. and Lew, A. A. 2011. Nature-based Tourism Resources Privatization in China: A System Dynamic Analysis of Opportunities and Risks. Tourism Recreation Research 36(2). 99–111.

Kerebungu, F. 2008. Pengembangan Industri Pariwisata Budaya dalam Meningkatkan Pendapatan Asli Daerah (PAD) Kota Manado. Jurnal Aplikasi Manajemen. Vol. 6, No. 1, April 2008.

Ministry of Economic Development, 2010. Tourism Development Strategy 2011–2016. Ministry of Economic Development of Sri Lanka. http://www.sltda.lk /sites/default/files/English.pdf.

Lagarense, B. E. S. and Walasendow, A. 2016. Developing Marine and Coastal-based Sport Tourism on the Waterfront: The Case of Manado Waterfront, Indonesia. Journal of Indonesian Tourism and Development Studies 4(3). 107–114.

Miles, M. B. and Huberman, A. M. 2014. Qualitative Data, Analysis and Design. Sage Publication, Inc.

Pangemanan, A., Maryunani, Hakim, L., Polii, B. 2012. Economic analysis of Bunaken National Park ecotourism area based on the capacity and visitation level. Asian Transactions on Basic and Applied Sciences 2(4). 34–40.

Primadany, S. R., Mardiyono, Riyanto. 2015. Analisis strategi pengembangan pariwisata daerah (studi pada Dinas Kebudayaan dan Pariwisata Daerah Kabupaten Nganjuk). Jurnal Administrasi Publik 1(4). 135–143.

Rares, J. J. 2015. Manajemen pengelolaan Taman Nasional Bunaken oleh pemerintah Provinsi Sulawesi Utara. Jurnal LPPM Bidang EkoSosBudKum 2(2). 40–43.

Salim, A. 2015. Pariwisata Bahari Berbasis Konservasi dan Keberlanjutan Ekonomi Masyarakat Raja Ampat. Available at: http://aguswi-kkp.com/pariwisata-bahari-konservasi-dan-keberlanjutan-ekonomi-masyarakat-raja-ampat/.

Scott, N. 2011. Tourism Policy: A Strategic Review. Goodfellow Publishers Ltd, Oxford.

Stange, J. and David Brown, 2011. Tourism Destination Management Achieving Sustainable and Competitive Results. https://www.usaid.gov/sites/default/files/documents/2151/DMOworkbook_130318.pdf.

UNWTO, 2012. UNWTO Tourism Highlights 2012 Edition. http://mkt.unwto.org/sites/all/files/docpdf/unwtohighlights12enhr.pdf.

A three-stage data envelopment analysis combined with artificial neural network model for measuring the efficiency of electric utilities

Pavala Malar Kannan & G. Marthandan
Multimedia University (MMU), Cyberjaya, Malaysia

ABSTRACT: The study of efficiency measurement was pioneered by Farnell in 1957. Since then, researchers in this field have developed many models to measure the efficiency of firms. However, there is no universally acceptable model to measure the efficiency of electric utilities due to differences in the selection of measures, countries' socioeconomic structures, as well as quality and availability of data.

The purpose of this study was to measure and rank the efficiency of electric utilities using a Three-Stage Data Envelopment Analysis-Artificial Neural Network (3S DEA-ANN) model. The model is developed using R programming and RStudio software.

Keywords: Data Envelopment Analysis, Artificial Neural Network, Efficiency

1 INTRODUCTION

The goal of this study was to measure and rank the efficiency of electric utilities using a three-stage data envelopment analysis-artificial neural network (3S DEA-ANN) hybrid model.

Since the early 1990s, regulators worldwide have been implementing incentive-based models to promote efficiency in infrastructure industries. Incentive-based models are devised to promote efficiency in industries, where competition among market players is low or nonexistent (Giannakis, Jamasb, & Pollitt, 2005).

The choice of efficiency measurement models and measures can significantly affect the efficiency of electric utilities. Therefore, inaccurate efficiency measurement models or measures will lead to imprecise results and cause the decision makers to take less than optimum decisions.

1.1 *Incentive-based regulation models*

Vogelsang (2002) stated that price cap model that includes incentives for cost reductions and incentives for more efficient pricing will facilitate the opening of public sectors such as electric utility sector for competition. However, the price cap model will also encourage the incumbent electric utilities to compete more aggressively and show anticompetitive behaviors.

Ter-Martirosyan and Kwoka investigated incentive-based regulation quality of service of the firms. They pointed out that any incentives schemes that encourage a firm to lower costs will also lead to lower quality. Such incentives scheme was partly responsible for the major blackout in Northwest and Midwest of the Unites States and parts of Canada in 2003 quality (Ter-Martirosyan & Kwoka, 2010).

In addition, Ter-Martirosyan and Kwoka (2010) also concluded that incentive-based regulation without strict quality benchmarks will lead to a significant increase in the duration of power outages.

Therefore, price cap model alone is not sufficient to improve the service quality of electric utilities. A more comprehensive model that includes cost and quality benchmarks may lead to more efficient electric utilities without sacrificing the service quality of the utilities.

1.2 Research gap

Haney and Pollitt mentioned that the choice of measures used in benchmarking affects the results of benchmarking exercise. If all measures are not considered and if regulators decide to choose or weight certain measures, the measured efficiency can change significantly (Haney & Pollitt, 2013). Alexandridis, Patrinos, Sarimveis, and Tsekouras also supported this view and stated that correct selection of measures is very important to eliminate insignificant measures, reduce the size of the model, and produce a more accurate prediction (Alexandridis, Patrinos, Sarimveis, & Tsekouras, 2005).

Nataraja & Johnson (2011a) cited Sexton et al. (1986) and stated that the position of frontier in data envelopment analysis (DEA) is affected by the selection of measures. The inclusion or exclusion of measures can alter the shape and position of the frontier.

The measures that are used in efficiency measurement depend on the choice of methodology and technical requirements of the selected model, as well as quality and availability of data. There are no universally applicable measures that can be used to measure the efficiency of electric utilities. However, measures or factors should reflect the electric supply objectives and service level (Thakur, Deshmukh, & Kaushik, 2006).

Table 1. Methodology for measuring electric utility efficiency.

No.	Authors/(year)	Research objective	Efficiency measurement methodology
1	Chen, Zhou, & Yang, (2017)	Evaluate China's electric energy efficiency	DEA based on game relationship
2	Sağlam, (2017)	Productive efficiency of wind farm in the USA	Two-stage DEA.
3	Omrani, Gharizadeh Beiragh, & Shafiei Kaleibari, (2015)	To measure the performance of Iranian electricity distribution companies.	Combination of game theory, principal component analysis (PCA), and DEA.
4	Gouveia, Dias, Antunes, Boucinha, & Inácio, (2015)	To benchmark the maintenance and outage repair activities in an electricity distribution company.	Value-based DEA.

Table 2. Methodology for measuring electric utility efficiency.

No.	Authors/(year)	Input measures	Output measures
1	Chen, Zhou, & (Yang, 2017)	Power consumption Employed population Fixed capital investment	Gross regional product SO_2 emission CO_2 emission Ammonia emission
2	Sağlam, (2017)	Installed capacity Wind turbines Wind power density	Generated electricity Value of production Homes powered
3	Omrani, Gharizadeh Beiragh, & Shafiei Kaleibari, (2015)	Transformer capacity No. of transformers Terrestrial network length Aerial network length No. of employees Areas	Energy delivery Energy consumption of other customers Industrial energy consumption Household energy consumption No of other customers No of industrial customers No of household customers No of street lights
4	Gouveia, Dias, Antunes, Boucinha, & Inácio, (2015)	Maintenance and outage repairing costs Supply interruptions Complaints per customer Number of incidents	Clients Network line length

Over the years, researchers have used many different measures to determine the performance of electric utilities. However, the measures that significantly influence the efficiency of electric utilities are not conclusive and need to be investigated further.

Haney & Pollitt (2013) stated that the primary objective of incentive-based regulation is to increase the efficiency of electric utilities. Because the utilities are rewarded on the basis of their performance, the benchmarking methods and techniques must be selected carefully in order to measure the performance accurately (T. Jamasb & Pollitt, 2001).

Nepal & Jamasb (2015) also stated that benchmarking can be a useful tool in accessing efficiency and performance of electric utilities. The results of the benchmarking can be used to compare the relative efficiency of the electric utilities with their peers.

The efficiency of electric utilities can be calculated using benchmarking methods such as DEA, Production Function, and DEA & Malmquist index (Haney & Pollitt, 2001). The selection of efficiency measurement model is also affected by the availability and quality of data (Nigam, Thakur, Sethi, & Singh, 2012). Furthermore, there are no benchmarking methods that can measure the efficiency of electric utilities accurately.

Researchers also have employed different models and measures to score the efficiency of electric utilities as shown in Tables 1 and 2. However, the models may not be perfect because the researchers may have accidently omitted critical measures as explained by Shuttleworth (2005) and need to be refined further.

2 LITERATURE REVIEW

2.1 *History of electric utility regulations*

Initially, the electric industry was not regulated by the government. However, this scenario changed following the 1877 Munn versus Illinois United States Supreme Court case.

The 1877, the Munn versus Illinois case was brought to the Supreme Court by a private company because Illinois introduced a legislature to regulate or control storage and transport charges of agricultural products. Initially, Munn and Scott were found guilty by a lower court after it violated the new law. However, the Munn and Scott appealed in the Supreme Court and lost ("Supreme Court Case: Munn v. Illinois 1877," 2015). However, the Supreme Court decided that the government can regulate private industries if the public can benefit from such regulations ("Munn v. Illinois," 2015).

Legislators enacted an act to curb the abuses of holding companies in 1935. The new act is known as PUHCA, which is an abbreviation for Public Utility Holding Company Act (Adams, 2008). The PUHCA reorganizes the large holding companies into many smaller utilizes. The reorganization enabled the States to curb the abuses of holding companies. PUHCA also allowed the smaller electric utilities to control generation, transmission, and distribution of electric supply in its region (Blacconiere et al., 2000).

The PUHCA was further fine-tuned in 1978 to allow new electric power generation utilities to sell their electricity output to other electric utilities instead of consumers (Adams, 2008). This lowers the barrier of entry for new entrant into the electric generation sector and encouraged competition among electric power generation utilities.

Adams (2008) also stated that the Energy Policy Act (EPACT) of 1992 restructured the electric utilities even further. This was followed by other improved laws such as EPACT 2005. EPACT encouraged a nationwide competition in the generation sector. In addition to that, it allowed corporations to sell electricity wholesale or retail anywhere in the world. Similarly, EPACT further encouraged competition by introducing incentives for certain types of fuel and also by giving authority to the Federal Energy Regulatory Commission to regulate wholesale power and gas markets (Federal Energy Regulatory Commission, 2006).

2.2 *Electricity market regulation models*

Traditionally, electricity market has been a monopolistic market. However, regulators worldwide are introducing market-oriented measures such as EPACT 1992 and EPACT 2005 into the electricity market to deregulate and improve the efficiency of the electric utilities. Currently, most electricity market regulators worldwide have adopted the incentive-based regulation model instead of the traditional rate of return regulation model. Some regulators have implemented price and revenue cap model. However, they are less in number compared to the more popular incentive-based regulation (Tooraj Jamasb & Pollitt, 2003).

The early electricity market deregulation and structural reforms were introduced in industrialized countries and worked well in those countries. However, this model failed in some countries such as the United States (California), Canada (Alberta and Ontario), and Brazil. The structure of the first model or 1G model is shown in Table 3.

The 1G regulation model evolved into 2G and 3G regulation models. Currently, most of the industrialized countries in America, the European Union, and Australia are using 2G and 3G models to regulate and encourage competition in the electricity markets. The structure of 2G and 3G models are shown in Table 4.

2.3 *Efficiency measurement tools*

Amado, Santos, and Sequeira (2013) cited the discussion of Kumbhakar and Hjalmarsson (1998) that it is important to evaluate the efficiency and productivity of electric utilities because their performance is greatly influenced by management. They also stressed that evaluating the efficiency of an electric utility from a technical viewpoint alone is not useful because electricity business also has a close relationship with customers. Amado, Santos, and Sequeira (2013) also cited Santos et al. (2013), who stated that some business entities such as electricity distribution or retail business have a closer relationship with the customer compared to generation.

Amado, Santos, and Sequeira (2013) also stressed that it is important to measure the performance of an electric utility so that the efficiency and effectiveness of the use of resources by the electric utility can be determined. This will allow the electric utility to use its resources efficiently and effectively and to deliver high-quality services to its customers.

The researchers also discussed that the performance of an electric utility can be evaluated internally or benchmarked against other similar electric utilities. Through benchmarking exercise, the electric utility will be able to compare itself with other similar utilities in the region or worldwide.

Table 3. 1G regulation models.

Unit	Regulation models
Generation	Long-term contracts or single-buyer model
Transmission	Hybrid model or price caps or rate of return model
Distribution	Monopoly with price caps or rate of return or hybrid model

Source: Gentzoglanis (2013).

Table 4. 2G and 3G regulation models.

Unit	2G regulation model	3G regulation model
Generation	"Independent System Operator" (ISO) and "Regional Transmission Organization" (RTO)	ISO, RTO, merchant investment, "Transmission Congestion Rights" (TCC), and/or "Financial Transmission Rights" (FTR)
Transmission	Monopoly, competition with price caps or rate of return or hybrid model	Competition
Distribution	Competition and "Ladder of Investment" (LoI)	Competition

Source: Gentzoglanis (2013).

However, benchmarking of vertically electric utilities is an under-researched subject, and there are very few references in this subject. However, benchmarking of the generation and distribution or customer service business entities is fairly researched, and there are a number of references regarding this subject.

For example, Giannakis et al. (2005) carried out a benchmarking study of electricity distribution utilities in the United Kingdom between 1991–1992 and 1998–1999. They also incorporated quality in the benchmarking study. The technical efficiency and productivity change of the electric utilities were calculated using hybrid DEA and Malmquist indices method. The researchers concluded that there is no significant relationship between cost-efficient firms and high-quality service. They also stressed that benchmarking with integrated quality of service will reveal a more accurate result as opposed to cost-only approaches.

Shrivastava, Sharma, and Chauhan (2012) carried out a study to assess the efficiency and benchmark thermal power plants in India. They carried out the study using the Data Envelopment Analysis mathematical tool. The performances of various sizes or capacity of electric power plants were evaluated and compared or benchmarked with each other. They also compared or benchmarked the performance of private and state-owned power plants. They found that small electric power plants perform poorly compared to medium-sized and large power plants. They also found that privately owned power plants perform better compared to the state-owned ones.

Similarly, Lowry and Getachew (2009) also discussed how statistical benchmarking is being utilized to regulate the electric utilities worldwide. They also stressed that benchmarking is important to measure the level of efficiency of electric utilities relative to the best-practice firms. They also suggested that benchmarking results should be reflected in the electricity rates or tariffs. They also stated that accurate benchmarking results will give more confidence to the regulators and assist them to set the electricity rates or tariffs.

Researchers worldwide have used various methods or tools to calculate the efficiency of electric utilities.

3 METHODOLOGY

3.1 Introduction

This study enabled us to examine the efficiency of electric utilities using a three-stage data envelopment analysis—artificial neural network (3S DEA-ANN) hybrid model.

3.2 Input and output measures

A total of 13 input measures and 5 output measures identified for this study are listed in Table 5.

Table 5. Input and output measures.

No.	Input measure	Output measure
1	Fuel cost	Net electrical energy output
2	Total manpower	Utilization of net capacity
3	Generation capacity	Availability
4	Total cost	Thermal efficiency
5	Operating expenses	Emissions
6	Capital expenditure	–
7	Number of customers	–
8	Energy loss	–
9	Network length	–
10	Electrical facilities	–
11	Area (km^2)	–
12	SAIDI (System Average Interruption Duration Index)	–
13	SAIFI (System Average Interruption Frequency Index)	–

3.3 Three-stage data envelopment analysis—artificial neural network model

Three-stage DEA was used by researchers to overcome limitations of traditional DEA model. The traditional DEA model is not able to overcome problems related to environmental measures. The three-stage DEA are able to overcome this problem by using stochastic frontier analysis (SFA) to break down random error term into efficiency and statistical noise (Iparraguirre & Ma, 2014).

Iparraguirre and Ma (2014) used three-stage DEA to examine the efficiency of social care for older people in 148 English Councils from 2009 to 2010. Initially, they found a high level of efficiency in social care services. However, after controlling the some of the environmental effects, they found that more stringent eligibility criteria and higher assessment costs affect the efficiency negatively.

Similarly, Shyu and Chiang (2012) used the three-stage DEA method to measure managerial efficiencies of branches of banks in Taiwan. They investigated 123 such branches and found that the efficiency measured by three-stage DEA varies significantly compared to traditional DEA. This suggests that the environmental measures have significant effect on bank branch efficiencies.

In the first stage of the three-stage DEA method, randomly selected data will be used to carry out traditional DEA. This is accomplished by measuring the efficiency of the selected electric utilities using traditional DEA. In the second stage, stage 1 slacks are decomposed using stochastic frontier analysis (SFA). In the third stage, the input data will be replaced by the input data that have been adjusted for the observable environmental measures and statistical noise. The adjusted input data will be used to measure the efficiency of the electric utilities using traditional DEAs.

The results of the three-stage DEA will then be used to train an artificial neural network (ANN) as shown in Figure 2. The trained ANN will then be used to measure the efficiency of the entire data set.

3.4 Analysis

The model was developed using R language and R Studio software. Because this study is still in the early stages, we were only able to carry out a limited analysis.

The efficiency score from the preliminary results was used to determine relative efficiency of the electric utilities and rank the utilities.

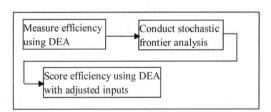

Figure 1. Three-stage DEA.

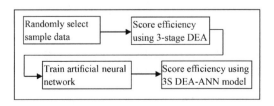

Figure 2. Three-stage DEA-ANN model.

Table 6. Existing and new hybrid models.

No.	Efficiency model	Type
1	Data envelopment analysis	Existing model
2	Data envelopment analysis—artificial neural network	Existing model
3	Three-stage virtual frontier—data envelopment analysis	New hybrid model
4	Three-stage virtual frontier—data envelopment analysis—artificial neural network	New hybrid model

4 FUTURE WORKS

The authors plan to compare the efficiency of electric utilities using several existing and new models. The existing models and planned hybrid models are listed in Table 6.

The models will be used to determine the efficiency score of the electric utilities and rank the utilities.

5 CONCLUSION

This study has enabled us to measure electric utilities efficiency score using the 3S DEA-ANN model and rank them accordingly. In future, the model will be compared against other existing and new models to determine the best efficiency model for electric utilities.

REFERENCES

Adams, A. (2008). Impact of Deregulation on Cost Efficiency, Financial Performance and Shareholder Wealth of Electric Utilities in the United States. Retrieved from http://etd.library.vanderbilt.edu/available/etd-09052008-143901/unrestricted/AsterAdamsDissertationFinal.pdf.

Alexandridis, A., Patrinos, P., Sarimveis, H., & Tsekouras, G. (2005). A two-stage evolutionary algorithm for variable selection in the development of RBF neural network models. *Chemometrics and Intelligent Laboratory Systems*, *75*(2), 149–162. https://doi.org/10.1016/j.chemolab.2004.06.004.

Amado, C. a. F., Santos, S. P., & Sequeira, J. F. C. (2013). Using Data Envelopment Analysis to support the design of process improvement interventions in electricity distribution. *European Journal of Operational Research*, *228*(1), 226–235. https://doi.org/10.1016/j.ejor.2013.01.015.

Blacconiere, W. G., Johnson, M. F., & Johnson, M. S. (2000). Market valuation and deregulation of electric utilities. *Journal of Accounting and Economics*, *29*(2), 231–260. https://doi.org/10.1016/S0165-4101(00)00021-5.

Chen, W., Zhou, K., & Yang, S. (2017). Evaluation of China's electric energy efficiency under environmental constraints: A DEA cross efficiency model based on game relationship. *Journal of Cleaner Production*, *164*, 38–44. https://doi.org/10.1016/j.jclepro.2017.06.178.

Federal Energy Regulatory Commission. (2006). *Energy Policy Act of 2005: Fact Sheet*. Washington, D.C.

Gentzoglanis, A. (2013). Regulation of the electricity industry in Africa. *African Journal of Economic and Management Studies*, *4*(1), 34–57. https://doi.org/10.1108/20400701311303140.

Giannakis, D., Jamasb, T., & Pollitt, M. (2005). Benchmarking and incentive regulation of quality of service: An application to the UK electricity distribution networks. *Energy Policy*, *33*(17), 2256–2271. https://doi.org/10.1016/j.enpol.2004.04.021.

Goto, M., & Tsutsui, M. (2008). Technical efficiency and impacts of deregulation: An analysis of three functions in U.S. electric power utilities during the period from 1992 through 2000. *Energy Economics*, *30*(1), 15–38. https://doi.org/10.1016/j.eneco.2006.05.020.

Gouveia, M. C., Dias, L. C., Antunes, C. H., Boucinha, J., & Inácio, C. F. (2015). Benchmarking of maintenance and outage repair in an electricity distribution company using the value-based DEA method. *Omega*, *53*, 104–114. https://doi.org/10.1016/j.omega.2014.12.003.

Haney, A. B., & Pollitt, M. G. (2013). International benchmarking of electricity transmission by regulators: A contrast between theory and practice? *Energy Policy*, *62*, 267–281. https://doi.org/10.1016/j.enpol.2013.07.042.

Iparraguirre, J. L., & Ma, R. (2014). Efficiency in the provision of social care for older people. A three-stage Data Envelopment Analysis using self-reported quality of life. *Socio-Economic Planning Sciences*, *1*(1), 1–14. https://doi.org/10.1016/j.seps.2014.10.001.

Jamasb, T., & Pollitt, M. (2001). Benchmarking and regulation : international electricity experience. *Utility Policies*, *9*(2001), 107–130.

Jamasb, T., & Pollitt, M. (2003). International benchmarking and regulation: An application to European electricity distribution utilities. *Energy Policy*, *31*(15), 1609–1622. https://doi.org/10.1016/S0301-4215(02)00226-4.

Lee, D. P. (1999). Information to users. *Arbor Ciencia Pensamiento Y Cultura*. https://doi.org/10.16953/deusbed.74839.

Lowry, M. N., & Getachew, L. (2009). Statistical benchmarking in utility regulation: Role, standards and methods. *Energy Policy*, *37*(4), 1323–1330. https://doi.org/10.1016/j.enpol.2008.11.027.

Munn v. Illinois. (2015). Retrieved December 13, 2015, from https://www.oyez.org/cases/1850-1900/94us113.

Nataraja, N. R., & Johnson, A. L. (2011). Guidelines for Using Variable Selection Techniques in Data Envelopment Analysis. *European Journal of Operational Research*, *412*(3), 662–669.

Nepal, R., & Jamasb, T. (2015). Incentive regulation and utility benchmarking for electricity network security. *Economic Analysis and Policy*, (1993), 1–11. https://doi.org/10.1016/j.eap.2015.11.001.

Nigam, V., Thakur, T., Sethi, V. K., & Singh, R. P. (2012). Benchmarking of Indian mobile telecom operators using DEA with sensitivity analysis. *Benchmarking: An International Journal*, *19*(2), 219–238. https://doi.org/10.1108/14635771211224545.

Omrani, H., Gharizadeh Beiragh, R., & Shafiei Kaleibari, S. (2015). Performance assessment of Iranian electricity distribution companies by an integrated cooperative game data envelopment analysis principal component analysis approach. *International Journal of Electrical Power & Energy Systems*, *64*, 617–625. https://doi.org/10.1016/j.ijepes.2014.07.045.

Sağlam, Ü. (2017). Assessment of the productive efficiency of large wind farms in the United States: An application of two-stage data envelopment analysis. *Energy Conversion and Management*, *153*(October), 188–214. https://doi.org/10.1016/j.enconman.2017.09.062.

Shrivastava, N., Sharma, S., & Chauhan, K. (2012). Efficiency assessment and benchmarking of thermal power plants in India. *Energy Policy*, *40*, 159–176. https://doi.org/10.1016/j.enpol.2011.09.020.

Shuttleworth, G. (2005). Benchmarking of electricity networks: Practical problems with its use for regulation. *Utilities Policy*, *13*(4), 310–317. https://doi.org/10.1016/j.jup.2005.01.002.

Shyu, J., & Chiang, T. (2012). Measuring the true managerial efficiency of bank branches in Taiwan: A three-stage DEA analysis. *Expert Systems with Applications*, *39*(13), 11494–11502. https://doi.org/10.1016/j.eswa.2012.04.005.

Supreme Court Case/: Munn v. Illinois 1877. (2015). Retrieved December 13, 2015, from http://www.let.rug.nl/usa/documents/1876-1900/supreme-court-case-munn-v-illinois-1877.php.

Ter-Martirosyan, A., & Kwoka, J. (2010). Incentive regulation, service quality, and standards in U.S. electricity distribution. *Journal of Regulatory Economics*, *38*(3), 258–273. https://doi.org/10.1007/s11149-010-9126-z.

Thakur, T., Deshmukh, S. G., & Kaushik, S. C. (2006). Efficiency evaluation of the state owned electric utilities in India. *Energy Policy*, *34*(17), 2788–2804. https://doi.org/10.1016/j.enpol.2005.03.022.

Vogelsang, I. (2002). Incentive regulation and competition in public utility markets: A 20-year perspective. *Journal of Regulatory Economics*, *22*(1), 5–27. https://doi.org/10.1023/A:1019992018453.

Zhao, X., & Ma, C. (2013). Deregulation, vertical unbundling and the performance of China's large coal-fired power plants. *Energy Economics*, *40*, 474–483. https://doi.org/10.1016/j.eneco.2013.08.003.

Emotional change of immigrant youths in Korea after collage art therapy: A case study

Geummi Wang
Institute of Eduin Art Therapy, Seoul, Korea

Youngsoon Kim & Youngsub Oh
Inha University, Incheon, Korea

ABSTRACT: The purpose of this study is to analyze the emotional changes of immigrant youths after collage art therapy. For this purpose, three immigrant youths were selected. A total of 10 sessions of collage group art therapy were conducted for 60 min each week from November 2016 to February 2017. As a qualitative research, this study analyzed their change of emotions, based on the meanings of the symbols and components in their artworks. Components like symbols, meanings, individual conflict, and desire have positively changed. Their early works showed negative symbols and passive actions as unconscious defense mechanism. Their mid-term works showed their concentration on art activities and dynamics with interest and creativity in art medium. Their late works showed positive symbols like achievement desire and adaptation to reality. Therefore, collage art therapy can be used as a suitable psychotherapy method for the emotional stability of immigrant youths.

1 INTRODUCTION

As international marriage and foreign workers have drastically increased in Korea since the end of the twentieth century, Korean society has been transformed into a multicultural one, which is composed of a variety of ethnic groups. This study focused on immigrant youths among immigrant groups. According to Ministry of Education in Korea (2017), 7418 immigrant youths have enrolled in the public education. However, at the same time, 3 out of 10 school-aged immigrant youths are not attending to school due to language problem, visa status, and socioeconomic status (Bae, 2016).

Thus, Korea needs to support immigrant youths with psychological counseling in systematic policy (Kim, 2011). They have advantages like dual cultural background and bilingual ability. If they are given educational opportunities, they can grow as valuable human resources in the global era. Currently, Korea is gradually expanding educational opportunities and psychotherapeutic system to immigrant youths as members of Korean society (Kim, 2015).

Accordingly, psychotherapeutic programs like collage art therapy can help them express their own inner side. However, studies on collage group art therapy for immigrant youths are not enough. Therefore, a case study of emotional changes of immigrant youths in the experience of collage group art therapy programs is inevitable.

The purpose of this study is to provide fundamental information to develop effective services and education programs that contribute to the emotional stability of immigrant youths. A specific research question to achieve the purpose of this study is as follows: How do the emotional changes of immigrant youths, who participated in the collage group art therapy program, appear in their artworks?

2 THEORETICAL BACKGROUND

2.1 Collage art therapy

Collage refers to the way in which the screen is made up of prints, cloth, iron, wood, sand, leaves, and so on (Sim, 2016).

Collage art therapy is a task of cutting and pasting existing images in mass media like magazines and in various industrial media. This work can deepen the understanding and insight of oneself. The process of selecting a favorite image from magazine is likely to be an expression of self-emotion and to facilitate spontaneity, self-expression, and self-opening (Wang, 2017; Lee, 1997).

The magazine picture collage technique, which is used as art therapy technique, was introduced in 1972 by Burk and Provancher as an evaluation technique in the US occupational therapy center. It was then developed as therapeutic means by Sugiura (1994).

Currently, art therapy field in Korea can select various collage media due to the development of the stationery industry. This means that it is possible to use various media according to symptoms and developmental stages from infant to adult. Also, the collage can be devised variously according to the intention of the therapist and the symptoms and complaints of the client, and it is most important to proceed with consideration of sex, age, and disability severity.

Especially, group art therapy helps discover the self within the group. In other words, it visualizes the image of oneself, so that participants can recognize his/her psychological state and actively express its emotional direction (Wang et al., 2016).

The process of collage production is interesting. The collage therapist must carefully consider the production process. Then, presentation medium should be selected according to the session progress. The client may be reminded of many past affairs, events, and trauma in the process of putting and pasting media, color, and magazine photos. Thus, the therapist can maximize the effectiveness of the treatment by knowing the detailed procedures.

Particularly, collage group art therapy goes beyond the limit of verbal communication and allows the immigrant youths to see their own image during the session of exchanging their mutual feelings through visual work. It is sincere and efficient for the adaptation process of society because it makes participants realize reality. In addition, this group therapy is more effective in self-management ability and social interaction by analyzing the internal and social conflict factors as well as being able to feel therapeutic power by promoting interrelationship through group works (Jeong, 2014). However, during the process, the therapist should actively intervene to help solve the problem when conflict between the participants or professional insight is needed (Park, 2007).

Therefore, in this study, a collage group art therapy program was constructed on the basis of individual diagnosis method, group diagnosis method, postcard collage method, media-mixed formative method, and circular collage method among several techniques.

2.2 Immigrant youths in Korea

Immigrant youth includes a child from an immigrant who remarried with Korean spouse, a child who entered the country with foreign parents, a foreign youth who entered the country in adolescence for the purpose of work or study purposes, a North Korean youth defector, or a North Korean child born in a third country (Yang & Cho, 2012).

According to Ministry of Education in Korea (2017), the number of children of multicultural families is 99,186, which account for more than 1.68% of all students, of whom 7418 are immigrant youths. Among these, 4583 are in elementary school, 1627 are in middle school, and 1208 are in high school students. However, the total number of immigrant youths, including preschoolers, is about 20,000.

The biggest difficulty of immigrant youths is Korean language. Although their academic needs are very high, there is no individualized, level-specific learning method, so they have difficulties in adapting to school life.

Previous studies have suggested the following characteristics of immigrant youths. They have grown up in other cultures and have acquired the language and culture of the mother country. Thus, after their entry into Korea, they experienced cultural confusion in a wide range of living areas. In addition, there are few cases of resentment and anger toward parents who were difficult in communication during the separation period with their parents. Immigrant youths are likely to enter Korea due to their parents' remarriage rather than their own will (Ryu & Oh, 2012). The processes of family disintegration, separation, reunification, and migration to other countries can bring emotional anxiety and crisis experiences to them. Unfortunately, immigrant youths in Korea are still in separation and neglect from the low level of socioeconomic situation of their parents. This family situation makes their sense of alienation worse and becomes an obstacle of their re-socialization (Kim et al., 2012).

Immigrant youths were born in a country other than Korea and entered Korea during their adolescence, so they have different levels of adaptation to Korea from Korea-born multicultural children.

In adolescence, social support in the form of love, protection, and acceptance is given parents, teachers, and friends. When the social support increases, self-efficacy also increases, and social adjustment and psychological adjustment are improved (Park et al., 2002). Therefore, family, peer, and teachers are important supporters in the social network of adolescence, and they have many preventive functions from various problems in social relations. This social support is an effective factor for adaptation of adolescence, and it maintains emotional stability in youth's developmental period.

In this sense, a program for emotional development is needed as an intervention for their emotional stability. However, linguistic intervention techniques, which require a large amount of language abilities, are limited to immigrant youths with limited abilities of the Korean language. Thus, other programs to overcome these limitations should be provided to them. Therefore, if they are properly educated, they can grow into valuable human resources in the global era. Therefore, their adaption to school and social life is a very important task for the Korean society.

3 RESEARCH METHOD

3.1 *Research participants*

The research participants were three immigrant youths, including two youths in school and a youth out of school. Although eight youths first participated in this program, finally three youths were selected because of continuous attendance, translation issue, and completeness of artworks. Their nationalities were China and Uzbekistan. Before coming to Korea, they had no experiences to go abroad, because of their low socioeconomic status. They had stayed for less than 1 year in Korea. They were able to speak and listen to the Korean language. Their profiles are shown in Table 1.

Participant A's family consists of father, stepmother, and a younger brother. He can speak and listen to the Korean language. His main issues are related to language, injury (burn), and disinterest in stepmother. Participant B's family members are maternal grandmother (cohabitation), mother, and stepfather (non-cohabitation). She is well of speaking Korean. Her main issues are family reunion and the making of friends. Participant C's family consists of Korean-Uzbek father and mother, and a sibling. She can speak both Korean and English. She wants to learn advanced Korean language. In general, as immigrant youths enter Korea, following their mother or father's remarriage with Korean spouse, they are not ready to speak Korean (except for participant C). However, their parents begin to teach them Korean after their entry into Korea.

Table 1. Profiles of participants.

Name	Age (gender)	Nationality	Entry to Korea	Attending school
A	16 (M)	China	2015.9	No
B	17 (F)	China	2015.8	Yes
C	15 (F)	Uzbekistan	2016.5	Yes

3.2 Collage art therapy program

The program was created and developed on the basis of previous works (Wang, 2017), and it was also structured in accordance with emotional, cognitive, and volitional levels.

This therapy was conducted from December 2016 to February 2017 throughout a series of 10 sessions. The process of art therapy is composed of three stages: early stage, middle stage, and late stage. The contents are shown in Table 2.

In the early stage, the research participants shared the individual psychological state and conflict for the purpose of rapport formation and relaxation of tension between the research participants and researchers.

In the middle stage, from the 4th session to the 7th session, the participants were trained to express their emotions and to recognize their self-understanding and realization of self-image.

In the late stage, from the 8th session to the 10th session, the goal was set as an insight into the positive self-image in the future and the reality adaptation.

3.3 Collage analysis

The collage analysis criteria of this study refer to the study of Sugiura (1994), the founder of collage art therapy, and Aoki (2005) and Keun-Mae Lee (2009). The criteria are shown in Table 3.

Table 2. Collage art therapy program.

Stage	Session	Theme	Activities
Early	1	Rapport	My image
	2	Desireseeking	Mandala
	3		My tree
Middle	4	Self & reality under-standing	Egg
	5		Clay
	6		Mask
	7		Gift
Late	8	Reality adaptation & future	Photo image
	9		Family
	10		Hope tree

Table 3. Collage analysis criteria.

Area	Section
Form analysis	Number of pieces
	Blank
	Deviation from drawing paper
	Use of back side
	Overlapped use
	Method of paper cutting
	Shape of piece
	Position of blank
	Position of drawing paper
	Balance of works
Content analysis	Photo
	Expression
Overall analysis	Integrity in sessions
	Theme in sessions
	Use of space in sessions

4 RESULT

4.1 Within-case analysis

4.1.1 Research participant A

A's artworks are shown in Table 4.

In the early stage, participant A did not have eye contact with therapist. He gazed only at the art medium and hesitated in doing activities. He did not answer the therapist's questions. He was reluctant to use scissors due to his hand with burns. His collage pictures or ornamental pieces were monotonous and symmetrical. Other persons did not appear in his collage works. He expressed only himself.

In the middle stage, he was interested in art medium. He expressed his desire, that is, repeatedly selecting medium. He started to answer therapist's questions and help organizing things around him. He sometimes laughed and revealed his pleasant feelings. He cut collage pieces with square shape rather than photograph shape. He constructed collage within the size of drawing paper. He expressed his own childhood memory and his expectation desire of family.

In the late stage, he gladly greeted with others and asked his own questions about session theme. While he repeatedly explained the same thing, he presented his own words to the end in front of members. He used almost same size of square. His collage pieces were positioned symmetrically and well centered. He naturally expressed desire for belonging to his family and their support. He also realistically expressed his own future career.

4.1.2 Research participant B

B's artworks are shown in Table 5.

In the early stage, participant B had an attitude to look bright. However, she hesitated in selecting art medium. She attached the given collage pieces with inclined angle. There was a big margin in the drawing paper. She expressed self-centered images in past and present tenses.

In the middle stage, participant B cooperated with members in selecting art medium. She was comfortable and relaxed and familiar with sessions. She cut and pasted a full size of square to fill the drawing paper. Thus, there was no margin in the paper. She expressed a community consisting of herself, parents, maternal grandmother, and even an animal.

Table 4. A's artworks.

Session	Work	Session	Work
1	Introducing myself	6	
2		7	
3		8	
4		9	
5		10	

Table 5. B's artworks.

Session	Work	Session	Work
1	Introducing myself	6	
2		7	
3		8	
4		9	
5		10	

In the late stage, she no longer hesitated by adding medium. She used proper size and content of collage. She constructed integrative contents of collage work and aligned those in the center of drawing paper. She expressed her own image as interpreter, which she hopes to become in the future.

4.1.3 *Research participant C*

C's artworks are shown in Table 6.

In the early stage, participant C often looked bright and naturally had eye contact. However, she controlled her younger brother. She used pieces to fill the entire size of drawing paper. She expressed her own superiority.

In the middle stage, she communicated with members with active actions. She was loud. She freely went to the toilet during the session, freely selected medium. She was good at concentration and tranquility. She sufficiently expressed her own desire.

In the late stage, her choice of medium was changed. She was still about to control her younger brother. She focused on collage composition with mature defensive expression. She expressed sociality and mature self-independence.

4.2 *Cross-case analysis*

Table 7 shows the emotional changes and common characteristics from the individual works of participants on the basis of the analysis of the symbol and testimony.

In the early stage, participants were homesick. However, they hesitated it because of their less ability to speak Korean. Furthermore, they expressed defensive and passive attitudes in the early intervention.

In the middle stage, participants showed interest in art medium and attempted to express specifically their own works. Their activities in the mid-term sessions became more serious than early session.

In the late stage, participants wanted to take picture of their works and show those to family and friends. They attempted to use various media. Their expression about future was specific. They expressed responsibility for family, interest in economic activities, academic desire, and so on.

Table 6. C's artworks.

Session	Work	Session	Work
1	Introducing myself	6	
2		7	
3		8	
4		9	
5		10	

Table 7. Emotional change of participants.

Stage	Common feature	Characteristics
Early	• A longing for a family left in their home country and a childhood memory • A passive attitude toward self-expression from the lack of the Korean language	• Defensive self-expression • Passive attitudes in early intervention • A hesitant language expression
Middle	• Repeated curiosity about the art medium, especially available for grain media • Trying to express one's work in a linguistic way	• Serious attitude to mandala activities • expressing mother language
Late	• Desire to show family or friends after taking picture of one's work • Introducing various art medium • Expression of one's dream or career change	• Specific job choices • Economic activity Questions • Family support Responsibility • Confidence in bilingualism • Academic needs

5 CONCLUSION AND DISCUSSION

In this study, we identified the immigrant youths' phased change of class atmosphere, expression ability, facial expression, completeness of artworks, participation, and so on. On the basis of these several evidences, we is evident that collage art therapy is effective to immigrant youths.

In the collage art therapy's early stage of intimacy formation and desire seeking, pictorial change appeared as a process of expressing inner desire by reminiscing childhood experiences in the mother country.

In the mid-term stage of self-understanding and reality recognition, their works showed an interest in art medium and a strong sense of belonging to the community through mutual cooperation. In addition, activities for creativity and sharing of works brought about dynamic within the group.

In the late stage of self-esteem and reality adaptation, they revealed the inner side of themselves in a way that the contents of artworks interacted within the group.

In conclusion, collage group art therapy provided an opportunity for immigrant youths to relieve emotional difficulties such as language limitations, unfamiliar environments, and stress experienced in different educational systems with their home countries. This therapy contributed to the emotional stability by interaction with other people through the formation of a protective and stable group. Researchers also hope that immigrant youths will grow as healthy members in Korean society through continuous collage art therapy programs.

The limitation of this study is that participants' verbal and nonverbal expression, including bodily and facial expression, cannot be fully analyzed. Furthermore, it did not fully deal with the results by participants' individual differences such as nationality, gender, age, and school attendance. For further study, a variety of art media can be attempted for collage art therapy program. And as immigrant youths with various backgrounds can participate in this program, their individual difference of emotional changes after this program can be studied.

ACKNOWLEDGEMENT

This study was supported by the Ministry of Education of the Republic of Korea and National Research of Korea (NRF-2017S1 A5B4055802).

REFERENCES

Bae, S.R. 2016. A study on the status of migrant youths and supporting their self-reliance. *National Youth Policy Institute Research Paper*.
Burk, R.E. & Provancher, M.A. 1972. Magazine picture collage as an evaluative technique. *American Journal of Occupational Therapy* 26(1): 36–39.
Jeong, Y.J. 2014. The characteristics of art and the therapeutic meaning of the creative process in art therapy. *Korean Journal of Art Therapy*, 23(5): 1221–1237.
Kim, M.J. 2011. Multicultural education for immigrant students. *Journal of Education & Culture* 17(2): 55–76.
Kim, N.Y. 2015. *A study on the legislation of art therapy: Approval of visual art therapy as a complementary and substitutional medical therapy and its institutionalization*. Unpublished doctoral Dissertation. Dongguk University.
Kim, Y.S. & Park, B.S. & Trang, P.T.H. 2012. A qualitative study on re-socialization experience of the immigrant youths with enculturation perspective. *The Language and Culture* 8(3): 37–63.
Lee, E.J. 1997. *A study on the breaking from the stagnancy of fine arts guidance for the higher grades in elementary School: Mainly through the collage's technique*. Unpublished master's thesis. Korea National University of Education.
Lee, K.M. 2009. A research on the formal characteristics of collage expression of elementary students grouped by grade. *Korean Journal of Art Therapy* 16(2): 171–188.
Ministry of Education (2017). *Plan of Multicultural Education Support*. Retrieved October 31, 2017, from http://blog.naver.com/moeblog/220909378876.
Park, S.H. 2007. *Group art treatment case study for broken family children's self-respect elevation*. Unpublished master's thesis. Jeonju University of Education.
Park, Y.S & Min, B.K. & Kim, U.C. 2002. Longitudinal study of adolescents' life-satisfaction: An analysis of social support and self-efficacy. *Journal of Education & Culture* 8: 161–202.
Ryu, B.R. & Oh, S.B. 2012. An analysis of educational opportunities for and adaptation of immigrant youth. *Multicultural Education Studies* 5(1): 29–50.
Sim, E.J. 2016. *The effects of collage group art therapy program on depression of the elderly in care facilities*. Unpublished master's thesis. Pyeongtaek University.
Sugiura. 1994. *Collage Therapy*, Tokyo: Mishima.
Wang, G.M. 2017, A case study on the collage art therapy for the migrant youth. Unpublished doctoral Dissertation. Inha University.
Wang, G.M. & Kim, Y.S. & Kim, M.S. 2016. A case study on the emotional changes of immigrant youth participated in the collage group art therapy. *The Journal of Learner-Centered Curriculum and Instruction* 16(11): 1017–1039.
Yang, K.M. & Jo, H.Y. 2012. The exploratory study of psycho-social adjustment of immigrant youth in Korea. *Korean Journal of Youth Studies* 19(11): 195–224.

A study on the "individual accountability" of undergraduate students in the cooperative learning-centered liberal education class

Youngsoon Kim, Gihwa Kim & Hee Choi
Inha University, Incheon, Korea

ABSTRACT: The purpose of this study is to examine the role-playing experience of college students participating in liberal arts course "Multicultural Society and Coexistence Humanities". Here, we deal with the personal accountability of undergraduate students in a role-playing, learning, applying, and cooperative learning model in the liberal arts class. Through the learning process, we had an interest about the individual accountability. This study concentrated on the issues of how the undergraduate students understand and interact with this accountability. In the class, four social interactions (conflict, exchange, cooperation, and competition) play an important role in understanding the personal accountability. Through these interactions, we tried to understand the content of the individual book experience of the research participants and its meaning. As a result of the study, three main topics of the students participating in the class are explained, namely responsibility of the role, positive interdependence among team members, and moral reflection.

1 INTRODUCTION

The purpose of this study is to examine the role-playing experience of college students participating in liberal arts course "Multicultural Society and Coexistence Humanities". This course was based on a cooperative learning role playing in blended learning during 3 h (an offline class for 1.5 h and an online class for 1.5 h a week).

Cooperative learning started from overseas in the 1970s, and domestic discussion started in the mid-1980s by Park (1985). Since then, models and methods have been developed, and studies have been conducted in various aspects in the early 1990s. There are also studies on teachers' perceptions of cooperative learning methods, models, strategies, and cooperative learning, including studies on various school classes, ranging from infants to elementary schools, middle and high schools, universities, and graduate schools.

Examples of studies of university students are Kim (2003), Ahn & Kim (2015), and Park (2010). In these studies, it was analyzed that cooperative learning influences academic achievement, peer relationship, and self-directed learning. Hong & Lee (2013) conducted linguistic interaction analysis through cooperative learning for university students. Park & Ko (2016) analyzed the mediating effect of communication ability in the context of cooperative learning. Recently, Lee (2017) has analyzed the degree of reflection of the basic elements of cooperative learning performance of college students and the difference of class satisfaction according to a cooperative learning style.

Park (2007) investigated collaborative learning experiences on the basis of a study on graduate students. Park (2014) conducted a study on the attitudes and behaviors of college students participating in an integrated presentation discussion using cooperative learning. However, a research on cooperative learning experience through role playing for college students has not been conducted.

The aim of this study is to investigate the experience of college students participating in liberal arts classes in the process of role playing and the meaning of experience through cooperative learning. The students started from the awareness of how to understand and interact with the learned knowledge through role playing while applying the cooperative learning model in the liberal arts class of "Multicultural Society and Coexistence Humanities".

The key social interactions in this role playing were conflict, exchange, cooperation, and competition. This study aims to examine the individual's role in the process of role playing, focusing on these four social interactions. There are two research questions in this study: (1) How do college students participating in the cooperative class role playing experience personal accountability? (2) What is the meaning of the experience of individual accountability?

2 THEORETICAL BACKGROUND

The concept of cooperative learning and the role-playing learning are important in this study. The philosophical presupposition of cooperative learning is as follows: Humans are social animals. This means that, in order for a human to pursue a human life, we can also interpret it in the sense of helping the members of society and receiving help from them. However, in order for individuals to adapt to society, it is common to think that I can succeed when sacrificing others for me. This is the concept of competition. But can all succeed together? This kind of thinking can always leave the human mind. This is the concept of collaboration. Likewise, cannot class members achieve common goals together in learning? Obviously, as members of society can live together well, learning can achieve the learning goals of the members together. If these ideas are positively recognized, motivation for cooperative learning can be induced.

According to Chiu, M. M. (2008), cooperative learning is an educational approach that aims to organize classroom activities into academic learning experiences. There are much more to cooperative learning than merely arranging students into groups, and it has been described as structuring positive interdependence. Students must work in groups to complete tasks collectively toward academic goals. Unlike individual learning, which can be competitive in nature, cooperative learning can capitalize on resources and skills themselves.

Furthermore, the teacher's role changes from giving information to facilitating students' learning. Everyone succeeds when the group does. Ross and Smyth (1995) describe successful cooperative learning tasks as intellectually demanding, creative, open-ended, and involve higher-order thinking tasks. Cooperative learning has also been linked to increased levels of student satisfaction.

Above, we examined the characteristics of cooperative learning directly. Now we will confirm the opinions of various scholars on the components of cooperative learning. Aronson (1978) defined collaborative learning as a way to achieve a given learning goal by working together in a group. However, Cohen (1994) found that all learners were small group forms that could participate in clearly assigned joint tasks. In other words, if the concept of cooperative learning is defined, the first cooperative learning is a small group form in which the students collaborate with one another to promote the learning effect by interacting with each other. The second is socialization training for coexisting social process.

The components of cooperative learning have various opinions from scholars. Johnson and Johnson & Johnson (1989) and Johnson et al. (1992) found positive interdependence, individual accountability, interpersonal interactions, interpersonal and social skills, and group processes. Slavin (1991) emphasized positive interdependence and personal accountability. Cohen (1986) concluded that open concept and discovery task emphasize high-order thinking skills, group tasks require participation of other members, multiple tasks are related to major topics, and role assigned to each group member are important.

Rottier & Ogan (1991) considered group cohesion, interpersonal interaction, personal accountability, social skills, collective responsibility, teacher supervision, and group evaluation. Ormrod (1995) addressed the interdependence of group members, personal accountability, teacher supervision, group evaluation, clear group goals, and small group size cooperative learning. Sharan (1990) emphasized positive interactions, interpersonal interactions, individual accountability, small group and interpersonal skills, and group evaluation.

In addition to analyzing prior research on cooperative learning, the concept of role playing is needed. This is because this study deals with role-play-based cooperative learning classes.

Role play is one of the learning methods to set scenes that take place in reality and to allow each person play his/her own role so that he/she can cope correctly when something through a similar experience. It is used in a wide range of education, such as learning conversation through words from other countries, how to cope with customers in companies, and so on, as well as piercing leadership. By experimenting with various scenes similar to real ones, we can achieve the same effect as the experience. Therefore, when confronted with the same scene in reality, it is possible to reduce the awkwardness and to cope quickly.

In the study of such a series of cooperative learning, common factors are interpersonal interchange and personal accountability, positive interdependence among group members, interpersonal relations and social skills, and group evaluation. This study focused on "individual accountability", which is one of the components of cooperative learning among college students participating in cooperative learning.

3 RESEARCH METHOD

In this study, we used case study and phenomenological description method to explore the meaning of individual experiences of undergraduate students in cooperative learning process. The reason for using the phenomenological methodology is that phenomenological description can structure the vivid appearance of human life.

We also applied the collective multiple case study method, which is an appropriate research method to find common points through cases experienced by several individuals rather than one case. In addition, the phenomenological study method (Shin et al., 2004) was applied to reveal the essence of experience in cooperative class.

3.1 Research participants

Research participants were grouped into four groups, with each group consisting of four to six participants, considering their diverse social and demographic backgrounds. Each group had a name matching the characteristics of the participants, and team activities were conducted one to two times a week centered on the facilitator. During a semester, researchers met participants through SNS online and interacted in the school library study room after class.

The key themes presented in role playing were the following four social interactions: "conflict", "exchange", "cooperation", and "competition". The learning method was to set up the story itself and to share the role playing so that the interaction between the team members became active. Table 1 shows information of the study participants who participated in the cooperative learning class with role playing.

The team consisted of various regions, majors, and grades. This is because the role-playing theme is "social interaction" in order to reflect its background and interaction with other groups. Participants will experience various interactions with team members through role playing, which is a way of cooperative teaching. In addition, there will be various dynamics within the team to form a collaborative structure and achieve the team's shared goals. In the process of collaborative learning, research participants conducted research with the premise that they would experience personal accountability in a natural way.

3.2 Data collection and analysis process

Data collection was conducted by reconstructing a group inquiry model of the six stages of Sharan & Sharan (1994). These stages are research group organization phase, planning inquiry and role sharing phase, inquiry execution phase, presentation preparation phase, presentation phase, and activity evaluation phase. The stage of role playing was conducted in the order of idea exchange and discussion, script writing and modification, rehearsal, and performance. After the class was completed, the questionnaires were distributed to and filled by the participating students. The researcher designed the lesson, observed the progress of the roll play with the teaching assistant, and wrote the observation log every lesson.

Table 1. Information of research participants.

Group	R.P	Gr.	Major	Home region
A Group	A 1	2	Electrical Engineering	Gyeonggi
	A 2	3	Chemical Engineering	Seoul
	A 3	1	Food & Nutrition	Gyeongnam
	A 4	2	Drama	Gyeonggi
	A 5	1	Media Informatics	Gyeongnam
B Group	B 1	2	Information Communication	Incheon
	B 2	3	Computer Information	Incheon
	B 3	1	Philosophy	Chungbuk
	B 4	4	Material Science	Gangwen
	B 5	1	International Trade	Gyeonggi
C Group	C 1	2	Business	Seoul
	C 2	2	Information & Communication	Guangju
	C 3	4	Electronic Engineering	Ulsan
	C 4	1	Food and Nutrition	Gyeonggi
	C 5	2	Consumer Studies	Gyeongbuk
	C 6	1	Business	Incheon
D Group	D 1	3	Information Communication	Gangwen
	D 2	4	Computer Information	Incheon
	D 3	2	Biology	Seoul
	D 4	4	English Education	Daegu

Data analysis was approached through phenomenological analysis (Giorgi, 1985). We analyzed the meanings of the participants' experiences by extracting the sentences containing meaningful experiences in journals and open questionnaires, which were collected by the individual in role playing. An analysis unit can be a text, a paragraph, a sentence, or a part of a sentence. In some cases, one paragraph in the text is proved to be meaningful as an analytical unit, but, in general, it is not possible to interpret it because of its small size (Gläser & Laudel, 2010). Therefore, in this study, the process of integrating the structure was carried out by extracting the meaning into the sentence with unit analysis, sentence unit, and paragraph unit to focus on the experience and clarify the relation with the meaning.

In order to ensure the reliability and validity of the research results, two seminars and colloquiums were held in June and August of 2017, and the research team, fellow researchers, and experts conducted various tests. Furthermore, it was verified that the data were collected according to IRB's research ethics regulations. We also conducted in-depth discussions on the appropriateness of data analysis and interpretation methods within the research team.

4 INDIVIDUAL ACCOUNTABILITY IN COOPERATIVE LEARNING

The purpose of this study is to analyze the experience and meaning of the individual accountability in cooperative learning for undergraduate students participating in liberal arts classes. As a result of analyzing the collected data, we classify the core subjects into the following three categories: (1) responsibility for role, (2) positive interdependence, and (3) moral reflection.

4.1 *Responsibility for roles*

Responsibility is a concept that refers to some kind of attitude being revealed in the relationship between the self and others. Levinas argued that all cannot be free from this responsibility because they are responsible for responding to the call of the other (Lee, 2008). The individual learning structures do not care about other learners, and individuals have an assessment and responsibility for completed assignments. However, the cooperative learning structure is responsible for the team members' learning, so it differs from the individual learning structure in sharing, helping each other, and interacting with each other.

The research participants in this study were aware of their responsibility for role-playing in the cooperative learning process, and performed continuous evaluation. They considered both the roles of others and the role of the whole team, as well as their role in evaluating roles and responsibilities. In the evaluation of individual roles and responsibilities, individual internal and external factors were in operation. In addition, they negatively assessed the fact that they were more active in their opinion or inaction because of their personality traits. Furthermore, they showed their willingness to realize and improve themselves that they lacked the ability to interact with others through their experiences in cooperative classes. They also recalled that they did not spend more time and effort to achieve the shared goals in terms of external factors.

> "I was disappointed that I was not actively presenting my opinion. I am afraid to express my opinion confidently in front of other people, and I think I should fix it." (Participant A5)

> "I was disappointed that I could not invest time in the role play part. I should have looked more often as a facilitator, but I could not do it because I was busy." (Participant C1)

In the interviews of the participants, it is noteworthy that the responsibility for the role was assessed not only for the reflection on themselves but also for the whole team. They were judging that the achievement of the goal was insufficient. The reason for this is not only the role of the individual but also the overall harmony. It is interpreted that the high responsibility and active role of each individual are important in order to coexist with a large number of others, but it is necessary to harmonize mutual members.

> "Overall, the role played in a sad atmosphere. The nature of acting required free time and occasionally a process, and we could get good results in a free atmosphere rather than a set frame, but we could not." (Participant C3)

4.2 *Positive interdependence*

The conception "positive interdependence" is an element of cooperative and collaborative learning where members of a group who share common goals perceive that working together is individually and collectively beneficial, and success depends on the participation of all the members.

The most important part in cooperative learning is to increase positive interdependence. But this is possible when personal responsibility is backed up. In other words, when individuals play their roles, all members who want to achieve the same goal will be able to achieve their goals. In collaborative learning, the responsibility and role of an individual is blurred and problems arise from it. Especially, a common problem appears in the situation of cooperative learning. This is the learner having a non-cooperative and invisible attitude toward cooperative learning activities. It is also a question of the "free rider" who takes the score together in the group evaluation without making any effort.

However, this study did not reveal any intrusive attitudes toward the roles and responsibilities to be performed as a team member during the cooperative learning. The reason is that the way of cooperative learning is role playing. Role playing is based on the proper role of all members. Thus, it can be seen that the role contribution is prevented from being concentrated in the minority within the team and naturally, the participation of all team members is promoted.

> "I did not fall all together and tried to do it often because I tried it often." (Participant B1)

> "Everyone worked hard without riding freely. I have not been bothered by age, grade, sex, etc." (Participant C1)

The research participant A3 described himself as "a part in the clock". If any one of the parts in the clock does not work, the watch will stop functioning. This is interpreted as meaning that any team member will have a negative impact on the entire team if he/she does not play his/her role. In other words, it was recognized that the role of an individual in cooperative learning is important for achieving common goals.

"Like the parts in the clock, I sincerely tried to do what I did to organize my team activities." (Participant A3)

4.3 *Moral reflection*

Individual learning is evaluated according to its own abilities and efforts. In addition, individual learners carry out their own roles as appropriate and assume responsibility for the results. Therefore, we do not have to consider others in our personal learning. However, in collaborative activities, the evaluation affects others and the team as a whole. In the course of cooperative learning, an individual is consciously concerned that another person is injured because of oneself. This can be viewed as a moral reflection, which has a positive effect on increasing individual learning activities and enhancing their capacity and ability. The results of this study showed that the research participants correctly perceived their inadequacy in the cooperative learning process. The subject of moral reflection is self, but it was targeted to others as a result.

"It was awkward because it was the first time that a proper role play was done by myself. I was good at taking on the role that suits each person ... But if I played more naturally and played the adverb ..." (Participant A4)

"It seems like I was shivering when I was acting poorly in front of people. I should not be wrong because I made a mistake." (Participant D1)

This perception of research participant D1 reveals the regret and self-reflection of the team's score. D1 thinks responsibility for the low score is due to his performance. This can be seen as self-reflection of participant D1.

5 CONCLUSION

In this study, we tried to find the experiences of undergraduate students performed in the role-playing process of the liberal arts classes and to understand the meaning of experience through cooperative learning. For this purpose, we applied a collective multiple case study method, which is an appropriate research method to find common points through cases experienced by several individuals rather than one case.

In particular, this study addressed only the experience of individual accountability among the characteristics and components of various cooperative learning. The results of the research categorized the following three main themes: (1) responsibility for roles; (2) positive interdependence; and (3) moral reflection.

First, research participants can find responsibility for their role. Participants were aware of their responsibility for role-playing in the cooperative learning process and performed a continuous evaluation.

Second, there was a positive interdependence in roles and responsibilities as team members during cooperative learning. Participants realized that if they were not dependent on each other, they could not complete the team's goals.

Third, we find moral reflection on the research participants. It was confirmed that it increases the learning activity of the moral individual and positively affects his/her ability and ability to improve.

As described in Introduction, previous studies focused on the effects of cooperative learning and the distribution of cooperative learning components. On the contrary, this study focused on the individual accountability among various cooperative learning factors. Especially, this study attempted to explore the meaning of the individual accountability of college students who participated in the role-playing class based on cooperative learning.

The results show that individual accountability plays an important role for the promoting global citizenship. Bauman (1992) indicated that the global citizenship is the rights, responsibilities, and duties that come with being a member of the global entity as a citizen of a particular country. This promoting global citizenship does not mean that such a person denounces or waives his/her nationality, more local identities, but such identities are given "second place" to their membership in a global community. Then, individual accountability can originate global citizenship if transnational migration is spreading.

It is very important that cooperative learning be practiced in schools in a society such as Korea where multicultural phenomenon has become prominent recently. In order to become a university student in Korea's educational reality, individual learning was more important than cooperative learning. In this respect, the introduction of cooperative learning is necessary for university students who will be leaders of future society. Particularly, the individual accountability among the components of cooperative learning will help fulfill his/her role as a citizen in a society where diversity is soaring.

In this context, the current generation should provide learning that will enhance individual's commitment to future generations for the sustainable Earth. This initiative is also relevant to the implementation of the United Nations goals for sustainable development. A future challenge is to devise a more sophisticated program that will increase the sense of individual accountability.

ACKNOWLEDGEMENT

This study was supported by the Ministry of Education of the Republic of Korea and National Research of Korea (NRF-2017S1A5B4055802).

REFERENCES

Ahn, Doe-Hee., & Lee, Yu-Ree. (2015). Influencing effects of collaborative learning and self-directed learning on life satisfaction in university students. *Korean journal of youth studies*, 22(7), 1–30.
Aronson, E. (1978). *The jigsaw classroom*. Sage.
Bauman, Z. (1992). *Intimations of Postmodernity*. London: Routledge.
Chiu, M.M. (2008). *Flowing toward correct contributions during groups' mathematics problem solving: A statistical discourse analysis. Journal of the Learning Sciences, 17(3), 415–463.*
Cohen, E.G. (1994). Restructuring the classroom: Conditions for productive small groups. *Review of Education Research*, 64(1), 1–35.
Giorgi, A.(1985). *Phenomenology and psychological research. Duquesne.* Shin, Kyung-Rim et al., (Tr) (2004). Seoul: Myunmosa.
Gläser, J., & Laudel, G.(2010). *Experteninterviews und qualitative Inhaltsanalyse. Springer-Verlag.* Woo, Sang-Su et al., (Tr) (2012). Seoul: Communication Books.
Hong, Soon-Tae., & Lee, Sang-Kook. (2013). An Analysis of Korean College Students' Verbal Interactions during their Cooperative Reading Activity. *Journal of English Language and Literature,* 55(3), 555–594.
Johnson, D.W., Holubec, E.J., Johnson, R.T., & Roy, P. (1992). Circles of learning. Alexandria, VA: Association for Supervision and Curriculum Development.
Johnson, D.W., & Johnson, R.T. (1989). Cooperation and competition: Theory and research. Interaction Book Company.
Kim, Kie-Jeong. (2003). The Effects of Cooperative Learning and Traditional Lecturing Strategies on the Cognitive and Affective Outcomes of College Students. *Journal of Educational advancement*, 22(1), 43–71.
Lee, Mi-A. (2017). University Students' Perceptions on the Level of Basic Elements Embedded in Cooperative Learning and Class Satisfaction. *Journal of educational innovation research,* 27(1), 115–132.
Lee, Yu-Taek. (2008). Eine philosophische Reflexion über den Begriff Verantwortung: Im Hinblick auf E. Levinas und H. *Jonas. Researches Heidegger,* 17, 63–94.

Ormrod, J.E.(1995). Educational psychology: Principles and applications. Englewood Cliffs, NJ: Prentice-Hall.

Park, Ji-Hoe., & Ko, Jang-Wan. (2016). Relationship between Collaborative Self-efficacy and Problem-solving Skills of University Students: Mediating Effect of Communication Skills. *Journal of educational innovation research,* 26(1), 169–192.

Park, Min-Jeong. (2007). Learning Experience of based Instruction. *Journal of Curriculum Studies,* 25(3), 265–288.

Park, Seong-Ik. (1985). Effects of Cooperative and Competitive Learning Strategy on Cognitive-Affective Outcomes. *Educational Research,* 7(1), 79–94.

Park, Sang-Min. (2014). The Cooperative Learning Based the Unificational Presentation Discussion Class Study. Ratio et Oratio, 7(1), 7–44.

Park, Yong-Han. (2010). Effects of Cooperative Learning on Goal Orientation, Motivation, and Achievement of College Students. *Asian journal of education,* 11(1), 91–119.

Rottier, J., & Ogan, B.J. (1991). *Cooperative learning in middle-level schools.* Natl Education Assn.

Ross, J., & Smythe, E. (1995). Differentiating cooperative learning to meet the needs of gifted learners: A case for transformational leadership. *Journal for the Education of the Gifted,* 19, 63–82.

Sharan, S. (1990). *Cooperative learning: Theory and research.* New York, NY: Praeger.

Sharan, Y., & Sharan, S. (1994). What do we want to study? How should we go about it? Group Investigation in the Cooperative Social Studies Classroom, In Cooperative Learning in Social Studies: A Handbook for Teachers, edited by R.J. Stahl, NY: Addison-Wesley Publishing Company, 257–276.

Shin, Kyung-Rim., Cho, Myung-Ok., & Yang, Jin-Hyang. (2004). *Qualitative Methodology.* Ewha Womans University Press.

Slavin, R.E. (1991). Synthesis of research of cooperative learning. *Educational leadership,* 48(5), 71–82.

Effect of competency on the job performance of frontline employees in Islamic banks: The moderator effect of religiosity

Mohd Fodli Bin Hamzah & Muhammad Nasri Bin Md. Hussain
University Utara Malaysia, Kedah, Malaysia

ABSTRACT: The rapid growth and development of Islamic banks should be aligned with the effective and efficient human capital development process. However, over time, there have been some problems that create a public hesitation toward the Islamic banking system. Several studies have been conducted on the Islamic banking system, especially in the areas of customer satisfaction, the selection factors of banking facilities among the consumers, analysis on the development and growth, and liquidity requirement. However, there is still a lack of research on employees of Islamic banks, especially in relation to their competency and work performance. Hence, this study is conducted to fill the gap that exists in this subject. Spencer & Spencer's theory states that competency must be a combination of technical competency and behavioral competency. The establishment of behavioral competency can be developed through a self-religious approach. In this regard, this study aims to explore the moderator effect of religiosity to determine the relationship between competency and the performance of frontline employees in Islamic banks. We hope to contribute to the Islamic banking industry by giving feedback on the effectiveness of the program conducted by the relevant parties by assessing into the real performance of the employees in the field.

Keywords: Islamic Banking, competency, human capital, job performance

1 INTRODUCTION

In an increasingly competitive and complex business environment along with the era of globalization, the organization should equip their workers with knowledge, proactive skills, attitudes, and positive values to ensure that the organization remains competitive and innovative in the industry (Bhatti, 2008). The growth and development of Islamic banks are so rapid and have gained global attention, which has become key competitor to the conventional banking system (Abduh & Alias, 2014). The Islamic banking system is one of the fastest growing segments in the global financial sector (Sayani & Miniaoui, 2013). Hence, the need for competence and committed employees is crucial and should be parallel with the development and growth of the Islamic banking industry (Lamba, 2012). One of the problems currently faced by Islamic banks is the lack of a comprehensive framework that is capable of handling the issues of the allocation of effective resources to maximize the return and values for the stockholders (Arshad & Tahir, 2016). As a result, Islamic banks are faced with various risks, both direct and indirect. Direct risks occur due to the noncompliance to the Islamic contract underlying the products and services offered by Islamic banks such as the failure to understand the basic concept like the tenet and requirement under the Islamic buy and sale contract. Meanwhile, indirect risks occur due to the interest rate, economic conditions, and liquidity, as well as governance and legal issues (Rafay & Sadiq, 2015). These issues lead to the continuous negative perception of the society toward Islamic banking. Therefore, provision of professional and competent employees to handle the abovementioned issues while ensuring that Islamic banking continues to grow and gain high trust in the community is needed. Thus, the employees who have the expertise are considered vital and a valuable asset to the organization as well as to the industry (Vasiliauskien, Stanik, Lipinskien, Nord, & Lietuva, 2005). Because the subject of this study is related to Islamic banking, the discussion relates only to the Islamic religion as the moderator effect to influence the relationship between the competency and job performance of the employees of Islamic banks.

2 GAP IN THE LITERATURE

Previous studies have stated that competent employees will show excellent work performance and there is a positive correlation between both variables (Ahmad & Schroeder, 2003; Mani, 2013). However, these two correlations have not been specially tested against the Islamic banking frontline employees who are responsible for dealing and liaising with the customers directly in their daily tasks in either Malaysia or other countries of the world. An effective and aggressive human capital development program should be carried out and given serious attention in line with the development of the Islamic banking industry in order to provide skilled and competent workforce to the industry and compete with the conventional banking system (Mohd Ali, Zakiah Mohamed, 2015). However, the effectiveness of the program can still be questioned in view of the acceptance and perception of the public on the Islamic banking system. These criticisms and negative perceptions lead to the question as to how far Islamic banking personnel have the ability in terms of knowledge and skills to explain and educate the public about the concept of Islamic banking. The negative criticism and argument illustrated that the performance of employees in the Islamic banking has not been at a satisfactory level despite various efforts being undertaken by the authority (Azizan, 2011). In addition, the employees should also exhibit a noble behavior and morals in accordance with the teachings of Islam. These positive skills and traits are very important and are very helpful in improving work performance as well as helping the Islamic banking industry to achieve the goals set forth (Azmi, 2015). Other studies related to the Islamic banking system were the factors affecting Islamic banking itself (Awan & Bukhari, 2011), level of customer satisfaction (Saad, 2012), quality of service (Kamariah et al., 2013; Abedniya & Zaeim, 2011; Razzaque, 2014), Islamic banking performance (Imam & Kpodar, 2010), human resources (Norhanim Dewa & Sabarudin Zakaria, 2012), and perception toward Islamic banking (Loo, 2010). Direct studies on the influence of competency on the performance of Islamic banking staff especially the frontline employees and the moderation factor of the religion are lacking and require further investigation.

3 THEORETICAL BACKGROUND

According to Campbell (1993), the performance of employees can be measured and monitored. The performance is the behavior and actions of every employee who will contribute to the achievement of the objectives of an organization. (McCloy, Campbell, & Cudeck, 1994) Campbell (1993) developed an outstanding model which was widely used to discuss the individual performance in the workforce (Razzaque, 2014). The model has emphasized that the determining factor of one's work performance depends on four factors, namely declarative knowledge, knowledge of procedures, skills, and motivation. In the aspect of a person's job, he/she needs to know what to do and how to do it and should have a desire to do it. David C. McClelland developed a theory which states that a person with good academic background is not necessarily excellent in his/her performance. McClelland deviated the understanding of competencies in terms of knowledge, skills, and attitudes and focused on a specific self-image, value, as well as traits (Li & Hongmei, 1973). According to Spenser & Spenser's theory, skills and knowledge competency can be viewed and valued, and are known as technical competency, while self-concept, attitude, and motives are more personal, and are known as behavioral competencies that are difficult to see and value (Spencer & Spencer, 1993). Hence, the conceptual research framework, as shown in Figure 1, was underpinned by the theory of performance and the theory of competency, taking into consideration the moderator effect of religiosity as follows:

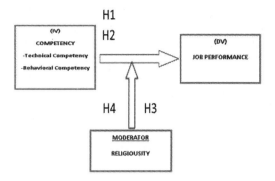

Figure 1.

3.1 *Job performance*

McCloy et al. (1994) defined work performance as a measure of habits or behaviors that contribute toward achieving the goals and objectives of an organization. In other words, performance indicates not only the outcome of the act but also the habits and actions of individuals contributing to the achievement of organizational goals. Habits and actions are under the control of the individual. It will determine and influence the actions and performance of workers (McCloy et al., 1994). Performance should be a combination of task performance, which is the performance in accordance with the task as described in the job descriptions and contextual performance, which is the willingness of the workers to perform the task outside the scope of their actual job description (Goodman & Svyantek, 1999). This combination of task and contextual performance leads to high performance and loyal employees in the organization (Michlitsch, 2000). Loyal employees will adapt to any changes in the organization. A previous study stated that the elements of organizational change have a positive impact on the job performance of employees (Khosa et al., 2015).

3.2 *Competency*

Competency is a set of knowledge, skills, behavior, attributes, and motives that must be owned, valued, mastered, and actualized by employees in carrying out their assigned duties and responsibilities (Vathanophas, Vichita; Thai-ngam, 2007). The term *competency* has been widely discussed by many researchers and it refers to the ability and capability to do a task (Vazirani, 2010). Competency is also a basic feature that recognizes the ability, efficiency, and skills possessed by an employee in performing his/her duties. It can be adapted in every job done to produce excellent work and contribute to the achievement of organizational goals (Vathanophas, Vichita; Thai-ngam, 2007). According to Spencer & Spencer (1993), competency is divided into two types: technical competency and behavioral competency. Technical competency is inclusive of knowledge and skills that are required in the field of work. Meanwhile, behavioral competency is inclusive of self-concept, nature, and motive. It will drive one's ability to improve performance in a field of work. The combination of these two types of competencies makes the worker more flexible in performing the assignment (Hamid et al., 2014).

3.3 *Relationship between competency and job performance*

Several studies have been conducted on the competency of the employees and the development of competency model and their relation to job performance. Norhanim Dewa and Sabarudin Zakaria, (2012) studied the importance of human capital in the Islamic banking industry in order to ensure the performance sustainability and competitiveness of the Islamic banking industry. The study highlighted the crucial need for education and knowledge development among the employees of the banks. It also highlighted that the employees should determine the need to undertake training and development courses as well as to identify the type of training required to enhance their job performance. Husna (2012) conducted a study to explore into the manpower competency of Bank Islam Malaysia Berhad. However, the focus was only on the selection process and training needed for BIMB. The study revealed that the employees of the bank need to undergo an effective training program and high level of competency in order to ensure the quality of services offered to the customer (Husna et al., 2012). The training and development program is considered an important factor to influence employee's performance despite other factors such as working environment and workplace conditions. The study proved that the employee who has undergone a proper training program has the competency and can thus perform better (Malik, Ahmad, Gomez, & Ali, 2011). Competencies will differentiate between well-performing workers and nonperforming workers (Irawan, 2011). Panggabean (2013) examined the influence of competency, including education level, job experience, and education background, on the work performance of the employees. The result showed that competency has significantly influenced the job performance of the employees. It has also been proven that the increment of competency level reflects the performance of the employees to perform their job (Panggabean, 2013). June, Kheng, and Mahmood, (2013) examined the relationship between competency, person-job fit, and the job performance among the employees of service SMEs in Malaysia. The study found that the employees need to have a better understanding of the job assigned to them in order for them to perform better in their job and to deliver high-quality services. In addition, the employees should possess a relevant competency level, which reflects their performance and quality of services to the customers (June, Kheng, & Mahmood, 2013). The majority of previous studies concern with the relationship between competency and job performance of the workers in several industries, but there was a lack of focus on the frontline employees of Islamic banks. As such, the first and second hypotheses for this study, as shown in Figure 1, are as follows:

Hypothesis 1: *There is a significant relationship between competency and the performance of frontline employees of Islamic banking.*

Hypothesis 2: *The element of competency has a positive relationship with the performance of frontline employees of Islamic banking.*

3.4 *Religiosity*

According to Western researchers, religiosity or religion is defined as a system of belief and practice in which a group of people interpret and respond to what they feel is supernatural and sacred. This interpretation was made by Johnstones (1975), as quoted from Khairam (2000). It is related to the system of belief and worship of the extraordinary creatures that control the power and the universe while spirituality is a personal quest to understand the answers to the ultimate questions about life and the purpose of life as well as the relationship with the most holy or transcendent (Ismail, 2013). Religion, from the Islamic perspective, can be measured, but the level of true religion is only known to God. Religion is important in any society and plays an important role in determining the pattern of life in the society. Islam is known as a religion that guides all aspects of life or ways of life. It is not just a ceremony, but more than that. Even individuals who practice true Islamic teachings will have different lifestyles and may have a religion but do not practice it (Dr. Rusnah Muhamad, 2008). A previous study highlighted the importance of religion and spirituality in human life. Studies were conducted on the unethical issues of religion and business practices (Dr. Rusnah Muhamad, 2008), religious influence on behavior (Dubinsky, 2014; Olowookere, 2014), religious influence on Islamic banking (Echchabi & Abd. Aziz, 2012), religious influence on work performance (Inasoria, 2014), and religious role as moderator (Ismail, 2013; Achour, Mohd Nor, & MohdYusoff, 2016).

3.5 *Relationship between religiosity and performance*

Extensive studies have been conducted to examine religious and spiritual influences in the workplace and important elements in influencing the behavior of the employees. Believing in the power of God who possesses the natural and spiritual power has a significant influence on the motivation and performance of the employees. The study found a positive correlation between prayer and motivation and the factors of commitment to the organization's commitment and self-esteem as well as a positive attitude toward the success of the organization (Zulkifli, 2013). Previous studies concerned with the effect of religiosity on the job performance of employees. On the basis of literature review (140 articles), Karakas (2009) identified three different perspectives regarding the spirituality and organizational performance. These three perspectives spirituality improve the welfare of workers and the quality of life and also bring employees closer to organizations and communities (Karakas, 2009). Religiosity has a significant influence on work ethic. Employees who have high level of religiosity tend to be more ethical in their manner at the workplace. They will have high level of awareness to differentiate between good and bad behaviors (Abdi, Fatimah, Wira, Nor, & Radzi, 2014). The study found that intrinsic religious orientation has a significant positive influence on emotional intelligence, but extrinsic religious orientation has a negative influence on the emotional intelligence (Krauss, Hamzah, & Idris, 2007). With intrinsic orientation, the employees will be keener to make extra effort to work harder and perform better in the workplace. The present study will explore in-depth the level of religiosity among the frontline employees in Islamic banks, which influences their job performance. In addition, the study would explore into the moderator effect of religiosity on the relationship between job competency and performance. As such, the third and fourth hypotheses, as shown in Figure 1, are as follows:

Hypothesis 3: *The level of religiosity has a significant relationship toward the job performance among the frontline employees of Islamic banking*

Hypothesis 4: *Religiosity moderates the relationship between competency and work performance of frontline employees of Islamic banking.*

4 METHODOLOGY

In this study, we used the survey method for data collection, which made use of a self-administered questionnaire. Data will be collected from 280 respondents. The unit of analysis is the frontline employees at the branch level of Islamic banks in Malaysia inclusive of the manager, marketing staff, front desk staff, and customer service officers. They have direct contact of service with the customers. The performance of frontline employees would be much complex within a service sector context, especially financial services such as banking. They are also the ones that sell services and also responsible for post sales services such as answering customers' questions or adjusting policies based on customer demands (Julia M, 2009). One of the measurements for the effectiveness of frontline employees is the loyalty of the customers to repeat their transactions with the bank once they are satisfied with the services offered. The previous study revealed that there is a positive correlation between the competency of employees and customer satisfaction (Mengesha, 2015). The study that adopted the measurement of religiosity was conducted by Eric and Krauss (2011). The scale *Muslim Religiosity-Personality Inventory (MRPI)* was developed to measure the accuracy and validity of religiosity of Muslims in Malaysia from two important aspects, namely Islamic world view, which reflects the divine doctrine of the belief over the oneness of god, and the elements of religiosity personality, which represent the man-to-man and man-to-God relationships. The score will be provided for each religiosity aspect in order to develop the religiosity profile for each of the respondents (Eric & Krauss, 2011).

5 RECOMMENDATIONS AND CONCLUSION

The study aims to offer itself as guidance and sheds light on its benefits to the policy makers and the Islamic banking institutions in either Malaysia or other countries to have a better plan on formulating a retaining program for their employees by looking into the competency level of employees, particularly the frontline employees. In fact, the success measurement of the Islamic banking industry depends on not only the monetary aspect but also the overall performance, which includes the human capital and the acceptance of the public toward the products and services (Vazirani, 2010). The study also gives some feedback to the authority, learning institutions as well as the industry to enhance their module and program by assessing the real performance of the employees in the field.

Previous studies have highlighted the important role of religiosity to develop attitudes and behavior of individuals (Osman-gani, Hashim, & Ismail, 2007). Thus, this study aimed to look into the effect of religiosity on the competency and performance of the employees and to ensure that the employees of the Islamic banking system, especially the frontline staff, portray the real image of Islam, especially in terms of distinguishing the Islamic banking system from the conventional banking system. Finally, the study contributed to the existing literature in predicting the influence of competency on the job performance of frontline employees in Islamic banking sector and focused on the religiosity as the moderator.

REFERENCES

Abdi, M.F., Fatimah, S., Wira, D., Nor, M., & Radzi, N.Z. (2014). The Impact of Islamic Work Ethics on Job Performance and Organizational Commitment, 1–12.
Abduh, M., & Alias, A. (2014). Factors Determine Islamic Banking Performance in Malaysia: A Multiple Regression Approach. *Journal of Islamic Banking and Finance*, Jan(March), 44–54.
Abedniya, A., & Zaeim, M.N. (2011). Measuring the perceive service Quality in the Islamic Banking System in, *2*(13), 122–135.
Achour, M., Mohd Nor, M.R., & MohdYusoff, M.Y.Z. (2016). Islamic Personal Religiosity as a Moderator of Job Strain and Employee's Well-Being: The Case of Malaysian Academic and Administrative Staff. *Journal of Religion and Health*, *55*(4), 1300–1311.
Ahmad, S., & Schroeder, R.G. (2003). Management practices on operational he impact of human resource performance: recognizing country and industry differences, *21*, 19–43.
Arshad, M.U., & Tahir, M.S. (2016). Issues in transformation from conventional banking to Islamic banking. *International Journal of Economics and Financial Issues*, *6*(3), 220–224.
Awan, H.M., & Bukhari, K.S. (2011). Customer's criteria for selecting an Islamic bank: evidence from Pakistan.
Azizan, A. (2011). An empirical study on banks' clients' sensitivity towards the adoption of Arabic terminology amongst Islamic banks.
Azmi, I.A.G. (2015). Islamic human resource practices and organizational performance: Some findings in a developing country.
Dr. Rusnah Muhamad, (2008). The Influence of Religiosity and Perceptions of Unethical Business Practices: An Empirical Investigation of the Malaysian Muslims Dr. Rusnah Muhamad, 1–12.
Dubinsky, M.K.O.O.A.J. (2014). Influence of religiosity on retail sales people's ethical perceptions: the case in Iran.
Echchabi, A., & Abd. Aziz, H. (2012). The Relationship between Religiosity and Customers' Adoption of Islamic Banking Services in Morocco. *Oman Chapter of Arabian Journal of Business and Management Review*, *1*(10), 89–94. doi:10.12816/0002190.
Eric, S., & Krauss. (2011). 2011 The Muslim Religiosity-Personality Inventory (MRPI) Scoring Manual, 0–20.
Goodman, S.A., & Svyantek, D.J. (1999). Person-organization fit and contextual performance: Do shared values matter. *Journal of Vocational Behavior*, *55*(2), 254–275.
Husna, N., Nor, S., Hj, N., Japar, M., Bakar, A.A., & Abdullah, S. (2012). Manpower Competency in Bank Islam Malaysia Berhad, *1*, 23–37.
IInasoria, R.D.C. (2014). Effects of Profile, Religiosity and Job Attitude on the Job Performance of the Philippine National Police: The Case of Bulacan Province, *2*(5), 22–32.
Irawan, I. (2011). The design of Spencer generic competency as a model for banking supervisors position specification in Surabaya. *Journal of Economics, Business, and Accountancy …*, *14*(3), 217–224.
Ismail, (2013) Establishing linkages between religiosity and spirituality on employee performance.
Julia M., David M., (2009) Competency-based Systems for Frontline Employees, *IPIC/IPMA-HR Conference, Nashville, TN*, 14–17.

June, S., Kheng, Y.K., & Mahmood, R. (2013). Determining the Importance of Competency and Person-Job Fit for the Job Performance of Service SMEs Employees in Malaysia, *9*(10). doi:10.5539/ass.v9n10p114.

Kamariah, N., Mat, N., Mujtaba, A.M., Al-refai, A.N., Badara, A.M., & Abubakar, F.M. (2013). Direct Effect of Service Quality Dimensions on Customer Satisfaction and Customer Loyalty in Nigerian Islamic, *3*(1), 6–11.

Karakas, F. (2009). Spirituality and Performance in Organizations: A Literature Review, *44*(0), 1–43.

Khattak, N.A. (2010). Customer satisfaction and awareness of Islamic banking system in Pakistan, *4*(May), 662–671.

Khosa, Z.A., Rehman, Z.U., Asad, A., Bilal, M.A., Hussain, N., & Scholars, M.P. (2015). The Impact of Organizational Change on the Employee's Performance in the Banking Sector of Pakistan. *IOSR Journal of Business and Management Ver. II*, *17*(3), 2319–7668. doi:10.9790/487X-17325461.

Lamba, G.K.J.S.T. (2012). Exploring the impact of total quality service on bank employees' organisational commitment.

Loo, M. (2010). Attitudes and Perceptions towards Islamic Banking among Muslims and Non-Muslims in Malaysia: Implications for Marketing to Baby, *3*(June 2005), 453–485.

Malik, M.I., Ahmad, A., Gomez, S.F., & Ali, M. (2011). A study of work environment and employees' performance in Pakistan. *African Journal of Business Management*, *5*(34), 13227–13232. doi:10.5897/AJBM11.1502.

McCloy, R.A., Campbell, J.P., & Cudeck, R. (1994). A confirmatory test of a model of performance determinants. *Journal of Applied Psychology*, *79*(4), 493–505. doi:10.1037/0021-9010.79.4.493.

Mengesha, A.H. (2015). Effects of marketing competency of frontline employees on customer satisfaction: A study on commercial bank of Ethiopia, *1*(4), 39–50.

Michlitsch, J.F. (2000). High-performing, loyal employees: the real way to implement strategy. *Strategy & Leadership*, *28*(6), 28–33.

Mohd Ali, Zakiah Mohamed, Shahida Shahmi. (2015). Competency of Shariah Auditors in Malaysia: Issues and Challenges. *Journal of Islamic Finance*, *4*(1), 22–30.

Norhanim Dewa, & Sabarudin Zakaria. (2012). Training and Development of Human Capital in Islamic Banking Industry. *Journal of Islamic Economics, Banking and Finance*, *8*(1), 95–108.

Olowookere, E.I. (2014). Influence of Religiosity and Organizational Commitment on Organizational Citizenship Behaviours: A Critical Review o f Literature, *1*(3), 48–63.

Osman-gani, A.M., Hashim, J., & Ismail, Y. (2007). Effects of Religiosity, Spirituality, and Personal Values on Employee Performance: A Conceptual Analysis, (2003).

Panggabean, N.N. (2013). Pengaruh Kompetensi Terhadap Prestasi, *1*(2), 104–113.

Rafay, A., & Sadiq, R. (2015). Problems and issues in transformation from conventional banking to Islamic banking: Literature review for the need of a comprehensive framework for a smooth change. *City University Research Journal*, *5*(02), 315–326.

Razzaque, J.K.F.M.A. (2014). Service quality and satisfaction in the banking sector. doi:10.1108/IJQRM-02-2013-0031.

Saad, N.M. (2012). Comparative analysis of customer satisfaction on Islamic and conventional banks in Malaysia. *Asian Social Science*, *8*(1), 73–80.

Sayani, H., & Miniaoui, H. (2013). Determinants of bank selection in the United Arab Emirates. *International Journal of Bank Marketing*, *31*(3), 206–228.

Sintok, U.U.M., Aman, K.D., & Alqahtani, A.H.M. (2012). Employees' Perception and Organizational Commitment: A Study on the Banking Sector in Gaza, Palestine Graduate School of Business, *3*(16), 299–312.

Spencer, L.M., & Spencer, S.M. (1993). Competence at Work: Models for Superior Performance. *John Wiley & Sons*, 1–372.

Vasiliauskien, L., Stanik, B., Lipinskien, D., Nord, a B., & Lietuva, L.B. (2005). The Employees' Competence Development inside Organization: Managerial Solutions, *5*(5).

Vathanophas, Vichita; Thai-ngam, J. (2007). Competency Requirements for Effective Job Performance in The Thai Public Sector. *Contemporary Management Research*, *3*(1), 45–70.

Zulkifli, R.M. (2013). Entrepreneurial Orientation and Business Success of Malay Entrepreneurs: Religiosity as Moderator, *3*(10), 264–275.

Role of communication in IT project success

K. Shasheela Devi, G. Marthandan & K. Rathimala
Multimedia University (MMU), Cyberjaya, Malaysia

ABSTRACT: Many organizations rely on appropriate project management, and communication management is a primary area for project realizations. There is consistently high project failure rate in the IT industry, and past studies mention communication as a key factor for project success or failure. Hence, this research on the role of communication in IT project success empirically endured. The Media Synchronicity Theory (MST) and the Triple C model are adapted in this study, where the characteristics of project communication and project success are identified, followed by the assessment of the relationship between project communication and project success. The impact of communication performance and project management on project success is established. Questionnaire data from 127 respondents were analyzed using the Partial Least Squares Structural Equation Modeling (PLS-SEM) method. Empirical evidence on the role of communication in IT project success contributes to the body of knowledge and industry of project management.

Keywords: Project management, communication, project success

1 INTRODUCTION

Project management is highly embraced in IT across industries (Schwalbe et al., 2011). Project management growth projected for project-intensive industries and IT is one of the project-intense industries (PMI Pulse, 2015, Thomas, 2014, PricewaterhouseCoopers, 2012). Organizations achieve project performance with project management in place, to meet strategic business objectives and financial savings (PMI Pulse, 2015).

On the contrary, reports still indicate high failure rates of IT projects (PMI Pulse, 2015, Verner et al, 2014, Bloch, Blumberg, & Laartz, 2012, Geneca, 2011). The Standish Chaos Report cited by many researchers reported that the success rate of IT projects is only 29% on average. It was reported that the current project failure rates are higher than those 5 years ago (The Standish Group Report, 2014).

Studies also reported that one of the key factors for project failures is poor communication (PMI Pulse, 2015, Bond-Barnard et al., 2013, Marnewick, 2012, Maidin, 2010). The Bull Survey in 1998 established the major causes for project failures, with poor communication in the first place (Bull Study, 1998). Ineffective communication between vendors and users is the main cause of IT project failures in Malaysia (Haslinda et al., 2011). The fact is that communication cuts across all the project management processes and knowledge areas, which emphasizes the importance of communication in project management framework by PMI and PRINCE2 (PMI, 2013 & de Carvalho, 2014). Communication is the heart of project management that drives project success (Sarhadi, 2016).

Prior studies established that the importance of communication in project management is very much exploratory (Marnewick, 2012, Hairul et al., 2011, Maidin, 2010). The consistently high project failure rates in the IT industry drives this study to explore further on project success. The existence of communication across all project management phases and being a key factor for project success triggers this study to unleash the role of communication in IT project success empirically. Furthermore, studies related to project management are scarce in Malaysia; thus, this study focuses on the IT industry in the Malaysian context. The MST and the Triple C model are adapted into this research of project management.

This explanatory study aims to (1) explore the characteristics of communication in project, (2) establish the relationship between the characteristics of communication and project success, (3) determine if communication performance mediates between the communication characteristics and project success, and (4) determine if project management mediates between communication performance and project success.

2 LITERATURE REVIEW

2.1 Media Synchronicity Theory (MST)

The MST established that media fit improves communication performance, which then leads to project success (Wang et al., 2016 & Dennis et al., 2008). The conceptual definitions and operational measures for media capabilities in the theory are still under development stage that provides opportunities for further empirical testing and validation (He & Yang, 2016).

2.2 Triple C model

The Triple C model in project management has communication as the first "C" that drives project success. Referring to the Triple C model, communication should take into account the what, why, who, when, where, and how? (Badiru, 2008). These are the characteristics of communication.

2.3 Communication characteristics

Referring to the Triple C model, as mentioned in Section 2.2, communication characteristics from different fields of studies were adopted into this research. The characteristics for this study are communication timing (An et al., 2015, Yean et al., 2014), communication frequency (Yean et al., 2014, Kennedy et al, 2011), and communication velocity (Boyanov, 2012, Leary, 2011, Dennis et al., 2008). The communication timing is the measure of the project information communicated in a timely manner (Ling & Ma, 2014, Christensen, 2014, Wang & Anumba, 2009). Communication frequency is defined as the number of times communication takes place (Ling & Ma, 2014, Muller, 2003). Communication velocity is the speed of information transferred or communicated within a project (Abdessameud et al., 2015, Boyanov, 2012).

2.4 Communication performance

Communication performance is about achieving a shared understanding when a sender transmits information and the receiver receives the information. The linkage between communication performance and project success requires further study (Dennis et al, 2008). Communication tied to business strategies, should be set with measurable targets to be evaluated (Zerfass & Sherzada, 2014). Communication performance can be measured based on the communication products (output), communication processes, stakeholder satisfaction (interaction), and the results (impact) (Vos, 2015).

2.5 Project management

Project management is attained by applying and integrating project management processes, which are categorized into five process groups: initiation, planning, execution, monitoring and controlling, and closing to meet project requirements throughout the project's life cycle. Project management is the first factor affecting project success (Ozguler, 2016). Past studies have shown some evidence that value sought from a high-performing project management system is associated with the success of projects (Joslin & Muller, 2015, Ahonen et al, 2015, Asad & Pinnington, 2014).

2.6 Project success

PMI has defined project success as product and project quality, meeting schedule, compliance to budget, and customer satisfaction. P2M has defined project success as physical product or service delivered with novelty, differentiation, and innovation. APM have defined project success as level of satisfaction by stakeholders and measured by criteria set and agreed at the initial stage of a project. PRINCE2 has stated why project objectives have to be achieved (Marnewick, 2012).

For decades, project success was measured by the solid triple constraints: cost, time, and scope/quality. Newer indicators like stakeholder satisfaction and business success were identified as project success indicators in several recent studies (Joslin & Müller, 2015, Eigbe et al., 2015, Marnewick, 2012, Tmeemy et al., 2011, Papke-Shields et al., 2010).

2.7 Research model and hypothesis development

Communication is the key to the success or failure of a project (Hodgkinson, 2009). This is mentioned in prior studies, but project failures are still highlighted by the industry practitioners, which triggered research on the role of communication in IT project success. This established the main gap for this research to endure. Considering this idea and observation into the communication theories, MST was the suitable theory measuring project success. However, the MST was addressing on a single characteristic, the communication media's effect on project success (Wang et al, 2016, Min, 2012). The Triple C model indicates several communication characteristics that are important for project management. Following further literature review, communication is vital for management functions (Dupe, 2015). Communication management is necessary for successful communication in projects (Muszynska, 2015, Caltrans, 2007). Further studies on the correlation between communication management and project performance are recommended (de Carvalho, 2014). Both the MST and the Triple C model were adapted to suit the objective of this research to study the role of communication in IT project success according to the research model in Figure 1, and the following hypotheses are proposed:

H1: Timely communication improves communication performance

H2: Frequent communication improves communication performance

H3: Faster communication velocity improves communication performance

A study comprising all the characteristics of communication against project management was a gap found during the literature review phase and adopted for this study. One new characteristic adopted was velocity; this is common in the communication studies in the engineering field but new for project management. The communication media capabilities in MST replaced with communication characteristics identified from various previous studies should be tested empirically in this research model.

The communication performance exists within the MST, and the relationship between communication performance and project success requires further study (Dennis et al., 2008). Thus, this gap considered, and the communication performance adopted as a mediator between communication characteristics and project success forms the following hypotheses:

H4: Communication performance mediates between communication characteristics and project success

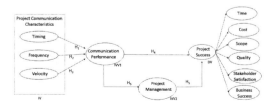

Figure 1. Research model.

Project management being the studied field and associated with project success in prior studies (Joslin & Muller, 2015, Ahonen et al, 2015) proposes the H5 hypotheses:

H5: Project management mediates between communication performance and project success

3 RESEARCH METHODOLOGY

Questionnaire is used as the main data collection instrument, which is designed based on past literature and modified accordingly to suit this research. A pilot study inclusive of validity and reliability test conducted with 20 respondents.

The questionnaire survey used ordinal scale for demographic data collection and interval scale (7-point Likert scale) for remaining data collection. The 7-point scale is the optimum rating scale and high in reliability. A scale higher than 7-point demands discrimination and a scale lower than 7-point suffers loss of reliability due to coarseness of grouping (Symonds, 1924). A recent study supported that 7-point Likert scale provides accuracy of measure and is appropriate for electronically distributed or unsupervised usability questionnaires (Finstad, 2010).

Data were collected from IT professionals in IT projects within Malaysia. Out of 200 questionnaires distributed online and manually to the IT industry-working professionals, 135 responses were obtained. Excluding the incomplete questionnaires, only 127 were found to be usable, which records a response rate of 63.5%.

4 RESEARCH FINDINGS

4.1 Data analysis

Structural equation modeling (SEM) has become a "must" statistical technique for researchers in the social sciences (Hooper et al, 2008). SEM analysis is the most effective way to test mediated effects. SEM can concurrently test all the relationships in the model, be it the loading of observed indicators to the latent constructs, the paths from independent variable to the mediator and the dependent variable, and the path from the mediator to the dependent variable. SEM also has an advantage of dealing with measurement errors. SEM is able to provide higher level of validity and reliability (Li, 2011, Gefen et al, 2011).

This research uses partial least square (PLS) SEM for data analysis. PLS-SEM is a second-generation multivariate statistical technique used to test and estimate causal relationship among multiple independent and dependent constructs (Urbach & Ahlemann, 2010).

4.2 Construct reliability and validity

The model was measured using the SmartPLS 3.0 tool. Composite reliability and the Cronbach's alpha were measured to determine the model's reliability. According to Table 1, the composite reliability for all the variables are greater than 0.70 indicating high reliability. The Cronbach's alpha values for all the variables are greater than 0.70, except for "timing", which stands at 0.672, which is an acceptable level, indicating that internal quality is achieved.

Followed by the validity test by measuring the average variance extracted (AVE) and the indicator loadings. The AVE for all the variables are greater than 0.50, which means that a good internal quality model is achieved (Hair et al., 2017).

Discriminant validity ensures that a reflective construct has the strongest relationship with its own indicators compared to other constructs in the PLS path model (Hair et al., 2017). In a recent study, the heterotrait–monotrait ratio of correlations (HTMT) approach shows superior performance to Fornell–Larcker or cross-loadings in assessing discriminant validity Discriminant validity is achieved if the HTMT values are lower than 0.9 (Henseler et al., 2015). This research model has achieved discriminant validity as all the HTMT results are lower than 0.90 according to Table 2.

4.3 Structural model testing results

The research model was tested with structural equation modeling (SEM). As shown in Figure 2, the estimations are significantly different for the structural model (*p < 0.05 and ***p < 0.001).

All the paths in the PLS analysis are positively significant at level 0.05, except for path "frequency – communication performance", with a t-value of 0.564, which is lower than 1.96. This "frequency – communication performance" path is not significant.

Thus, all hypotheses are supported except for H2 is rejected. The model fit (R2) shows 48.9%, 47.9%, and 63.4% of the communication performance, project management, and project success.

Table 1. Cronbach's alpha, composite reliability, and AVE results.

	Cronbach's alpha	Composite reliability	AVE
Business success	0.924	0.941	0.727
Communication performance	0.787	0.862	0.61
Cost	0.76	0.849	0.587
Frequency	0.806	0.868	0.572
Project management	0.86	0.893	0.545
Project success	0.96	0.964	0.511
Quality	0.905	0.93	0.726
Scope	0.734	0.853	0.669
Stakeholder Satisfaction	0.877	0.917	0.735
Time	0.903	0.934	0.781
Timing	0.672	0.803	0.513
Velocity	0.838	0.883	0.561

Table 2. HTMT results.

Communication performance	Frequency	Project management	Project success	Timing	Velocity
Frequency	0.596				
Project management	0.822	0.629			
Project success	0.74	0.503	0.855		
Timing	0.718	0.853	0.784	0.601	
Velocity	0.814	0.67	0.783	0.621	0.706

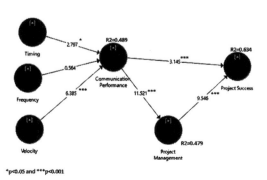

Figure 2. Structural model results.

5 MANAGERIAL IMPLICATIONS

The findings of this research offer some insights into the theoretical and practical implications in the field of project management. The empirical evidence of this research supports four key findings. As hypothesized, the findings show that (1) timely communication improves communication performance; (2) faster communication velocity improves communication performance; (3) communication performance mediates between communication characteristics and project success; and (4) project management mediates between communication performance and project success. However, one hypotheses is not supported, which indicates that frequent communication does not improve communication performance. This is a contradiction to the initial expectation of this research. This could be due to different work culture and environment in project settings in Malaysia. Supporting this statement in a study on communication in distributed projects with different time zones, noted communication occurs more often in small time zone range for distributed software projects. The reply time for emails in projects with small time zone range are shorter than with large time zone (Nordio et al., 2011). Too frequent communication might also impact the project timeline negatively and becomes a waste of time as it needs much effort for communication to take place in projects. A study suggested keeping a goal of reducing average time spent on communicating without compromising the quality to reduce ineffectiveness (Hodgkinson, 2009).

Looking into project postmortems, a reference to communication issue will be found in some form or manner. As the project team expands, the line of communication also increases, which requires extra effort within the team to communicate for project effectiveness (Hodgkinson, 2009). The communication characteristics have to be managed well to achieve common understanding to achieve project success. Despite the existence of many principles and methodologies of communication management with supporting tools, project failures still occur due to improper communication management. A good relationship and personal contacts within team members is required within communication management to ensure project success (Muszynska, 2015). Understanding the skills and techniques to communicate according to a person's emotional condition is essential for project success (Greenberger, 2016). A wide spectrum of prior project management studies exists in academic and industrial practice, inclusive but not limited to the tools and concepts of project management, the project management maturity, control factors such as industry, project complexity, country and organization culture vs. project management components: enablers, trainings, and leadership. From the communication perspective, its importance in organization change, locally and geographically distributed team, and project management does exist. A proper communication structure ensures successful communication management and project management (Lundberg & Seglert, 2011).

6 CONCLUSION AND FUTURE WORKS

The research objective was achieved, that is, communication does play a role in the success of IT projects. This research has also confirmed communication velocity as a new significant characteristic in communication management for successful project management. The findings of this research contribute to the body of knowledge in project management and can be applied in the industrial project management practices to improve project success.

Although the empirical findings contribute to significant theoretical and practical insights, several limitations should be considered. First, the communication characteristics were tested to be limited to communication timing, communication frequency, and communication velocity. Other characteristics such as communication method and communication technology can be adopted into this model for future studies. Second, this research focused on the success rate of IT projects only. Perhaps future studies using this model could be on other project-intense fields like construction. Third, the subject of this study only covered the geographical context of Malaysia. However, this study should be conducted in other countries to determine how this model applies in different cultural and environment settings as the future direction of this research.

REFERENCES

Abdessameud, A., Polushin, I. G., & Tayebi, A. (2015). Synchronization of nonlinear systems with communication delays and intermittent information exchange. *Automatica*, 59, 1–8. doi:10.1016/j.automatica.2015.05.020.

Ahonen, J. J., Savolainen, P., Merikoski, H., & Nevalainen, J. (2015). Reported project management effort, project size, and contract type. *Journal of Systems and Software*, 109, 205–213. doi:10.1016/j.jss.2015.08.008.

Al-Tmeemy, S. M. H. M., Abdul-Rahman, H., & Harun, Z. (2011). Future criteria for success of building projects in Malaysia. *International Journal of Project Management*, 29(3), 337–348. doi:10.1016/j.ijproman.2010.03.003.

An, Z., Li, D., & Yu, J. (2015). Firm crash risk, information environment, and speed of leverage adjustment ☆. *Journal of Corporate Finance*, 31, 132–151. doi:10.1016/j.jcorpfin.2015.01.015.

Asad, F., & Pinnington, A. H. (2014). ScienceDirect Exploring the value of project management: Linking Project Management Performance and Project Success. *JPMA*, 32(2), 202–217. doi:10.1016/j.ijproman.2013.05.012.

Badiru, (2008). Triple C Model of Project Management; Communication, Cooperation, and Coordination. CRC Press, Taylor & Francis Group, NW.

Bloch, M., Blumberg, S., & Laartz, J. (2012). Delivering large-scale IT projects on time, on budget, and on value. *McKinsey & Company*, (June), 1–6.

Bond-Barnard, T. J., Steyn, H., & Fabris-Rotelli, I. (2013). The impact of a call centre on communication in a programme and its projects. *International Journal of Project Management*, 31(7), 1006–1016. doi:10.1016/j.ijproman.2012.12.012.

Boyanov (2012). On the Measurement of Quantity of Information on Speech, Sound and Image and their Link with the Information Processing. ACM: HCCE'12, March 8–13, 2012, Aizu-Wakamatsu, Fukushima, Japan.

Bull Survey 1998. Business Review, Vol. 27 Iss 1 pp. 80–99.

Caltran (2007). Project Communication Handbook Second Edition. Sacremanto: Office of Project Management Process Improvement.

Christensen, M. (2014). Communication as a Strategic Tool in Change Processes. *International Journal of Business Communication*, 51(4), 359–385. doi:10.1177/2329488414525442.

De Carvalho, M. M. (2014). An investigation of the role of communication in IT projects. *International Journal of Operations & Production Management*, 34(1), 36–64. doi:10.1108/IJOPM-11-2011-0439.

Dennis, A. R., Fuller, R. M., & Valacich, J. S. (2008). Media, Tasks, and Communication Processes: A Theory of Media Synchronicity. *MIS Quarterly*, 32(3), 575–600. doi:10.2307/25148857.

Dupe Adesubomi Abolade (2015). Organisational Communication, The Panacea for Improved Labour Relations. *DE Gruyter, Studies in Business and Economics no. 10(2)/2015*. Doi:10.1515/sbe-2015-0016.

Eigbe, A. P., Sauser, B. J., & Felder, W. (2015). Systemic analysis of the critical dimensions of project management that impact test and evaluation program outcomes. *International Journal of Project Management*, 33(4), 747–759. doi:10.1016/j.ijproman.2014.09.008.

Finstad, K. (2010). Response Interpolation and Scale Sensitivity: Evidence Against 5-Point Scales. *Journal of Usability Studies*, 5(3), 104–110.

Gefen, D., Rigdon, E. E., & Straub, D. (2011). An Update and Extension to SEM Guidelines for Administrative and Social Science Research. *MIS Quarterly*, 35(2), iii–A7.

Geneca, (2011). Industry Report: Doomed from the Start? Geneca LLC.

Greenberger, L. S. (2016), Effective Communications for Project Success. Remediation, 26: 121–128. doi:10.1002/rem.21463.

Hairul, M., Nasir, N., & Sahibuddin, S. (2011). Critical success factors for software projects: A comparative study, 6(10), 2174–2186. doi:10.5897/SRE10.1171.

Haslinda et al. (2011). Government ICT Project Failure Factors: Project Stakeholders' Views. Journal of Information Systems Research and Innovation.

He & Yang (2016). Using Wikis in Team Collaboration: A Media Capability Perspective. Elsevier: Information & Management (2016).

Henseler et al. (2015). A New Criterion For Assessing Discriminant Validity in Variance-Based Structural Equation Modeling. *J. of the Acad. Mark. Sci. (2015) 43:115–135*. doi: 10.1007/s11747-014-0403-8.

Hodgkinson Jeff (2009). Communication is the Key to Project Success. *IPMA USA*.

Hooper, D., Coughlan, J., & Mullen, M. (2008). Structural Equation Modelling: Guidelines for Determining Model Fit Structural equation modelling: guidelines for determining model fit, 6(1), 53–60.

Joslin, R., & Müller, R. (2015). Relationships between a project management methodology and project success in different project governance contexts. *International Journal of Project Management*, *33*(6), 1377–1392. doi:10.1016/j.ijproman.2015.03.005.

Kennedy D.M. et al. (2011). An Investigation of Project Complexity's Influence on Team Communication Using Monte Carlo Simulation. *Journal of Engineering & Technology Management* 28 (2011) 109–127.

Leary, D. E. O. (2011). The Use of Social Media in the Supply Chain: Survey and Extensions, *144*, 121–144. doi:10.1002/isaf.

Li, Spencer. (2011). Testing Mediation Using Multiple Regression and Structural Equation Modeling Analyses in Secondary Data. *Sage Publications Ltd., London.*

Ling Yean Yng & Ma Yuan. (2014). Effect of Competency and Communication on Project Outcomes in Cities in China. *Elsevier-Habitat International 44(2014) 324–331.*

Lundberg & Seglert. (2011). Project Communication for Successful Product Development—Developing a Project Overview at ITT W&WW. *Master of Science Thesis—KTH Industrial Engineering and Management.*

Maidin, S. S. (2010). IT Governance Practices Model in IT Project Approval and Implementation in Malaysian Public Sector, *1*(Iceie 2010), 532–536.

Marnewick, C. (2012). A Longitudinal Analysis of {ICT} Project Success.

Min, Q. (2012). Social Communication Technology, (71072108), 360–363. doi:10.1109/ICMeCG.2012.38.

Muszynska Karolina. (2015). Communication Management in Project Team—Practices and Patterns. *TIIM—Joint Conference 2015.*

Nordio, M., Estler, H. C., Meyer, B., Tschannen, J., Ghezzi, C., & Nitto, E. Di. (2011). How Do Distribution and Time Zones Affect Software Development? A Case Study on Communication. *2011 IEEE Sixth International Conference on Global Software Engineering*, 176–184. doi:10.1109/ICGSE.2011.22.

Ozguler (2016). Increase the projects' success rate through developing multi-cultural project management process. Procedia—Social and Behavioral Sciences 226 (2016) 236–242. doi:10.1016/j.sbspro.2016.06.184.

Papke-Shields, K. E., Beise, C., & Quan, J. (2010). Do project managers practice what they preach, and does it matter to project success? *International Journal of Project Management*, *28*(7), 650–662. doi:10.1016/j.ijproman.2009.11.002.

PMI, (2013). A Guide to the project management body of knowledge (PMBOK) 5th Edition. Project Management Institute Inc.

PMI Pulse, (2015). Capturing the Value of Project Management. Project Management Institute Inc.

PricewaterhouseCoopers. (2012). Insights and Trends: Current Portfolio, Programme, and Project Management Practices, 1–40.

Sarhadi (2016). Comparing Communication Style Within Project Teams of Three Project-Oriented Organizations in Iran. *Procedia-Social Behavioral Sciences* 226 (2016) 226–235.

Schwalbe, K., Hulbert, J., Calhoun, J., Sabatino, P. J., Shipp, L., Art, S., & Jenkins, S. (2011). Information Technology Project Management, REVISED Sixth Edition. Information Technology Project Management, REVISED Sixth Edition.

Tai, S., Wang, Y., & Anumba, C. J. (2009). A survey on communications in large-scale construction projects in China. *Engineering, Construction and Architectural Management*, *16*(2), 136–149. doi:10.1108/09699980910938019.

The Standish Group Report. (2014). *The Standish Group Report—Chaos.*

Thomas, P. (2014). *The state of the Project Management Office (PMO)*. Project Management Solutions Inc.

Urbach, N., & Ahlemann, F. (2010). Structural Equation Modeling in Information Systems Research Using Partial Least Squares. *Journal of Information Technology Theory and Application (JITTA)*, *11*(2), 5–40.

Verner, J. M., Babar, M. a., Cerpa, N., Hall, T., & Beecham, S. (2014). Factors that motivate software engineering teams: A four country empirical study. *Journal of Systems and Software*, *92*(1), 115–127. doi:10.1016/j.jss.2014.01.008.

Verčič, D., Zerfass, A., & Wiesenberg, M. (2015). Global public relations and communication management: A European perspective. *Public Relations Review*, (July). doi:10.1016/j.pubrev.2015.06.017.

Vos. (2015). Communication Health Check—Measuring Corporate Communication Performance. *Journal of Business Studies Quarterly* 2015, Volume 7, Number 1.

Wang et al. (2016). How Social Media Applications Affect B2B Communication and Improve Business Performance in SMEs. *Industrial Marketing Management* 54 (2016) 4–14.

Yean, F., Ling, Y., & Ma, Y. (2014). Effect of competency and communication on project outcomes in cities in China. *Habitat International*, 44, 324–331. doi:10.1016/j.habitatint.2014.07.002.

Pupils' self-regulated skills in relation to their perceived position in formal and informal school life processes

K. Hrbackova
Department of Pedagogical Sciences, Faculty of Humanities, Tomas Bata University in Zlín, Zlín, Czech Republic

ABSTRACT: The paper focuses on assessing the extent to which pupils in the upper-primary school level use self-regulation skills related to interpersonal cognitive problem-solving. The paper also aims to identify the extent to which self-regulation in pupils is dependent on their perceived position in formal and informal school life processes; such processes may play a supportive or endangering role in their learning. A multiple linear regression was employed to determine to what extent the perceived position in formal and informal processes of school life predict self-regulation in pupils (N = 1,133). The results showed that experiencing success and opportunity, pupils' social inclusion in peer groups, and teacher's positive attitude towards pupils are the most relevant predictor of how pupils apply their self-regulation skills. Negative experiencing did not explain any additional variance in the final scores.

1 INTRODUCTION

1.1 *The meaning of self-regulation*

Understanding behavioural self-regulation in pupils represents an important way of getting to know and understanding the causes, processes and consequences of this behaviour. We regard the social context of self-regulation as significant. The school environment, in particular, plays an important role in the development of pupils' self-regulation, because it is in this very environment that they encounter many life situations (arising from interpersonal relationships) to which they must react. We regard the ability to regulate own behaviour as one of the key prerequisites of a relationship with oneself and with others.

1.2 *Pupils' self-regulation*

Self-regulation is the internally controlled ability to regulate emotions, attention and behaviour in response to the demands of the external and internal environment. These abilities emerge and develop during childhood and adolescence, and clearly contribute to the individual's higher quality of life (Raffaelli, Crockett & Shen, 2005). It is a highly individual concept, based on goal-setting, planning, persistence, and influencing the environment, behaviour, emotions and attention (Rothbart & Posner, 2005).

Self-regulation relates mainly to cognitive and behavioural processes, through which the individual maintains (and if possible raises) the level of their emotional, motivational and cognitive reactions, which lead to a positive adaptation in the area of social relationships, success and a positive perception of oneself (Blair & Diamond, 2008). It can also be said that self-regulation affects the individual's ability to make a correct assessment related to other people (Gilbert, Krull & Pelham, 1988).

Self-regulation requires the ability to observe and control own behaviour, emotions and thoughts, and change them in accordance with the requirements of the specific situation. It is the internal process of controlling oneself, during which a person transforms their abilities into the skills necessary to manage specific situations (Vávrová, Hrbáčková & Hladík, 2015).

The self-regulation process aims to maintain control over oneself, i.e. it includes the ability to maintain inner well-being, focus one's attention in a certain direction, and be able to concentrate completely on the achievement of own goals (Riva & Ryan, 2015).

We regard behavioural self-regulation as a deeper internal own control mechanism, which leads to considered, deliberate and conscious behaviour. Self-regulation requires the ability to control own behaviour, regulate one's own emotional energy, attention and conduct, cope with situations which disrupt the internal system, bring about stress or are frustrating, and manage challenges or difficult situations (Bodrova & Leong, 2008; Lowry, 2014).

Self-regulation manifests itself in two possible dimensions: (1) a situation where a person controls their impulses results in the prevention of certain behaviour (the person stops taking action), (2) a case where a person starts with the primary need to realise their own goals results in the activation of certain behaviour (the person begins to take action).

The classic concept of human development presumes a tendency towards psychological growth and integration. Active growth is connected with a tendency towards synthesis and organisation. This general view of a person as an active, integrative being with the potential to act in accordance with themselves is contained in the psychodynamic and humanistic personality theories (Stuchlíková, 2010).

We can regard the self-determination theory, which deals with the need for the determination of oneself, as one of these theories. It is based on the fact that a person has the tendency to satisfy three basic psychological needs: the need for competence, the need for relationships with other people (relatedness) and the need for autonomy. It presumes that, in the fulfilment of these needs, the person is increasingly more differentiated and holistic during self-perception. This tendency leads to greater autonomy and overall intrapersonal and interpersonal organisedness (Deci & Ryan, 2002; Hrešan, 2014).

These needs are perceived as universal life needs, which are innate and unchanging over time, regardless of gender or the culture in which the person grows up (Chirkov, Ryan, Kim & Kaplan, 2003).

The realisation of own potential requires the support of the external environment. The social environment can support the saturation of these needs, or it can prevent their fulfilment. The reaction of the environment to the individual's undesirable behaviour, which indicates a low level of self-regulation, can cause either the start of the development of self-regulation abilities or, on the contrary, stagnation and the inability to develop self-regulation. If the social context is positive, supportive and stimulating, there is a presumption that the individual will be able to develop their self-regulation to a greater degree (Blair & Diamond, 2008).

The aim of our research is to clarify the degree of use of self-regulation skills related to interpersonal cognitive problem-solving in upper-primary school children. First and foremost, we want to ascertain the relationship between self-regulation skills and the perceived position of pupils in formal and informal school life processes, and the degree to which this perceived position predicts pupils' self-regulation skills related to interpersonal cognitive problem-solving.

2 METHOD

2.1 *Participants*

The research group includes upper-primary school pupils (N = 1,133 valid). The primary schools were chosen by random selection (random number generator) from all the primary schools in the Czech Republic.[1]

A total of 11 primary schools were selected for the group. 1,133 6th – 9th class pupils participated in the research, comprised of 596 boys and 537 girls of between 11 and 16 years of age (M = 13.29, SD = 1.293).

---

[1]Ministry of Education, Youth and Sport of the Czech Republic register of school and school facilities, situation as on 26/04/2017 (incomplete, mixed-grade, special and practical schools were eliminated from the list).

2.2 Measures

Self-regulation was evaluated using the *Means-Ends Problem Solving technique* (MEPS) (Platt & Spivack, 1989) to measure the degree to which children were and were not self-regulated possessed skills related to interpersonal cognitive problem-solving. The MEPS uses a story-based format where children are provided the beginning and ending of a story. The beginning poses a problem, the ending reports the outcome. Children were instructed to tell what happened in the middle of the story that connects the two. In essence, children are asked to generate the means by which the outcome was achieved, given the problem. Children are written and can be as long or short as necessary. Because children require additional time the stories and respond only 5 of the 10 scenarios were selected for administration (Appendix 1). The MEPS allows children to generate as many means as they possibly can, and these are scored as being relevant or irrelevant. To provide some standardization in the process, we have asked children to generate only the best answer for the middle of the story. This answer is then evaluated along a scale of 0 to 3, with 0 being no means or completely irrelevant means and 3 being a relevant means. The pupils were able to obtain a maximum of 15 points, whereby a higher score corresponded to a greater degree of use of self-regulation skills during interpersonal cognitive problem-solving.

On the basis of a factor analysis (analysis of the main components), we discovered that the questionnaire is one-dimensional and explains 56.69% of the variance. The internal consistency of the questionnaire in all five items, measured using Cronbach's coefficient, attains a value of $\alpha = .81$, which represents an acceptable degree of reliability.

We measured formal and informal school life processes using the *Pupils' Attitudes to School Life* questionnaire (Vojtová, 2009). This questionnaire evaluates how pupils perceive their own position in formal and informal school life processes, and defines the areas of school life which are supportive, or on the contrary high-risk, for their learning. The original instrument contains 38 items and is based on the *The Quality of School Life Scale – School Life Quality Questionnaire* (Williams & Batten, 1981) and the authorial collective theory of Binkley, Rust & Williams (1996). The questionnaire differentiates between 7 dimensions of school life: (1) Success and Opportunity, (2) Overall Satisfaction with School, (3) Formation of Identity, (4) Negative Experience, (5) School Status, (6) Teacher-Pupil Relationships. One extra dimension was added to the questionnaire – (7) Social Inclusion in Peer Groups (according to Vojtová & Fučík, 2012).

On the basis of a factor analysis, we have identified 5 factors which explain 48.04% of the variance. We identified the factor related to Success and Opportunity (F1), as well as the factors of Social Inclusion in Peer Groups (F2), Negative Experience (F3), Teacher Approach to Pupils (F4) and School Status (F5).

On the basis of a low factor score (below .40) and the significant saturation of more than one factor, we eliminated a total of 5 items, with the final version of the questionnaire including a total of 33 items. The pupils evaluate each statement with one choice on a four-point scale expressing the degree of agreement or disagreement. In individual dimensions of school life, we work with the arithmetical mean of all of the items appertaining to it. A higher score represents a more positive attitude to school life.

Factor 1, *Success and Opportunity*, explains 28.17% of the variance, and includes 10 items with a factor weight from .54 to .75. This factor reflects the pupils' opinion of their own position in learning processes, and of the opportunities which they receive in them. The positive results in all the items show that the schooling and learning correspond to their expectations. They regard school as a place which positively stimulates them to learn. They sense their ability to achieve good results in school work.

Factor 2, *Social Inclusion in Peer Groups*, expresses how pupils perceive their involvement in informal activities and pupil groups in school life. Positive results in this dimension indicate pupils' good feelings from their own involvement in informal peer social networks (relationships among pupils). This factor includes 6 items with a factor weight of .62 to .75, and explains 7.28% of the variance.

Factor 3, *Negative Experience* of school life, is connected with negative feelings such as loneliness and worry. The items in this factor are inverse, and have been transcoded (renumbered) in order to ascertain overall attitude to school life. Factor 3 includes a total of 5 items with a factor weight of .53 to .66, and explains 4.95% of the variance.

Factor 4, *Teacher Approach*, contains 9 items with a factor weight of .52 to .75, and explains 4.6% of the variance. If the results of this evaluation reach the positive parts of the spectrum, then they perceive the teacher as a person who provides them with interest, support and assistance, who is fair, on whom they can rely, and who respects the pupils' needs.

Factor 5, *School Status*, includes 3 items with a factor weight of .43 to .58, and explains 3.03% of the variance. This factor expresses how pupils perceive their position in school life processes, with an awareness of their own value and importance. The positive results indicate that others (both fellow pupils and teachers) value their person and regard them as important. They can therefore regard school as a safe social environment in which they are firmly anchored, and which provides them with support in case of need.

The overall evaluation of school life is expressed by the mean of the evaluation of all the items and the items in the individual dimensions. At the same time, we also worked with the distribution of the pupils into subgroups according to the character of the prevailing nominations (pupils with a prevailing positive evaluation of school life, pupils with ambivalent attitudes and pupils with negative attitudes). We defined these subgroups within a range of ±1 of a standard deviation from the mean.[2]

2.3 Procedure

The pupils filled in the questionnaires using the "paper – pencil" method. The data was processed using the **IBM SPSS** programme version 24. The one-way ANOVA, Pearson's correlation coefficient and multiple linear regression (the stepwise method) were used for the data analysis. At the same time, the prerequisites for using of the chosen test were examined; i.e. normality and homoscedasticity (Levene's test) were verified.

3 RESULTS

3.1 Degree of use of the pupils' self-regulation skills

The degree of self-regulation of the pupils' learning attains a mean score of $M = 6.93$ points ($SD = 3.38$) from a total number of 15 points (Table 1). The results of the analysis show that the degree of use of self-regulation skills connected with cognitive problem-solving by upper-primary school pupils is not very high. At the same time, it is apparent that there are significant differences in the degree of use of self-regulation skills depending on the perceived position in formal and informal school life processes ($p < .001$).

The highest degrees of self-regulation are achieved by pupils with a prevailing positive evaluation of school life of $M = 8.11$ ($SD = 2.99$). Pupils with ambivalent attitudes achieve a mean score of $M = 7.05$ ($SD = 3.27$). Pupils with negative attitudes achieve significantly lower self-regulation levels of $M = 4.96$ ($SD = 3.46$) compared to pupils with ambivalent attitudes ($p < .001$) and pupils with positive attitudes to school life ($p < .001$). It is apparent that the perceived position in formal and informal school life processes can play a significant role in the development of the self-regulation of pupils' learning, and most importantly in the use of pupils' self-regulation skills related to interpersonal cognitive problem-solving.

3.2 Relationship between self-regulation and pupils' attitudes to school life

We have discovered that the degree of use of self-regulation skills related to interpersonal cognitive problem-solving is very strongly connected with the perceived position of pupils in school life processes (Table 2).

---

[2] A range of ±1 of a standard deviation from the mean is defined as 2.4 to 3.3 points ($M = 2.8$, $SD = .44$). Pupils with negative attitudes fall under a value of 2.4 points, while pupils with a prevailing positive evaluation of school life fall above a value of 3.3 points.

The degree of pupils' self-regulation related to interpersonal cognitive problem-solving correlates very strongly with perception of success and opportunity (r = .260, p < .001), with social inclusion in peer groups (r = .227, p < .001), with a teacher approach (r = .260, p < .001) and with school status (r = .163, p < .001). On the other hand, the application of self-regulation skills is not connected with a negative experience of school life (r = −.043, p = .149).

It means that the more schooling and learning corresponds to pupils' expectations, and the more pupils regard school as a place which stimulates their learning and in which they have the opportunity to experience success, the greater the degree of application of their self-regulation skills related to interpersonal cognitive problem-solving. At the same time, it is evident that if the pupil in the class experiences good feelings from their own inclusion in informal peer groups, and if the pupil perceives the teacher's approach as supportive and respectful, then they also apply self-regulation skills to a greater degree when resolving problematic situations. Similarly, it also applies that, with an increasing awareness of own value and importance in the school environment, the pupils' ability to apply self-regulation skills during interpersonal cognitive problem-solving also grows. On the other hand, it is evident that negative experience does not play a role in the degree to which pupils apply self-regulation skills.

3.3 *Attitude to school life as a predictor of pupils' self-regulation*

To ascertain the degree to which the assessed school life processes play a part in the degree of self-regulation of upper-primary school pupils, we looked for a model which would best explain the degree of variability of the dependent variable, i.e. the degree of self-regulation skills related to interpersonal cognitive problem-solving (Table 3).

Table 1. Degree of the pupils' self-regulation depending on their attitude to school life.

Perceived school life	N	Mean	Std. deviation	Std. Error
Negative	167	4.96**	3.46	0.27
Ambivalent	769	7.05**	3.27	0.12
Positive	197	8.11**	2.99	0.21
Total	1133	6.93	3.38	0.10

**The results are significant at a significance level of .01.

Table 2. Relationship between self-regulation and pupils' attitudes to school life.

	Self-regulation	M	SD
Success and opportunity	0.260**	2.63	0.59
Social inclusion	0.227**	3.26	0.59
Negative experience	−0.043	2.38	0.56
Teacher approach	0.260**	2.76	0.58
School status	0.163**	2.54	0.58

**Correlation is significant at the .01 level (2-tailed).

Table 3. School life processes which play a part in the utilization of pupils' self-regulation skills.

	B	SE_B	β	Sig.
Success and opportunity	0.758	0.230	0.133	0.001
Social inclusion	0.769	0.187	0.134	0.001
Teacher approach	0.794	0.239	0.136	0.001
Negative experience	0.268	0.185	0.044	0.148
School status	0.098	0.214	0.017	0.648

To explain the degree of use of the pupils' self-regulation skills related to interpersonal cognitive problem-solving, the relevant variables are the perception of success and opportunity ($p = .001$), social inclusion in peer groups ($p < .001$) and a teacher approach ($p = .001$).

The created model, with three independent variables, explains 9.4% of the variability of the dependent variable, i.e. pupils' self-regulation (R Square = .094, $p = .001$).

There are mainly factors which we can describe as strengthening that play a part in the degree of use of pupils' self-regulation skills in connection with the perceived position in formal and informal school life processes. The more school corresponds to pupils' expectations (the enjoy learning, like going to school and experience success there), the greater the degree of use of self-regulation skills. The more the teacher is perceived as a person who provides interest, support and assistance, the greater the degree of use of pupils' self-regulation skills; also, if pupils perceive their involvement in informal activities and pupil groups in school life positively, they apply their self-regulation skills to a greater degree.

Given the fact that neither negative experience nor school status play any role in the degree of use of the pupils' self-regulation skills, we can conclude from the analysis that the application of self-regulation skills related to cognitive interpersonal problem-solving is influenced primarily by strengthening environmental attributes (experience of success, inclusion in peer groups and teacher approach); high-risk attributes connected with a feeling of negative experience do not play a very large role in the use of self-regulation skills.

3.4 *Conclusions*

From the results of the research, it follows that upper-primary school pupils do not use self-regulation skills during cognitive interpersonal problem-solving very much. At the same time, we can observe that the degree of use of self-regulation skills is dependent on the perceived position in formal and informal school life processes. Pupils with a prevailing positive evaluation of their own position in formal and informal school life processes apply self-regulation skills to a greater degree than pupils with ambivalent attitudes and pupils with negative attitudes.

The degree of use of self-regulation skills related to interpersonal cognitive problem-solving correlates very strongly with perception of success and opportunity, with social inclusion in peer groups, with teacher approach and with school status. On the other hand, the application of self-regulation skills is not connected with a negative experience of school life. Whether the pupil experiences negative feelings such as loneliness or worry in school is not connected with the degree of application of their self-regulation skills during interpersonal cognitive problem-solving.

When we looked for a model which would explain the effect of the perceived position in formal and informal school life processes on the degree of use of self-regulation skills, we identified three significant factors which we can describe as strengthening environmental attributes. The more schooling and learning corresponds to pupils' expectations, and the more they regard school as a place where they experience success, and which positively stimulates them to learn, the more they apply self-regulation skills during interpersonal cognitive problem-solving. An important role is also played by the perception of the teacher, and social inclusion in peer groups. If the pupil perceives the teacher as a person who displays interest and offers support and assistance, and if pupils perceive their involvement in peer groups positively, then they apply their self-regulation skills to a greater degree during interpersonal cognitive problem-solving. We can thus indirectly conclude that the development of self-regulation is influenced by strengthening environmental attributes, while high-risk attributes connected with a feeling of negative experience do not have such an effect on the use of self-regulation skills.

4 DISCUSSION

A positive perception of success and opportunity, teacher approach and inclusion in peer groups increase the degree of use of self-regulation skills. On the other hand, pupils who do not think they experience success in school and do not see school as an opportunity, who do not perceive the teacher's approach as respectful and supportive and do not think they are very involved in peer groups, do not use self-regulation skills during interpersonal cognitive problem-solving very much. However, this does not have to mean that, as a consequence of this attitude, they necessarily experience negative feelings (loneliness, worry etc.).

The question is why some pupils who experience feelings of worry, loneliness and loss of own value in school use self-regulation skills, while others do not. It is possible that a fundamental role is played in this respect by the cause of negative experience (and the pupils' attitudes to this experience), which can indirectly affect the pupils' use of their self-regulation skills. From our analysis, it does not follow that high-risk factors (negative experience connected with feelings of loneliness, worry or loss of awareness of own value) weaken self-regulation skills. However, we should mention that we only measured the degree of use/non-use of self-regulation skills. We did not measure the degree of failure in pupils' self-regulation (e.g. regulation of emotions), so we do not know whether negative experience can cause a self-regulation failure. We merely discovered that it is not connected with whether pupils use self-regulation skills or not. In further research, it would be interesting to focus on the role that negative experience (connected with the experience of so-called social pain) plays in the regulation of emotions, or the degree to which it plays a part in the failure of self-regulation, and in what direction it works.

We can regard as a significant discovery the fact that the positive perception of the position in formal and informal school life processes, specifically success and opportunity, as well as teacher approach and inclusion in peer groups, strengthens the degree of use of pupils' self-regulation skills during interpersonal cognitive problem-solving. We can say that a supportive environment plays a significant role in the development of self-regulation. However, how the pupils themselves perceive this environment, whether they experience success in it, is important. If they perceive learning as an opportunity, experience good feelings from their own inclusion in informal peer groups and perceive the teacher's approach as supportive and respectful, then they are also able to use their self-regulation skills to a greater degree in this environment. These self-regulation skills can help them manage and overcome problematic situations which they may encounter in school. If they use self-regulation skills, then they are aware of what they can do to manage conflict situations, and focus their attention on their actions (self-regulation), so that they can resolve this situation effectively according to their expectations.

ACKNOWLEDGEMENT

The article was created within the grant project GA CR 17–04816S *The Dynamics of Self-Regulation in Socially Excluded Pupils*.

REFERENCES

Binkley, M., Rust, K., & Williams, T. (eds.). 1996. *Reading literacy in an international perspective: Collected papers from the IEA Reading Literacy Study.* Washington, DC: U.S. Department of Education.

Blair, C., & Diamond, A. 2008. Biological processes in prevention and intervention: The promotion of self-regulation as a means of preventing school failure. *Development and Psychopathology* 20(3): 899–911.

Bodrova, E., & Leong, D. J. 2008. Developing self-regulation in young children: Can we keep all the crickets in the basket? *Young Children* 63(2): 56–58.

Chirkov, V. I., Ryan, R. M., Kim, Y., & Kaplan, U. 2003. Differentiating autonomy from individualism and independence: A self-determination theory perspective on internalization of cultural orientations and well-being. *Journal of Personality and Social Psychology* 84: 97–110.

Deci, E. L., & Ryan, R. M. 2002. *Handbook of self-determination research.* Rochester, NY: University of Rochester Press.

Gilbert, D. T., Krull, D. S., & Pelham, B. W. 1988. Of thoughts unspoken: Social inference and the self-regulation of behavior. *Journal of Personality and Social Psychology* 55(5): 685–694.

Hrešan, A. 2014. *Dynamika základních psychických potřeb (podle teorie sebedeterminace) v běžných životních podmínkách a v podmínkách dlouhodobé izolace*. Bakalářská práce. České Budějovice: Jihočeská univerzita v Českých Budějovicích.

Lowry, L. 2014. *What Is Behaviour Regulation? And What Does It Have To Do With Language Development?* The Hanen Centre. Available in: <http://www.hanen.org/SiteAssets/Helpful-Info/Articles/what-is-behaviour-and-what-does-it-have-to-do-with.aspx>.

Platt, J., & Spivack, G. 1989. *The MEPS procedure manual*. Philadelphia: Department of Mental Health Sciences, Hahnemann University.

Raffaelli, M., Crockett, L. J., & Shen, Y. L. 2005. Developmental stability and change in self-regulation from childhood to adolescence. *The Journal of Genetic Psychology* 166(1): 54–76.

Riva de La, S., Ryan, T. G. 2015. Effect of Self-Regulating Behaviour on Young Children's Academic Success. *International Journal of Early Childhood Special Education* 7(1): 69–96.

Rothbart, M. K., & Posner, M. I. 2005. Genes and experience in the development of executive attention and effortful control. *New Directions for Child and Adolescent Development* 2005(109): 101–108.

Stuchlíková, I. 2010. Motivace a osobnost. In: Blatný a kol. *Psychologie osobnosti*: 137–164. Praha: Grada.

Vojtová, V. & Fučík, P. 2012. Předcházení problémům v chování žáků. Dotazník pro žáky. Praha: Národní ústav pro vzdělávání.

Vojtová, V. 2009. Škola pro všechny—vyhledávání žáků v riziku poruch chování ve školním prostředí. *Orbis Scholae* 3(1): 79–97.

Vávrová, S., Hrbáčková, K. & Hladík, J. 2015. *Porozumění procesu autoregulace u dětí a mladistvých v institucionální péči*. Zlín: Univerzita Tomáše Bati ve Zlíně.

Williams, T. & Batten, M. 1981. *The quality of school life*. Melbourne: Australian Council for Educational Research.

APPENDIX 1

Example of items from *The Means-Ends Problem Solving Technique* questionnaire (MEPS).

1. *Beginning:* You and your parents moved to a different city and you don't know anybody there. You are going to new school for the first time. You wish to get to know new friends.
   Midpoint: ....
   End: The end of your story is that you have many new friends.
2. *Beginning*: You were the most successful student from your class in the final test, but some of your classmates don't speak to you because of this.
   Midpoint: ....
   End: The end of your story is that the classmates speak to you again.
3. *Beginning:* Your classmate accuses you falsely of breaking a window in a classroom. You view it as injustice.
   Midpoint: ....
   End: The end of your story is that your classmate apologized for accusing you falsely.
4. *Beginning:* You and your classmate had agreed on working on a collective task together, but s/he chose to work with someone else in the end.
   Midpoint: ....
   End: The end of your story is that you work with the classmate you agreed to work with at the beginning.
5. *Beginning*: Your classmates made fun of you again because of the clothes which you wore to school.
   Midpoint: ....
   End: The end of your story is that your classmates stopped making fun of your clothes.

Author index

Ablaev, I.M. 97
Ahmad, A.R. 89
Andreeva, I.N. 145
Artamonova, E.G. 139
Auysheeva, N.N. 103
Awang, M.M. 89

Badyina, K.Y. 185
Badyina, K.Yu. 159
Bakhtin, A.G. 165
Bakhtina, V.V. 165
Baklanova, T.Y. 103
Baranovich, D.-L. 131
Bazhenova, N.G. 219
Biryukova, N.A. 275
Blinova, M.L. 269
Borisova, E.Y. 171, 175
Bosova, L.L. 257
Boyarinova, G.N. 211

Carnia, E. 65
Chaldyshkina, N.N. 245
Chetthamrongchai, P. 1
Chew, F.P. 25, 33, 89
Chin, K.-Y. 57
Choi, H. 313

Danilova, O.V. 171
Daryakin, A.A. 81
Demchenko, M.V. 193
Dorogova, J.A. 283
Dremina, I.E. 181

Egoshina, R.A. 219

Fliginskikh, E.E. 185

Gavrilov, A.M. 265
Gavrilova, M.N. 197
Ghanieah, N. 75
Glukhova, N.N. 203, 211
Golovina, N.N. 159

Grunina, S.O. 245
Guseva, E.V. 203

Hamad, Z.H. 25
Hamzah, M.F.B. 321
Han, C.C. 131
Hong, Z.-W. 57
Hrbackova, K. 337
Hsiao, H.-Y. 57
Hussain, M.N.B.Md. 321
Hutagalung, F. 25, 75, 131

Ivanova, I.G. 219

Jatmika, H.M. 149

Kadykova, G.N. 211
Kairupan, S.B. 289
Kannan, P.M. 297
Karandaeva, T.A. 171
Kazakovtseva, M.V. 225
Kazarenkov, V.I. 275
Kazyro, G.N. 251
Khaptakhaeva, N.B. 103
Khydyrova, A.V. 139
Kim, G. 313
Kim, Y. 305, 313
Kolesova, T.V. 237
Kolomiets, D.L. 275
Kom, M. 41
Korableva, S.V. 181
Korotkov, S.G. 231
Kozina, I.B. 171, 175
Kozlova, N.N. 117
Krasnova, N.M. 203
Krylov, D.A. 231
Kurapova, I.A. 181
Kuzminykh, Zh.O. 165

Lavina, T.A. 257
Lavrentiev, S.Y. 231
Leng, C.H. 75

Lezhnina, L.V. 181
Li, L.P. 131
Lim, C.T. 33
Loskutovan, E.A. 283
Lumban Gaol, F. 41

Maksimov, V.N. 203
Mamedova, N.A. 139
Marthandan, G. 297, 329
Martínez-Arroyo, J.A. 7, 15
Mironova, M.D. 111
Mitrofanova, T.A. 237
Mukhina, S.A. 197
Munastiwi, E. 155

Naykhanova, L.V. 103
Ndii, M.Z. 65
Nesterov, V.N. 117

Oh, Y. 305

Petukhova, O.A. 245
Polozova, O.V. 197

Rahman, A.A.A. 89
Rathimala, K. 329
Romanova, E.V. 237
Romanova, M.S. 237
Rosli, N.A. 75
Rouiller, N. 193
Ruchkina, G.F. 193
Rusinova, N.V. 219
Rybakov, A.V. 159

Sabitova, N.M. 125
Semenova, G.N. 185
Sendouw, R.H.E. 289
Shalaeva, S.L. 245
Shamtieva, M.R. 145
Shasheela Devi, K. 329
Shavaleyeva, C.M. 125
Shcheglova, N.N. 145

345

Shishkina, O.V. 269
Shkalina, G.E. 211
Si, S. 41
Sinagatullin, I.M. 275
Sirazetdinov, R.M. 111
Sokolova, G.L. 251
Soldatkina, T.A. 251
Streltsov, R.V. 257
Sudharatna, Y. 1
Sujarwo 149
Sumaryanti 149
Supriatna, A.K. 65
Susanto, B. 41
Svetlova, V.A. 159

Timerkhanov, R.S. 125
Timofeev, V.V. 265
Toktarova, V.I. 261
Tomoliyus 149
Tsaregorodtsev, E.I. 225, 257

Ustymenko, O.N. 283
Utama, B. 149

Valenzo-Jiménez, M.A. 7, 15
Vengerovskiy, E.L. 193
Villa-Hernández, L.A. 15
Virtsev, M.Y. 111

Vissarov, A.V. 265
Volkova, T.A. 145

Wang, G. 305

Yagdarova, O.A. 165
Yakovleva, S.L. 185, 251
Yalalova, N.F. 81
Yambaeva, N.V. 283
Yarygin, A.A. 265

Zabolotskikh, O.P. 269
Zagainov, I.A. 269
Zimina, I.S. 197